Economic Trends and Sustainable Environmental Assessment

Editors

Andrew Adewale Alola
CREDS-Centre for Research on Digitalization and Sustainability
Inland Norway University of Applied Sciences, Norway

Festus Victor Bekun
Faculty of Economics, Administrative and Social Sciences
Department of International Logistics and Transportation
İstanbul Gelisim University, İstanbul, Turkey

Uju Violet Alola
Department of Tourism Guidance, Istanbul Gelisim University
Istanbul, Turkey

CRC Press
Taylor & Francis Group
Boca Raton London New York

CRC Press is an imprint of the
Taylor & Francis Group, an **informa** business
A SCIENCE PUBLISHERS BOOK

First edition published 2024
by CRC Press
2385 NW Executive Center Drive, Suite 320, Boca Raton FL 33431

and by CRC Press
4 Park Square, Milton Park, Abingdon, Oxon, OX14 4RN

CRC Press is an imprint of Taylor & Francis Group, LLC

Library of Congress Cataloging-in-Publication Data (applied for)

ISBN: 978-1-032-30675-9 (hbk)
ISBN: 978-1-032-30676-6 (pbk)
ISBN: 978-1-003-30621-4 (ebk)

DOI: 10.1201/9781003306214

Typeset in Times New Roman
by Prime Publishing Services

Preface

The world economic outlook over the past years outlines alarming statistics in terms of macroeconomic indices such as high inflation, soaring unemployment rate and increased inequality, and a most obvious slowdown in economic activities due to global pandemic COVID-19, alongside other country-specific economic turmoil. These economic trends come with inherent implications for the world economy given increasing globalization. From an environmental perspective, there is also a connection between the economy and energy consumption as has been highlighted through the text. Thus, there is a dire need to present a holistic compilation of economic agents and their adverse effects, especially as they relate to countries and regions of the world.

This book, *Economic Trends and Sustainable Environmental Assessment*, is meant to serve as a guide for academics, graduate students, PhD candidates, researchers, and government administrators. The reason being that it outlines economic trends and sustainable environmental assessment issues in contemporary times that encompass macro-economic issues and their spillover effect on developing, transition, and developed economies. *Economic Trends and Sustainable Environmental Assessment* also considers one of the contemporary issues outlined by the United Nations Development Goals—specifically, UNSDG 8, which highlights sustainable and decent work and UNSDG 13, which focuses on climate action. [Editor Note: please rephrase the explanation on UNSDG 8 and UNSDG 13 to make them clearer. What is meant by "decent work"? How does UNSDG focus on climate action?]

The book is divided into four sections: 'Technology and Innovation', 'Energy', 'Environment', and 'Economy'). The text contains seventeen chapters that outline trends, assessments, and interlinkages among the mentioned sections. Specifically, the chapters cover the impact of economic trends and their impact on the environment across diverse regions. They explore the need for energy diversification and clean or alternative energy source development to ensure a sustainable environment and sustainable development, especially from the perspectives of clean technology and innovation. Additionally, the current book also uses mathematical and econometrics approaches in several chapters to provide specific insights into the topic. Thus, with a blend of a comprehensive quantitative (econometric) approach and mix-methods application to the subject, this book provides ample insights to the target audience such as graduate students, PhD students, energy and environmental specialist, macroeconomic consultant, and government administrators, to name a few.

To further give motivation to our targeted audience, it worth stating that this book compilation was carefully chosen by experienced editors of the book with cumulative experience of almost a decade that informed the selection of elite top-tier scholars to contribute to this text. To this end, we make the claim that this book is timely for the target audience, and much more so for policymakers. Lastly, if the reader aspires to learn about a blueprint for cutting-edge aspects of technology and innovation, energy, environment, and the economy, we assure you that *Economic Trends and Sustainable Environmental Assessment* will surely help.

Contents

Environment

Economy

Technology & Innovation

Chapter 1

Innovation Diffusion and Technology Acceptance Theory

Emmanuel Akaiso

1. Introduction

Innovation can be defined as the transformation of ideas, goals, visions, and dreams of invention into valuable goods and services for people to exchange for an amount of money. Something everyone on earth must experience is changing, and many people are too afraid to experience this. As a result, they never move ahead in life because they have allowed themselves to be captives of fear, allowing themselves to be unable to alter their circumstances. After the difficult experience of the past twenty-four months due to COVID-19, the eyes of everyone have been opened to how important change, adaptability, and flexibility are. Looking back, we can see that change is particularly important for everyone. That is exactly where innovation comes in (Stenberg 2016). To paraphrase Steve Jobs, innovation differentiates a leader from a follower.

Furthermore, innovation is also defined as bringing in new ideas to modify something that is already in place to make it better, more unique, trendier, and more useful to the present circumstances of the customer (Cui et al. 2011). A study by Kogabayev and Maziliauskas (2017) explains the importance of innovation for the following reasons:

- To make us prepared and flexible when changes occur at distinct stages of our lives.
- To provide relatable solutions and answers to the problems customers are facing.
- To enhance growth in all business areas.
- To give businesses higher advantage ahead of their competitors.
- To help businesses make beneficial use of innovative technologies as they arrive.

Department of Contemporary Management, Isik University, Maslak, Istanbul, Turkey.
Email: akaisoe@gmail.com

1.1 Diffusion

In science, diffusion is defined as the movement of gaseous particles from a region of higher concentration to a region of lower concentration till the gaseous particles saturate the entire region. Relating this definition to marketing, diffusion in marketing is defined as the movement of marketing practices through a social space or social culture till the marketing practices saturate that social space (Scott and Mcguire 2017).

1.1.1 Business model innovation

Rogério et al. (2017) defines business model innovation (BMI) as a new flexible setup that consists of both internal and external aspects of an organizations leading to a competitive advantage where value creation and value delivery are achieved. "*Business model innovation is the art of changing a company or organization's structures so as to serve a more specific value and critical thinking skills to customers to have a lasting competitive advantage.*" BMI occurs when new ways of organizing businesses are put in place to maximize profit, thereby creating great customer value (Rogério et al. 2017).

There are diverse types of business model innovation, and they are summarized in Figure 1 (Teece 2010).

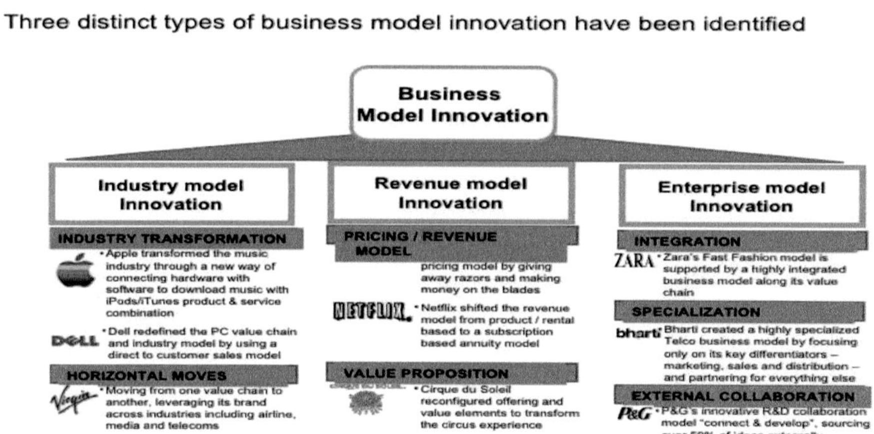

Figure 1: Explains business model innovation.

A remarkably interesting BMI story that can makes us further understand how BMI works is the story of Don Corning explained by Chesbrough (2010).

A chemical company named Dow Corning is a respectable supplier of silicone products for a substantial number of industries and businesses. The model of business with which they operated involved selling goods with huge consulting services and high gap. Despite this, the sale of some goods was redundant due to how the customers commonized the goods. Dow Corning found out that the growth of their business would only be further possible by twitching the business model and analyzing the cause of the redundant sales; the result of the analyzes turned out to be that some product buyers needed no support (no consulting service after the first consulting service they got) as they were seeking to purchase huge amounts of the goods at a very affordable price.

After finding this out, Dow Corning created a new business model to ride on the previous business model they have been using before (adding a cost-based business model to the previously existing value-based business model); which was to give products to the general market at a very affordable price even though there are no extra services attached to it. The establishment of this new business concept was done through a new Xia meter brand that offers affordable sales through internet services. This model was the exact opposite of the initial business model, but it worked, and the sales went up again (Dorin 2018).

1.1.2 Organizational innovation

Organizational innovation is the innovation that involves the use of interactive innovation, relational innovation, and organizational and technological innovation to bring together people who have unique skills, resources, and talents to achieve maximum success. "*Organizational innovation occurs when an organization makes use of new ways, new insight, new approaches, new process, new ideas, new opportunities to change the way the organization operates thereby achieving much success*" (Demircioglu 2016).

According to Da Silva and Silva Cirani (2020) organization innovation can be categorized into three aspects: (a) **organization structure** explains the structure put in place by the organization for their value creation and growth process, (b) **organization analysis** involves the constant analysis of the organization to check to see the growth process and analyze how well the company is doing and to see if the company needs to make any adjustments, and (c) **organizational change** is how adjustments or changes are made in an organization to increase the value creation and maximize profit through customer satisfaction (Steiber and Alänge 2015).

The company Carrier has a wonderful narrative. The consulting engineer at Sackett and Wilhelm's, Walter Timmis, tried solving the challenge the company faced with the printing machine which could produce multiple colors. The problem resulted from the fact that moisture affected the ink used for the printing and made it not to last long on the paper that was being printed upon, leading to reduction in the quality of multicolored materials produced, wastage of materials used, and loss in the productivity of the company in 1902. Walter Timmis then visited New York to see the sales manager of Buffalo Forge to ask for help and collaboration on the way forward for them to solve the problems they were going through at that moment. The sales manager advised Timmis on how the ink could be preserved even in a moist environment. So, Timmis started conducting research strategies to see if they could still maintain a moist monitored environment for the exact thing, they needed to do so that the ink and paper would not mismatch. In 1903, these demands were fulfilled, and the right moist environment was maintained, increasing the company's productivity. This process led to the invention of the new generation air-cooling systems (Cunha and Verhallen 1998).

1.1.3 Process innovation

Process innovation occurs when there is the establishment of new ways and new methods for production and service delivery to be done. This includes significant improvements in technique, equipment, and/or software. Process innovation strategies are usually constructed of capabilities generated from access to and recombination of external sources of knowledge (embodied knowledge, industry sources, and so on), as well as the innovation strategies to augment and reinforce such creativity skills (Cristina Scafuto et al. 2018).

Some aspects of process innovation involve using technology and production machines that help in designing products and constructing software, sales, and maintenance appliances, supply system and product delivery set ups as well as uplifting the accounting credibility and maximizing customer experiences (Dorin 2018). While your customers will see and value product innovation, a process change will probably be observed and valued solely by your employees. Changes in processes on average lower production costs more often than they enhance revenue. Therefore, process innovation is especially important in the business sector of the world we live in today (Huang and Rice 2012).

One interesting process innovation story is that of Differential, which had just completed a mobile sales dashboard for Grupo Bimbo. The bakery has sixty-five manufacturing sites and 2.5 million sales locations spread throughout 22 countries and three continents. As a result, those who were part of the managerial teams tended to meet their assistants all around the globe, online. The dashboards, which are mobile, enabled the managerial teams to make use of the records for the sales in all the country, provinces, and offices connected to them, eliminating doubts in the sale records and helping in time management (Widya-Hasuti et al. 2018).

1.2 Aims and objectives of this research

a) To understand what diffusion of business ideas is about.
b) To find out the various theoretical concepts that are used in the diffusion of goods and services.
c) To get deep insights into diffusion innovation theory and the technology acceptance model.
d) Analyzing and investigating the advantages, disadvantages, and limitations of each of the models.
e) Recommend the best model to be used by businesses.

2. Methodology

This paper is focused on analyzing theoretical models used in investigating the perspective of businesses in making their new goods or products to be diffused across a group of people and to be accepted by these people. For this research to be successful, this research focused on secondary sources, which includes research literature and theoretical published literature. The published literature was analyzed for the research models used to adopt innovations in businesses and to finally identify the advantages and disadvantages of the models, thereby showcasing which is more preferable or how they can both be combined to produce greatness in the business world.

3. Diffusion Innovation Theory

Diffusion innovation theory is the theory that explains how quickly a new idea or innovation spreads across a social space. The term diffusion is a marketing phenomenon that takes place among people in a social space in reaction to a new idea, new product, new item, or new way of life that has been introduced to them so as to work on their previously used methods or to improve how they do certain things in a better way (Dearing 2014). Making diffusion innovation theory (DIT) a reality has to be done through different strategic channels, that is, the ways through which the new innovation can be passed across through the social space from one person to another. There is a very important variable in DIT which is the time of adoption; we should note early enough that just because a new idea can be introduced does not necessarily mean it would be implemented by the people it was introduced to, so the time of adoption helps to explain how long it would take for the new idea to be implemented by people living in a social space or culture (Peres et al. 2009).

There can be an assessment of diffusion amongst people who are leaders in government positions, health organizations, and large tightly knitted communities in different cities and states across the globe. Figure 2 properly explains the correlations between the time of adoption and the properties of diffusion in unique situations, which also includes when a new idea is introduced, and also when a new idea does not diffuse due to certain factors or reasons applicable (Syahadiyanti and Subriadi 2018).

The figure explains how diffusion works in some social spaces. You can see the slow diffusion rate at the beginning because the idea was just introduced and people are getting to know what the idea is, After some time you see an increasing time of adoption on the graph, which means diffusion is increasing at this point. Next you see a reduction in the rate of diffusion, and this can be due to varied factors that affected the rate of diffusion in the social space (Huang et al. 2020).

So, an S-shaped graph is produced, but this is not always the case for all diffusion processes. There are cases where the time of adoption of certain ideas is strategically shorter, especially the spread of a new policy across a social space. The reason for the short adoption time is due to strategical measures taken by the government by creating an urgent national concern, which leads to rapid attention to the policy to be implemented and this makes people in that social space fully aware of the policy and adopt it quicker than expected because they have to be convinced to perceive that the new innovation would have a positive effect on them and their social environment at large (Weidman et al. 2018).

Looking at Figure 2 you will recognize that there are some parts of diffusion that actually goes uninvestigated because there could be other competing innovations as well that are more important, thereby hindering a new idea from being implemented

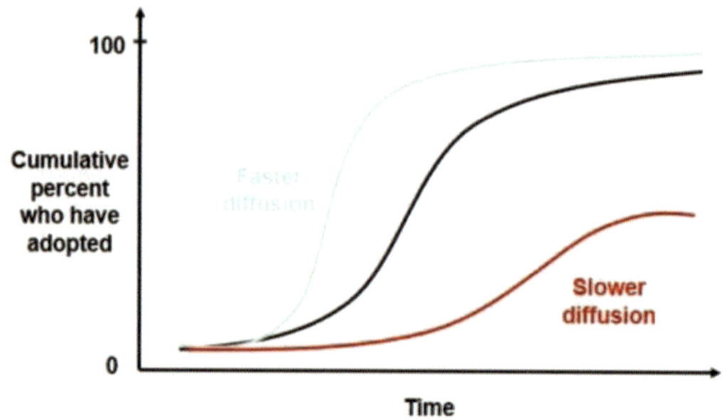

Figure 2: The context of diffusion. Source: (Dearing and Cox 2018).

or conducted (Peixoto et al. 2017). It is also worthy of note that failures of innovation diffusions also take place, and this proves that an innovation idea introduced to a social space may be adopted or rejected and the result of the diffusion is dependent on potential adopters (people that adopt the new innovations or people that allow the innovations to be adopted).

In the diffusion innovation theory, there is a resulting factor called acceleration and this occurs when a new innovation is quickly accepted or adopted by a social space because prominent members of that social space have seen the good in that innovation and they use their influence to pass that innovation across to other people in that social space. So, from tax policies to child welfare down to tobacco use reduction and many other policies introduced into the society, influential members of the society can help impact how quickly the new idea is adopted. Also, when influential members of society do not adopt an innovation or help to influence a policy adoption, this does not necessarily change the diffusion system as the diffusion theory is either adopted or rejected.

Over the ages, diffusion takes place across different social spaces. Innovation spreads through different spheres of life and helps to change people, policies, and laws. It helps achieve the best probable future for every society.

There are varied factors that affects the diffusion of latest ideas and some of them are:

a) ***Great advantage***: Persuading the target group that a new idea or innovation is much better than the idea or innovation currently used by people in certain social space. If they can see the advantage in using the newly proposed idea and the benefits and also how the new idea can create a better society for them to live in, then there is a huge probability for the new innovation or idea to be adopted and implemented (Tolba and Mourad 2018).

b) ***Deep connection***: The ability for a new idea or innovation to be connected/compactible with finding solutions to the problems or when a new idea/innovation easily solves the problems people in a certain social space are facing without changing their personal values makes it unproblematic for them to implement the new idea (Tolba and Mourad 2018).

c) ***Perceived use of ease***: Can people in a social space easily use the new idea and innovation introduced to them? Do they easily understand how beneficial the new idea would be to them and can it be easily implemented? (Zhang et al. 2015).

d) ***Probation***: Can the new idea be evaluated under probation basis before it has been implemented so people can see, analyze, and decide for themselves if they want to use certain innovations and ideas (Weigel et al. 2014).

e) ***Analytical result***: If the people that should implement a new idea or innovation can see a practical example of a test run and they can properly observe it for themselves to see the likely results they will get if they implement the innovation, there is a high probability that they will accept the idea (Khan et al. 2016).

One especially key factor to consider is that the success of the diffusion innovation theory is partly reliant on the modes and channels of communications because you need to inform and educate the people who need the ideas so they can adopt it. This process of communication could be through different channels, influential individuals, groups, and organizations, etc. Also, do not forget the value of time in the acceptance and the diffusion of an idea or an innovation.

3.1 Why or how do people adopt new ideas or innovations

The needs of people and the reasons why people do things are uniquely different from one another based on their acceptability of innovative ideas. Most times the first set of people to accept and adopt a new idea did so out of personal excitement without societal constraints; they are also called opinion leaders. The second set of people (early majority) adopt innovative ideas/innovations after analyzing and checking the advantages and disadvantages of the new idea, if the advantages outweigh the advantages, they then implement it. The last set of people to adopt an idea (late majority) do so because they feel societal pressure and they do so out of compulsion. The best set of adopters is the second set, where the new idea has been analyzed and investigated to be sure of its impact on society before it is adopted and implemented. The laggards are also like the innovators in that they are not under any societal constraints, but they take their time before adopting a new idea/innovation (Stenberg n.d.-b).

Dissemination occurs when a new idea/innovation is distributed to other members of the social space, so they are informed and updated as regards the new idea they need to implement. New innovations and ideas can be transmitted through special media channels, which include social media, mass gatherings, newspapers, etc., thereby making the information about the new idea readily available. So, every individual or industry that comes up with a new idea or innovation for the betterment of the lives of people within that social space needs to come up with mind-blowing ideas on how that particular idea can spread across a social space to the targeted audience that needs to be informed as regards the new innovation (Kee 2017).

After dissemination comes *implementation.* Implementation is when the new idea/innovation received is used. When a new idea has been disseminated to a certain group of people, the next question is whether the idea should be accepted/adopted or

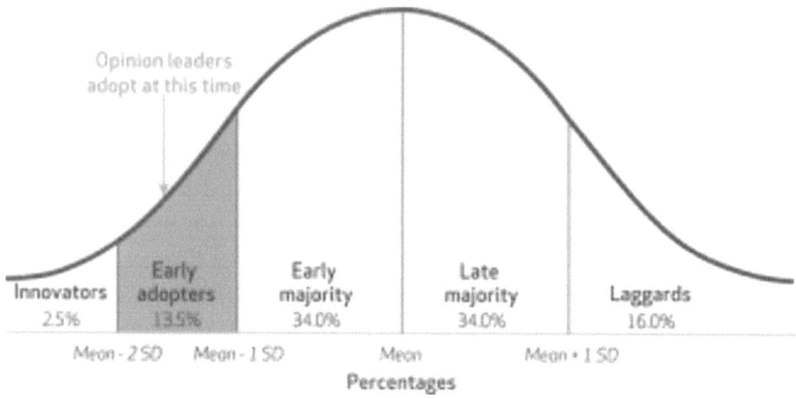

Figure 3: The graph explaining the innovation of the adopters with respect to the time of adoption (Dearing 2014, Dibra 2015).

rejected. Once the idea gets adopted, then it gets implemented, that is, the idea gets put to work. An *implementer* is an individual who receives a new idea and changes their way of life to adopt this new innovation. Most times the implementers may not necessarily be the chooser of the implementation, but the implementer has been influenced to use the new innovation presented (Owen et al. 2006).

Fidelity is the degree to which a new idea or innovation is implemented based on the way the originator of the idea planned it to be. Fidelity is usually determined by measuring how well people would accept the idea on a smaller scale and then extending it to the larger audience so that the progress can be properly tracked (Luiz Dias da Silva and Da 2017). The term fidelity could come exceedingly early before dissemination and implementation, and it could also be done after dissemination and implementation. Fidelity helps the business analyst know how individuals would react to innovative ideas and innovations. Fidelity in the diffusion process can be affected by two factors: reinvention and adaptation (Cam and Tran 2017). *Reinvention* is defined as the modifications the innovation originator makes to the innovation so it can be easily adopted and implemented by people in a certain social space. So, when an innovation is assessed on a small group of people during the fidelity process, certain results received help modify the innovation itself; that is what reinvention is about. *Adaptation* on the other hand is made by the implementers; they adjust their way of life to fit the new innovations in. They may need to change certain things just to make the innovation work for them based on how they are and where they live. So, in summary, reinvention is made by the innovation originator while adaptation is made by the implementers (Nutley et al. 2002).

Diffusion innovation theory gets implemented really fast if it is extremely easy to understand and implement (see Figure 4). So, every innovator should see the importance of making the new innovations relatable and extremely easy to use. Another factor t is when a new innovation rides on a previous innovation or is compatible with a previous innovation; it makes it easier for implementers to use this new innovation as they do not have to throw their previous innovations away since both the old and the new are compatible (Boyle 2010, Dibra 2015).

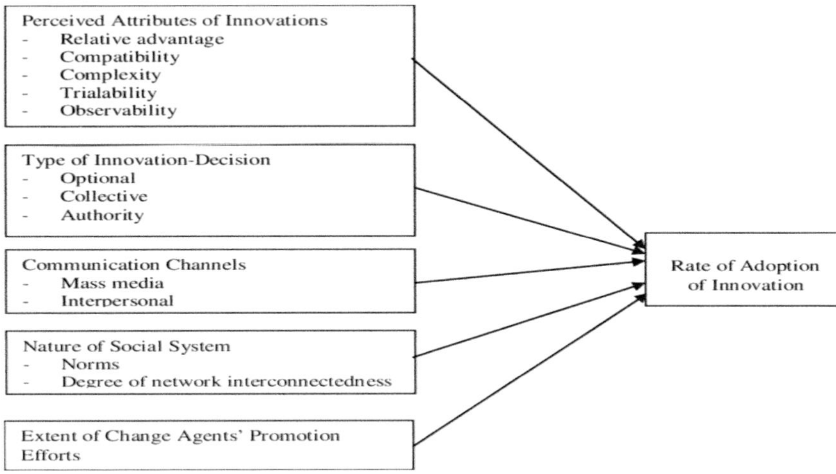

Figure 4: Rogers model on the rate of adoption of innovation (Dibra 2015).

3.2 Advantages of diffusion innovation theory (DIT)

First, the sustainability of DIT is an interesting narrative, and this is especially useful for every marketing practice. Marketing practices need to be sustainable, that is, they can be used for an extended period of time, and they can produce the right intent developed by the original innovators for the exact purposes the innovation was created for. The DIT is a sustainable theory for innovations in tourism, health and other government policies (Sai 2018).

Secondly, diffusion innovation theory involves the spread of innovations. So, the spread of this innovation supersedes the problems attached with a simple cost benefit analysis economic model. This spread analysis can also explain why or how people in certain social spaces adopt an innovation; it helps to provide a psychological and social interpretation as to why people adopt certain innovations. This theory is used in various countries across the globe at macro and micro levels of establishment on all types and sizes of innovations. It also gives a cost/benefit analysis on what it would cost people in a social space who adopt certain innovations (Weidman et al. 2018).

Thirdly, innovation disseminations in the diffusion innovation theory follows a very unpredictable direction. Instead of focusing on an individual's acceptance of a certain innovation, it showcases the innovation as a product of evolution. That is, there is great room for reinvention if the implementers have great issues with implementing the innovation. This is very much applicable in tourism and government policies, where the behavioral response of people to a certain social space is considered and evaluated so that the innovations can undergo innovation changes. An atmosphere for improving innovations increases competition and rejects stagnation. The continuous and constant need for improvements brings about the spread of innovations.

Fourthly, diffusion innovation theory is a well-thought-out theory which has developed and evolved from year to year and has been used in various disciplines, businesses, and areas of life. The DIT research has been assessed by thousands of researchers and has proven to be an amazing theory useful for the spread of innovations and ideas (Murray 2009, Weidman et al. 2018).

4. Technology Acceptance Model

The technology acceptance model (TAM) is an information systems theory that explains and shows cases how people in different socio-economical spaces accept, adopt, and use technology. Technology acceptance model is a model that describes why and how users accept and use a newly introduced technological model. It also considers the decisions individuals make before they accept certain technologies, considering how useful the technology is to them, and how easy the technology is to be used. TAM relies on two key factors, how useful the technology is, and how easy the technology is to use. TAM is a big and immensely powerful model for determining how different users would accept certain technological models (Deslonde and Becerra 2018).

4.1 Brief history of technology acceptance model (TAM)

TAM was first proposed by Fred Davis in 1985 for the sole aim of finding out the factors that influence the decisions of individuals to use certain technology. These factors include how useful the technology is, the ease of use for the technology, and the attitude of the individual towards the technological model (Surendran 2015).

This is the original model postulated by Davis in 1985 (Chuttur 2017). From Figure 5 above you can see that the actual system usage is dependent on the totality of the user motivation. The user motivation summarizes of how the user thinks the technology would be useful for them, how the user thinks the technology would be easy for them to use, and lastly their entire attitude towards the technology itself. These are enormously powerful factors to consider when using the technology acceptance model according to Fred Davis (Chuttur 2017).

This model in 1989 was later modified to give another model because of the reinvention and the evolving usage. This new modification contained information on

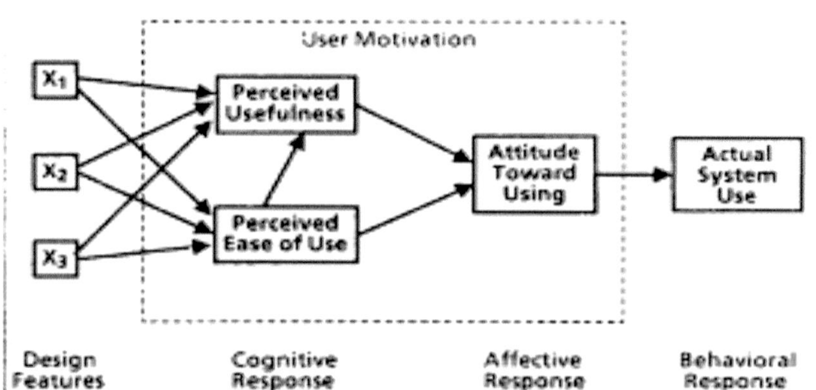

Figure 5: Original TAM proposed by Fred Davis (Chuttur 2017).

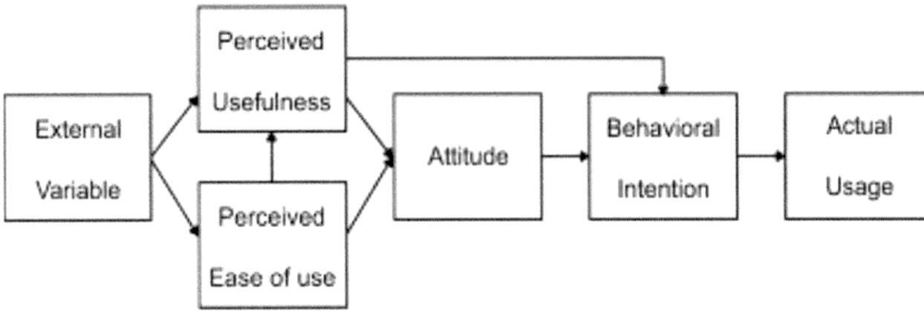

Figure 6: The first modified version of TAM (Chuttur 2017).

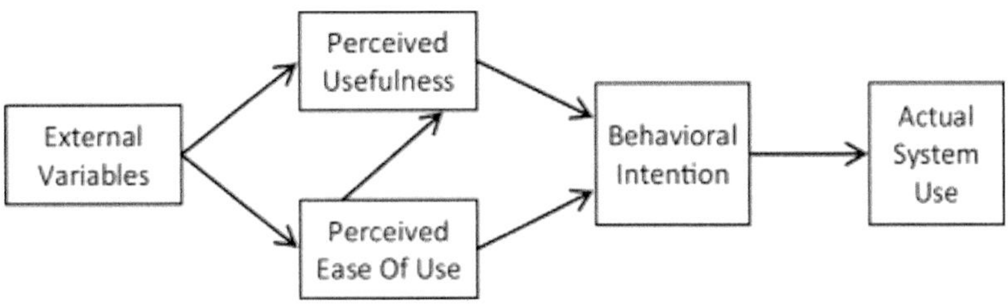

Figure 7: Final modified version of TAM (Chuttur 2017).

external variables as there are variables which are external that affect the adoption of technology.

Looking at Figure 6 and 7, you will see that in Figure 5 attitude is a factor standing on its own but in Figure 7, attitude has been fused together with behavioral intentions to help categorize all the factors affecting the decisions every individual would take before they accept the usage of certain technological models introduced to them. Also, the perceived usefulness and use of ease was seen to have a direct adverse effect on the usage of the technological system so the attitude was merged with the behavioral intention (Jamous et al. 2017).

Perceived use of ease can be described as how easily individuals can use a model. The individuals that the model is introduced to should be able to use the model easily; if they personally perceive, feel, or think that the model would be difficult for them to use, then they would reject the model and not have anything to do with it . That is why it is particularly important for models to be properly thought out and explained to the users before they are given for implementation.

Another interesting point to note here is that perceived ease of use is influenced by two factors called anchors and adjustments (see Figure 8). The anchors describe how well an individual can use the computer to help them adopt technology, the general beliefs that the individual has about computers while adjustments describe the enjoyment that comes from using the technology model in particular.

Figure 9 above explains how useful the individual finds the TAM model to be. The outlook of the model, is it beautiful and acceptable to the eyes? is it relevant to their betterment? Is the model producing high quality? Are the end results amazing? This and many more aspects factor individual look at before considering if a technology is useful or not.

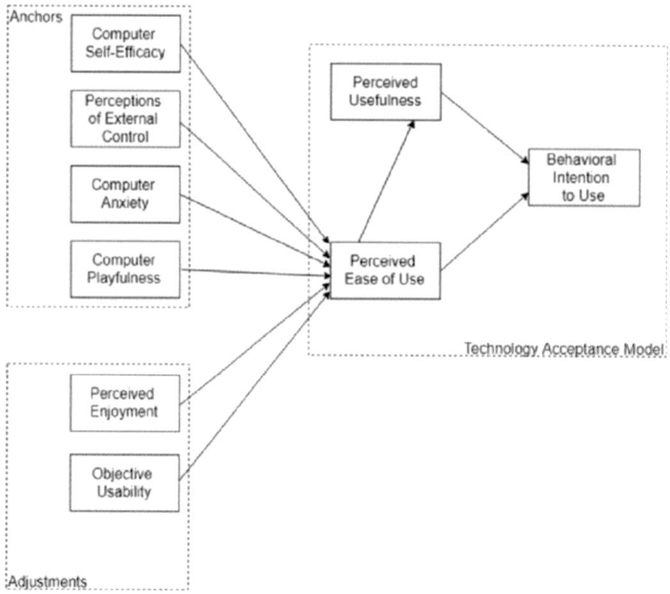

Figure 8: A summary of the factors that influence ease of use (Chuttur 2017).

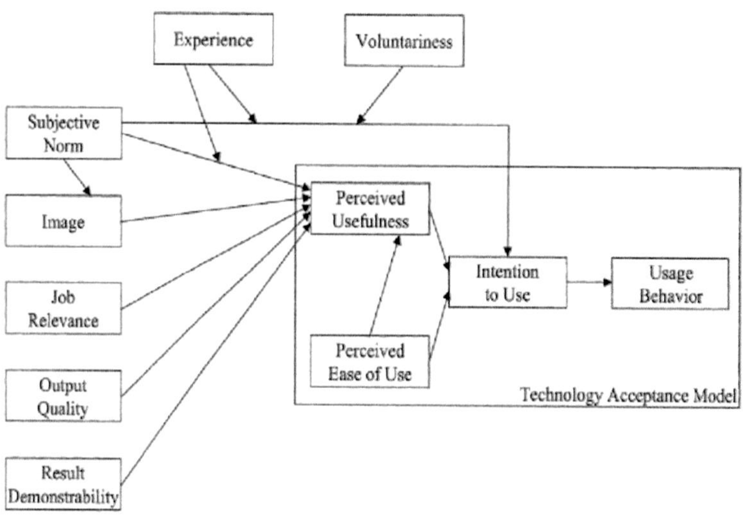

Figure 9: TAM 2 (Chuttur 2017).

4.2 Advantages of TAM

a) TAM is a relatively simple model and can be implemented easily, thereby making this model more useful and beautiful to use, compared to the theory of reasoned action or the theory of planned behavior.

b) TAM is simpler to use and less expensive to use because the methodology used in this model are direct and specifically related variables.

c) All previously done research points towards the fact that TAM is a verified model to predict the reasons why an individual would want to use certain technological models (Olushola et al. 2017).

4.3 Disadvantage of TAM

a) The summarized disadvantage of the TAM is the fact that TAM only measure usefulness of a model and the ease of using a model but cannot explain the real reasons behind an individual deciding to use a particular model (Marikyan 2016).

b) Also, the environmental settings for most TAM research were only done in voluntary environments and not mandatory or tough environments to measure distinct factors that enable TAM (Ajibade 2018).

4.4 Criticisms of TAM

According to Chitungo and Munongo (2013), the criticism of TAM is categorized into three parts:

a) The kind of methods used for the analysis of TAM.

It has been reported that most data used for the analysis of the TAM model were real self-reported data, which has not been subjected to further confirmation before using the TAM model, as it is known self-reported data are subjective and easily manipulated. Also, most research made use of students for their analysis, and this did not give a complete representative of all spheres of life so we can have a balanced model. Lastly, the TAM model only explained the voluntary use of the model and did not include explanations for the mandatory use.

b) The indicators and relationships existing in TAM.

There are some other factors that were not put into consideration in TAM. The original TAM was only analyzed in the voluntary environment and not done under mandatory conditions, as analysis under mandatory conditions produced other shocking variables as well as leading us to see that cognitive and affective aspects also plays a role in the perceived ease of use and perceived usefulness.

c) The foundational deep-rooted values of TAM.

TAM did not take into consideration the fact that an individual may decide to use a particular model and then stop using that model or uses an entirely different model. This could be maybe after the individual has reflected on their previous decision (Al-Rahmi et al. 2019).

5. Conclusion

In this research work both models were properly analyzed to showcase that every business needs to understand how their business ideas can be diffused across certain socioeconomic zones.

Diffusion innovation theory (DIT) can be affected by how easy such technology is, the deep connection people have with the technology, and also the analytical results people get when they use the technology. There are also different adopters that can determine how fast certain technology will diffuse.

The technology acceptance model (TAM) is affected by how useful people find the technology to be, how easy the technology is, the attitude of people towards the technology, and also the entire acceptance behavioral response. The limitation of the technology acceptance model is that the data used was self-reported data, which is subjective and can be manipulated. It also does not take into consideration that an individual can decide to stop using certain models and it does not take into consideration why this happens.

This diffusion innovation theory and technology acceptance model can be operationalized to increase the speed at which an idea is adopted and also enlarge the target market for any idea or policy.

References

Ajibade, P. (2018). *Technology Acceptance Model Limitations and Criticisms: Exploring Practical Applications and Use in Technology-related Studies, Mixed-method, and Qualitative Research.* http://digitalcommons.unl.edu/libphilprac/1941.

Al-Rahmi, W. M., Yahaya, N., Aldraiweesh, A. A., Alamri, M. M., Aljarboa, N. A., Alturki, U. and Aljeraiwi, A. A. (2019). Integrating technology acceptance model with innovation diffusion theory: An empirical investigation on students' intention to use e-learning systems. *IEEE Access*, 7: 26797–26809. https://doi.org/10.1109/ACCESS.2019.2899368.

Boyle, A. (2010). Some forecasts of the diffusion of e-assessment using a model. In *The Innovation Journal: The Public Sector Innovation Journal* (Vol. 15, Issue 1).

Cam, T. and Tran, T. (2017). Adding innovation diffusion theory to technology acceptance model: Understanding consumers' intention to use biofuels in Viet Nam Man Shin Cheng. In *International Review of Management and Business Research* (Vol. 6, Issue 2). www.irmbrjournal.com.

Chesbrough, H. (2010). Business model innovation: Opportunities and barriers. *Long Range Planning*, 43(2-3): 354–363. https://doi.org/10.1016/j.lrp.2009.07.010.

Chitungo, S. K. and Munongo, S. (2013). *Extending the Technology Acceptance Model to Mobile Banking Adoption in Rural Zimbabwe* (Vol. 3, Issue 1).

Christine E. Murray. (2009). Advantages of using DIT. *Journal of Counselling and Development*, 87(1): 108–116.

Chuttur, M. (2017). *Overview of the Technology Acceptance Model: Origins, Developments and Future Directions Household Energy Savings using live Monitoring View Project School Leadership in Challenging Times View Project*. https://www.researchgate.net/publication/277766395.

Cristina Scafuto, I., Claudia Belfort, A. and Teixeira, G. (2018). *Process Innovation: A Surfing Manufacturer Case Study*. https://www.redalyc.org/articulo.oa?id=499155729002.

Cui, T., Ping, J. W. and Pan, S. L. (2011). *Strategies of Crisis Management From Contingent Perspective. Digital Sustainability View project Social Innovation via Social Media View Project Strategies of Crisis Management from Contingent Perspective*. https://www.researchgate.net/publication/221229426.

da Silva, J. J. and Silva Cirani, C. B. (2020). The capability of organizational innovation: Systematic review of literature and research proposals. In *Gestao e Producao* (Vol. 27, Issue 4). Brazilian Institute for Information in Science and Technology. https://doi.org/10.1590/0104–530X4819-20.

Dearing, J. (2014). *Social Marketing and the Diffusion of Innovations City to City Diffusion of Carbon Mitigation Innovations View project Designing for Diffusion (D4D) View Project*. https://www.researchgate.net/publication/266613618.

Dearing, J. W. and Cox, J. G. (2018). Diffusion of innovations theory, principles, and practice. *Health Affairs*, 37(2): 183–190. https://doi.org/10.1377/hlthaff.2017.1104.

Demircioglu, M. A. (2016). Organizational Innovation. In *Global Encyclopedia of Public Administration, Public Policy, and Governance* (pp. 1–5). Springer International Publishing. https://doi.org/10.1007/978-3-319-31816-5_3017-1.

Deslonde, V. and Becerra, M. (2018). The Technology Acceptance Model (TAM): Exploring school counselors' acceptance and use of naviance. *The Professional Counselor*, 8(4): 369–382. https://doi.org/10.15241/vd.8.4.369.

Dibra, M. (2015). Rogers theory on diffusion of innovation-the most appropriate theoretical model in the study of factors influencing the integration of sustainability in tourism businesses. *Procedia - Social and Behavioral Sciences*, 195: 1453–1462. https://doi.org/10.1016/j.sbspro.2015.06.443.

Dorin, M. (2018). Product and process innovation: A new perspective on the organizational development. In *Article in International Journal of Advance Research and Innovative*. www.ijaemr.com.

Huang, C. Y., Wang, H. Y., Yang, C. L. and Shiau, S. J. H. (2020). A derivation of factors influencing the diffusion and adoption of an open-source learning platform. *Sustainability (Switzerland)*, 12(18). https://doi.org/10.3390/su12187532.

Huang, F. and Rice, J. (2012). Openness in product and process innovation. *International Journal of Innovation Management*, 16(4). https://doi.org/10.1142/S1363919612003812.

Jamous, M., Momani, A. M. and Jamous, M. M. (2017). The evolution of technology acceptance theories. In *International Journal of Contemporary Computer Research (IJCCR)* (Vol. 1, Issue 1). https://www.researchgate.net/publication/316644779.

Kee, K. F. (2017). *Adoption and Diffusion*.

Khan, K., Pathan, Z. H., Hussain, S., Latif, Z. and Lei, W. (2016). Factors affecting diffusion and adoption of information and communication technology among rural users in Khyber Pakhtunkhwa province, Pakistan. *International Journal of Information Technology and Management*, 1(1): 1. https://doi.org/10.1504/ijitm.2016.10003630.

Kogabayev, T. and Maziliauskas, A. (2017). The definition and classification of innovation. *Holistica – Journal of Business and Public Administration*, 8(1): 59–72. https://doi.org/10.1515/hjbpa-2017-0005.

Luiz Dias da Silva, M. and Da, D. (2017). *Diffusion and Adoption of Technology Amongst Engineering and Business Management Students*. https://www.redalyc.org/articulo.oa?id=499151081002.

Marikyan, D. (2016). *Technology Acceptance Model Theory Factsheet*.

Nutley, S., Davies, H. and Walter, I. (2002). *ESRC UK Centre for Evidence Based Policy and Practice: Working Paper 10 Conceptual Synthesis 1: Learning from the Diffusion of Innovations*.

Olushola, T. and Abiola, J. O. (2017). The efficacy of technology acceptance model: A review of applicable theoretical models in information technology research. In *Quest Journals Journal of Research in Business and Management* (Vol. 4). www.questjournals.org.

Owen, N., Glanz, K., Sallis, J. F. and Kelder, S. H. (2006). Evidence-based approaches to dissemination and diffusion of physical activity interventions. *American Journal of Preventive Medicine*, 31(4 SUPPL.): 35–44. https://doi.org/10.1016/j.amepre.2006.06.008.

Peixoto, M., Castro, A. C. and Nascimento, P. P. (2017). *Roger's Theory of Diffusion of Innovations and Institutional Changes in Brazilian Rural Advisory Services*.

Peres, R., Mahajan, V., Muller, E. and Recanati, L. (2009). *Innovation Diffusion and New Product Growth Models: A Critical Review and Research Directions*.

Pina Cunha, M. and Verhallen, T. M. (1998). Organizational Innovation: An overview of topics, models and research directions. *Comportamento Organizacional e Gestao*, 4(2): 5–34.

Rogério, M., Viegas, G. and Rodrigues Da, A. (n.d.). Business model innovation: A bibliometric review. *Bibliometric Review International Journal of Innovation*, 5: 2017. https://www.redalyc.org/articulo.oa?id=499151573003.

Sai, A. A. (2018). An exploratory study of innovation adoption in estonia. *Open Journal of Business and Management*, 06(04): 857–889. https://doi.org/10.4236/ojbm.2018.64064.

Scott, S. and Mcguire, J. (2017). Using diffusion of innovation theory to promote universally designed college instruction. *International Journal of Teaching*, 29(1): 119–128. http://www.isetl.org/ijtlhe/.

Steiber, A. and Alänge, S. (2015). Organizational innovation: A comprehensive model for catalyzing organizational development and change in a rapidly changing world. *Triple Helix*, 2(1). https://doi.org/10.1186/s40604-015-0021-6.

Surendran, P. (2015). *Technology Acceptance Model: A Survey of Literature*.

Syahadiyanti, L. and Subriadi, A. P. (2018). International journal of economics and financial issues diffusion of innovation theory utilization online financial transaction: Literature review. *International Journal of Economics and Financial Issues*, 8(3): 219–226. http:www.econjournals.com.

Teece, D. J. (2010). Business models, business strategy and innovation. *Long Range Planning*, 43(2-3): 172–194. https://doi.org/10.1016/j.lrp.2009.07.003.

Tolba, A. H. and Mourad, M. (2018). *Individual and Cultural Factors Affecting Diffusion of Innovation.*

Weidman, J., Myers-Lawson, C., Young-Corbett, D., Lawson, M., Fiori, C., Koebel, T. and Montague, E. (2018). *Prevention through Design: Use of the Diffusion of Innovation Model to Predict Adoption.*

Weigel, F. K., Hazen, B. T., Cegielski, C. G. and Hall, D. J. (2014). Diffusion of innovations and the theory of planned behavior in information systems research: A meta analysis. *Communications of the Association for Information Systems*, 34(1): 619–636. https://doi.org/10.17705/1cais.03431.

Widya-Hasuti, A., Mardani, A., Streimikiene, D., Sharifara, A. and Cavallaro, F. (2018). The role of process innovation between firm-specific capabilities and sustainable innovation in SMEs: Empirical evidence from Indonesia. *Sustainability (Switzerland)*, 10(7). https://doi.org/10.3390/su10072244.

Zhang, X., Yu, P., Yan, J., Ton, A. M. and Spil, I. (2015). Using diffusion of innovation theory to understand the factors impacting patient acceptance and use of consumer e-health innovations: A case study in a primary care clinic Healthcare needs and demand. *BMC Health Services Research*, 15(1). https://doi.org/10.1186/s12913–015–0726–2.

Chapter 2

Design and Innovation Performance

An Analysis of the Multifaceted Concept of Design

Carter Bloch[1] and *Carita Eklund*[1,2,*]

1. Introduction

Design is recognized to be fundamental for business performance (D'Ippolito 2014, Sanderson and Uzumeri 1995). Traditionally, design has been linked to the shaping (appearance) of end-products. However, design is now increasingly seen as an integrated part of different stages of the product development processes, from ideation to development to final design and marketing. The focus itself on product appearance can be seen as also influencing product function by changing how customers experience product use. Design can also be seen as a tool to help expand product portfolios and business approaches, for example, through the design of new services (Korper et al. 2020). Hence, the relation between design and innovation may be complex and multifaceted (Roper et al. 2016, Santamaría et al. 2009), encompassing many ways in which design can influence innovation.

Design thinking as an approach underlies many of these applications of design for innovation. Design thinking *can be defined as* "bringing designers' principles, approaches, methods, and tools to problem solving" (Brown 2008). And, while design thinking as a practice can be implemented in a number of fashions, some key elements are: gaining a deep understanding of user needs, incorporating knowledge of user needs through product development processes, and the use of multifunctional teams (Liedtka 2015).

[1] Aarhus University, Department of Political Science, Danish Centre for Studies in Research and Research Policy, Aarhus, Denmark.
[2] University of Vaasa, School of Accounting and Finance, Economics, Vaasa, Finland.
* Corresponding author: ceklund@uwasa.fi

Previous empirical work finds that design supports innovation. Czarnitzki and Thorwarth (2012) found that inhouse designers support both imitation and market novelty. Marsili and Salter (2006) showed that firms with design expenditures are more innovative than the firms without. Both articles stress design activities as one of many innovation capabilities. Furthermore, Filippetti (2011) found that design complements R&D, is more effective in dynamic firms with complex innovation strategies, and is associated with greater economic performance.

Existing quantitative analyses of how design affects innovation are however often unable to distinguish between different types of design activities, with an exception of Moultrie and Livesey (2014) and Roper et al. (2016). Most studies have either limited their focus to end-product design or use a single, broad conceptualization of design that encompasses all aspects (Filippetti 2011). Roper et al. (2016) examine different degrees of involvement of design in product development and its relation to innovation performance. To our knowledge, their research is the only one able to achieve a comprehensive quantitative picture of the role of design in innovation.

While the design notion is viewed to comprise a number of aspects, relating to appearance, perception of function, development processes, and management strategy, little systematic work has been done to examine how, or which these aspects of design, influence innovation. A related question is to what extent the integration of design across product development and other activities leads to better performance than, for example, a narrower use of design in appearance and shaping. While there is broad agreement that design can be beneficial, a strong focus on design can increase the complexity of development processes and potentially diffuse emphasis on technological development. The latter thus suggests that the role of design in relation to radical innovation may be unclear.

The purpose of this analysis is to examine the relation between the different aspects of design and innovation performance. This includes the use of design in the shaping of product innovations, the integration of design with other activities in the product development process, the use of design as a strategic management approach or to develop new business approaches and finally, the use of advanced methods to examine user needs and to integrate users in the development process.

We utilize unique data on design activities from the Danish Community Innovation Survey (CIS) for 2010 and 2012 to analyze how design relates to the propensity for innovation and innovation performance. For product innovation, we examine three levels of innovative novelty: new to the firm, the market, or the world.

We find that firms that use design systematically are much more likely to develop and implement a product innovation than firms that do not use design, both for imitation and for novel product innovations. However, there is little difference between firms that primarily use design for shaping and appearance and firms that also integrate design across development processes.

In contrast, when looking instead at the performance of these product innovations, we do not find that shaping is positively associated with innovation performance, while integrated design is positive and significant for both product innovations overall and for novel innovations. Similar results are found for the use of design to develop new business approaches. Advanced user methods are particularly important for the performance of novel product innovations.

We also find that the role of design differs for services and manufacturing. Within services, integrated design is most important for new to firm innovation, while it is most important for novel innovation in manufacturing. In addition, both new business approaches and advanced user methods appear to have a more important role for innovation performance within services than manufacturing.

The next section reviews literature on the role of design for innovation, which motivates the research questions presented in Section 3. Section 4 describes the data and method used in the analysis, while Section 5 presents the results of the analysis. The final section discusses the results of the analysis.

2. Literature Review

Design has been conceptualized in a number of ways in the literature (Creusen and Schoormans 2005, Galindo-Rueda and Millot 2015, Mortati 2015, Na et al. 2017, Norman and Verganti 2014, Walsh 1996). Design has traditionally been connected with work on the appearance of products, which can in some cases be viewed as an aesthetics communicator: it makes sense of things and creates artifacts that influence value-perceptions of products (Berkowitz 1987, Creusen and Schoormans 2005). Eisenman (2013) argues that visible design attributes are linked to the perceived functionality, and hence, value of products, with some attributes helping to explain what their products do, or to excite users and extend basic functions by highlighting their symbolic meanings (Radford and Bloch 2011, Rindova and Petkova 2007). This implies a close connection between aesthetic design and the success of technological innovations, where design can help to determine the cognitive and emotional responses to product innovations.

Design activities can also relate to business processes and management strategies. Na et al. (2017) develop the Design Innovation Spectrum concept, dividing design into the design of a product or a service, the management of design and the corporate level design thinking. Service design involves the use of design principles for service innovation and the development of new possibilities for service provision (Korper et al. 2020). The approach is related to design thinking in its use of design principles and co-creative user focus. While design-driven innovation is seen most closely in relation to the product development process, strategic and managerial aspects are also central elements. Landoni et al. (2016) highlights in particular the capability to manage design as an overall process that is closely linked with planning and business strategies.

The user is central to the integration of design in innovation processes, where observation and ethnography at the starting point can be used to gain a deep understanding of user needs and other co-creative approaches are used to involve users over the course of the design and development process (Dell'Era et al. 2020, Liedtka 2015).

The relation between innovation and design has gained attention in recent years (Roy and Riedel 1997, Utterback et al. 2006). Roy and Riedel (1997) compare successful and unsuccessful projects with survey data on over 220 design and innovation processes. They find design tasks related to features, performance, and quality relate to commercially successful projects, whereas style, price, and range tasks are related to commercially unsuccessful projects. Hernández et al. (2018) review the literature on the relation of design and innovation. Their review identifies a number of different roles that are attributed to design in innovation processes, reflecting both views of design as aesthetic characteristics and as a process or technique to develop innovations. These roles include: product differentiation, market introduction, transforming ideas into concepts, design as a creative thinking process, and as a technique to integrate concepts, people, and functions.

Norman and Verganti (2014) discuss how a user-centric approach affects the novelty and significance of innovation. Norman and Verganti (2014) attribute the biggest technology changes to technology-push innovations in which technology changes radically. Their main point is that human-centric design is unlikely to produce radical innovations.

Along similar lines, Mugge and Dahl (2013) examine the relation between design newness and innovation novelty for a sample of 130 consumers. They find that a low degree of design newness has a positive, explanatory effect on consumer evaluations of radical innovations, as it is easier for consumer to understand the novel products. However, they found no relation between design newness and consumer evaluation of incremental innovations.

Marsili and Salter (2006) examine the effect of shaping on innovation with the Dutch CIS and find a positive effect from design on the performance of new-to-the-firm and new-to-the-market products but not for improved existing products. Filippetti (2011), utilized 2009 Innobarometer data for all EU countries, and worked with a broader definition of design, which included graphic, packaging, process, product, service, and industrial design. Filippetti (2011) found that design is often not an alternative to a technology-based approach to innovation but typically a complement. While this definition of design is broader, the survey does not allow any distinction between the different types of design activities involved in innovation.

Roper et al. (2016) is the only study to our knowledge that empirically tested how different design activities affect innovation. They approached the relation between design and innovation by considering the three roles of designers: functional specialists, a bridging role, continuous involvement (Roper et al. 2016: 320). Functional specialism means that the designer is involved in one stage of product development, the bridging role has designers in diverse teams (defined as involvement in at least one of the three first stages, and at least one of the latter stages), and in continuous involvement, the designers are involved in the process from the start to the commercialization of the innovation. Roper et al. (2016) find that plants with design involvement have 9% higher share of innovative sales than innovative plants without the involvement of design staff. There was no difference in shares for functional specialists and bridging roles, while those with continuous involvement had 11% higher shares than other plants with design involvement.

The following key points can be summarized from this review of the literature on design and innovation. First, the integration of design principles in development, and also managerial processes are core elements in the broadened notion of design. Second, design also includes the creation of new business approaches, among these service design. Third, the use of advanced approaches to gain a deep understanding of user needs is an important part of many design approaches. Finally, it is not clear what role integrated design activities can have for technological development and radical innovation. In the analysis below, we seek to shed light on all these points, concerning both the use of design and how it is related to innovation performance. In particular, we will seek to measure design as shaping, the integration of design in processes, the strategic use of design, and the use of advanced methods to uncover user needs.

3. Research Questions

The purpose of this analysis is to examine the relation between the different aspects of design and innovation performance, thereby empirically testing many of the arguments presented above in the literature review. Our analysis is based on the Danish CIS with an expanded module on the design activities of Danish firms. As with Roper et al. (2016), we were able to distinguish different types of design approaches, though our design data, and hence also our analysis, is somewhat different than theirs. Whereas the design variables in Roper et al. (2016) represent designers' participation in different innovation stages, our measurement also includes the strategic use of design and the use of advanced methods to uncover user needs. Our data is not restricted to activities of designers, but concerns the nature of design activities in the firm overall. Hence, our measures may be a somewhat broader in this sense that they can relate to activities in other types of occupations.

In this analysis, we differentiate between three types of design: aesthetic design that focused on appearance (shaping), the integration of design throughout the product development process, and strategic design approaches. Hence, our first research question concerns the relation between design and innovation development.

RQ1-How is design related to the propensity to innovate?

Design can not only influence the development and introduction of innovations, but also their success, motivating our second research question:

RQ2-How is design related to innovation performance?

An additional question is whether the role of design differs across sectors, and in particular, for the manufacture of goods in comparison with the provision of services. While aspects concerning aesthetic design and appearance are most naturally connected with tangible goods, other aspects concerning the application of design principles to development processes or strategic decision making are equally relevant for both goods and services. Many of these aspects are included in the notion of service design (Korper et al. 2020). At the same time, it is important to keep in mind that most firms are active in both the production of goods and services.

RQ3-How does the role of design differ across manufacturing and services industries?

The user is a key element design approaches, where in particular, design thinking approaches involve the use of advanced observation and ethnographic methods to gain a deep understanding of user needs, and co-creative activities to incorporate the user perspective over the course of development processes (Liedtka 2015). Drawing on this, our fourth research question is:

RQ4-Is the use of advanced techniques to understand user needs linked to greater innovation performance?

Drawing on Verganti (2008), Roper et al. (2016) hypothesized that continuous design involvement should improve firms' capability of achieving radical innovation. However, opposing arguments can also be made that design detracts from technological work, thereby reducing the ability to generate radical innovation. In addition, Norman and Verganti (2014) argue that design can be viewed as the identification of market potential and usability research that they see as only likely to lead to small developments in technology, that is, only incremental innovations. Motivated by these discussions, our final research question is:

RQ5-How is design related to innovative novelty?

4. Methodology

The quantitative analysis uses firm-level data from two waves of the Danish CIS for 2010 and 2012. Each wave covers innovations and innovation activities for a three-year period, 2008–2010 and 2010–2012. These waves of the Danish CIS include an additional module covering design activities. These data offer a unique opportunity to model different design approaches and examine their relationship with innovation. The dataset comprises a representative sample of Danish firms in manufacturing and offering a broad range of services. We combine the waves to form a pooled sample with 6,906 observations and merge it with linked rich employer–employee data (IDA) for the control variables.

The standard CIS collects data on marketing innovations involving changes in aesthetic design (i.e., shaping), but not in design in general, or other forms of design activities such as design thinking or the use of design as a problem-solving approach. The module on design and innovation in the Danish CIS for 2010 and 2012 consisted of two questions: the first asks how much focus the firm has on design in operations, and the second covers specific design-based approaches. The latter include using design in product development problem-solving tools; the inclusion of designers in cross-functional teams or throughout development processes; design-based branding strategies that establish a common link between products, services, and concepts; the use of design thinking for product development strategies; and involving designers in the development of new business areas.

Table 1 describes the design questions included in the Danish CIS, while Table 2 includes descriptions for the variables used in the analysis below.

The data also includes measures of advanced approaches to capture user needs. Innovative firms are asked whether they utilize any of five approaches to understand user needs: basic customer contact, systematic methods (such as market analysis and focus groups), and three advanced methods (observation, co-creation with general users, and co-creation with lead users).

Table 1: Description of design questions and variables from the Danish CIS for 2010 and 2012.

Which statement reflects best your company's work with design?
– The company does not work with design in a systematic way
– Design is used as a final finish, forming or styling when something new is developed
– Design is an integrated – but not determining – element in the company's development work
– Design is a central, determining element in the company's development work
– Not relevant/Don't know
(If integrated or central) How does your company work with design?
– Design is used to **solve problems** related to the development of new concepts, products or services (problem solving)
– Designers are involved in **multidisciplinary teams** for the development of new concepts, products or services (teams)
– Designers are **involved from the beginning** with the development of new concepts, products or services (initiate)
– The company has a design policy that ensures visible **connection between products, services, concepts, brands, graphic design and sales venues (branding)**
– The company has a **design policy** for the development of new concepts, products or services
– Designers are involved in the **defining of new business approaches**

Source: Danish R&D and innovation questionnaire 2012, Statistics Denmark. Own translation from Danish to English.

Table 2: Variable descriptions.

New-to-firm innovation	= 1 if new-to-the-firm product innovation; 0 otherwise
New-to-market innovation	=1 if new-to-the-market product innovation; 0 otherwise
New-to-world innovation	= 1 if new-to-the-world product innovation; 0 otherwise
Share sales due to product innovations	= share of total sales from product innovations
Share sales due to novel innovations	= share of total sales from new-to-the-market or new-to-the-world product innovations
Share sales due to new to world innovations	= share of total sales from new-to-the-world product innovations
Shaping	= 1 if design as final shaping or if marketing innovation (and not integrated or central element); 0 otherwise
Integrated design	= 1 if design is (integrated or central element) and (problem solving or teams or initiate or branding); 0 otherwise
Design policy	= 1 if design policy; 0 otherwise
New business approach	= 1 if design used in development of new business approaches; 0 otherwise
Advanced user method	= 1 if at least one advanced method used to study user needs (observation or co-creation with general or lead users; 0 otherwise
R&D dummy	=1 if in-house R&D activities; 0 otherwise
Share master's degree	Share employees with master's degree or higher

Shaping is measured here as firms either using design as a last finish or having introduced a marketing innovation. Both final design and marketing changes involve specific functions, and this definition excludes all firms that use design across different elements. *Integrated design* is defined as those firms that have at least one of the first four design-related activities: problem solving, multidisciplinary teams, initiating innovations, and branding. The last two measures cover strategic, managerial activities that applies a design-based approach to different functions. *Design policy* here is measured as firms with either a design policy for concept, product or service development. *New business approaches* is the use of design in the definition of new business areas or approaches. New business can thus be related to the notion of service design.

The description of all variables used in the analysis is shown in Table 2. The analysis uses several measures of innovation and innovation performance as dependent variables. For product innovations, CIS contains data on their novelty level: new to the firm, new to the marketm or new to the world. Innovative performance is measured as the share of innovation sales in the last year of the survey period, where shares are given for each level of novelty.

Table 3 describes the data. Over the two observation years, we have 6,029 observations from the years 2010 and 2012. This number is after the exclusion of 877 observations where respondents did not answer the question on design. Some firms are included in both surveys. In the data, 28% of firms have a new-to-the-

Table 3: Descriptive statistics.

	All firms'	Innovative firms'
	mean	mean
New-to-firm innovation	0.284	0.977
New-to-market innovation	0.185	0.637
New-to-world innovation	0.088	0.301
Share sales due to new-to-firm innovation*		0.370
Share sales due to new-to-market innovation*		0.202
Share sales due to new-to-world innovation*		0.068
Shaping	0.237	0.385
Integrated design	0.201	0.370
Advanced user method*		0.446
Design policy	0.081	0.167
New business approach	0.045	0.090
Share masters	0.100	0.133
Employees	113	148
R&D (dummy)	0.249	0.576
Small (< 50 employees)	0.552	0.502
Service firm	0.629	0.515
Number of obs.	6 029	1 886

* survey question only answered by firms with an innovation.

firm product innovation, 19% have new-to-the-market, and 9% reported a new-to-the-world product innovation, where a world novelty is also a market and a firm novelty. Average sales from new products range from 37% (firm) to 20% (market) to 7% (world novelties). Table 3 presents five different measures of design activity. Around 24% use design primarily for shaping and for 20% design is integrated across development activities.

We test whether these variables explain the propensity to innovate and the share of innovative sales. Equation 1 presents the estimated equation of propensity to innovate, which is also called the innovation production function, and Equation 2 presents the estimated equation of innovation performance. Subscript n represents the novelty level of the product innovation. We have three novelty levels: new to the firm (imitation), new to the market (market novelty) and new to the world (world novelty). As control variables (X), we use the log of number employees, the share of employees with at least master's degree, export intensity, a year dummy for 2010, a dummy for R&D activity, and industry dummies. The main explanatory variables are four measures of design involvement:[1] shaping, integrated design, design policy, and new business approach. The three respective dependent variables in Equation 1 are the dummy variables for new-to-firm, new-to-market, and new-to-world innovations. Equation 1 is estimated using a probit model for each type.

$$P(inno_n) = \beta_0 + \beta_1\,SHAPING + \beta_2\,INTEGRATED + \beta_3\,DESIGN\_POLICY + \beta_4\,NEW\_BUSINESS\_APPROACH + X_i \quad (1)$$

The model of innovation performance, Equation 2, is much like Equation 1, but the dependent variables are the shares of sales from the three levels of novelty for product innovations. In addition to the four design measures listed above, we also include a measure for the use of advanced user needs approaches. In each case, the measures include the share of sales for the respective level and higher levels of novelty. The estimation is performed with a Tobit model.

$$innosales_n = \beta_0 + \beta_1\,SHAPING + \beta_2\,INTEGRATED + \beta_3\,DESIGN\_POLICY + \beta_4\,NEW\_BUSINESS\_APPROACH + \beta_5\,ADVANCED\_USER + X_i \quad (2)$$

5. Results

This section reports the results of the estimations. Table 4 presents the results for the propensity to innovate. Note that the results are reported as marginal effects. We can see that both shaping and integrated design are associated with a much higher propensity to produce a product innovation, and with similar coefficients for the two types of design roles. The results, for example, imply that firms that use design mainly for appearance (shaping) have over four times higher propensity to introduce a new to the firm innovation than firms without any systematic design activities. Coefficients for shaping and integrated design are generally declining in the novelty

[1] Data on advanced user methods only available for innovative firms.

Table 4: Design and the propensity to introduce a product innovation.

Variables	**(1) New to firm**	**(2) New to market**	**(3) New to world**
Shaping	4.768***	3.633***	2.641***
	(0.000)	(0.000)	(0.000)
Integrated design	4.090***	3.659***	2.946***
	(0.000)	(0.000)	(0.000)
y2010	1.104	1.230***	1.216*
	(0.157)	(0.007)	(0.055)
Design policy	1.801***	1.318**	1.033
	(0.000)	(0.049)	(0.848)
New business approach	1.216	1.334*	1.848***
	(0.245)	(0.086)	(0.002)
Log(emp)	1.076**	1.057*	1.009
	(0.012)	(0.080)	(0.828)
Share masters	0.729	1.862**	5.435***
	(0.191)	(0.0145)	(3.44e-08)
R&D dummy	6.845***	5.540***	6.001***
	(0.000)	(0.000)	(0.000)
Age	1.004	1.006**	1.008**
	(0.135)	(0.019)	(0.016)
Log (export/sales)	1.086**	1.127***	1.355***
	(0.027)	(0.010)	(0.000)
Constant	0.067***	0.031***	0.011***
	(0.000)	(0.000)	(0.000)
Observations	6,029	6,029	6,029
Log likelihood	–3599	–2887	–1792
Degrees of freedom	17	17	17
LR chi2	1983	1368	881.8
Prob > chi2	0.000	0.000	0.000
Pseudo R2	0.275	0.237	0.246

Marginal effects, p-values in parenthesis. *** $p < 0.01$, ** $p < 0.05$, * $p < 0.1$.

of innovations, but are still large for new to market and new to world product innovations. Hence, the results do not give any indication that focus on design comes at the cost of novel technological innovations.

The coefficient for design policy is positive and highly significant for new to firm innovation, but much less so for novel innovations. It is weakly significant for new-to-market innovation, and insignificant for new-to-world innovations. The opposite is the case for new business areas, which is only weakly significant for new market innovation but highly significant for new to world innovations. Both design policy and new business areas are associated with smaller increases in innovation propensities than for shaping and integrated design.

Tables 5 and 6 show the results for innovation performance, measured as the share of sales due to product innovations. Table 5 shows results for the full sample, while Table 6 shows results separately for manufacturing and services. Results are

Table 5: Share of sales from innovations, design. Tobit regression.

Variables	**(1) All product innovations**	**(3) Novel innovations**	**(5) New-to-world innovations**
Shaping	4.755	5.617	2.979
	(0.163)	(0.142)	(0.522)
Integrated design	9.123**	9.865**	8.187
	(0.0198)	(0.0232)	(0.113)
y2010	–6.036**	1.197	5.121
	(0.0203)	(0.677)	(0.135)
Design policy	–3.344	–9.001*	–3.232
	(0.437)	(0.0575)	(0.555)
New business approach	11.18**	9.562*	9.606
	(0.0328)	(0.0936)	(0.139)
Advanced user method	2.701	10.03***	6.127*
	(0.327)	(0.000941)	(0.0874)
Log(emp)	–2.529**	–2.368**	–3.066**
	(0.0157)	(0.0395)	(0.0295)
Share masters	7.479	45.59***	68.68***
	(0.401)	(2.53e-06)	(3.62e-10)
R&D dummy	15.29***	15.55***	21.54***
	(4.49e-07)	(3.66e-06)	(2.30e-07)
Age	–0.101	0.00217	0.0379
	(0.241)	(0.982)	(0.743)
Log (export/sales)	–0.795	0.810	10.07***
	(0.649)	(0.677)	(0.000454)
Constant	41.43***	–14.87**	–61.34***
	(0)	(0.0146)	(0)
Observations	1,753	1,753	1,753
LR chi2	–7075	–5423	–2851
Prob > chi2	0.000	0.000	0.000
Log likelihood	1735	1735	1735
df	18	18	18
Pseudo R2	0.00620	0.0112	0.0368

Robust p-values in parentheses, *** $p < 0.01$, ** $p < 0.05$, * $p < 0.1$. Novel innovations includes new to market and new to world innovative sales.

Table 6: Innovation performance explained by design position. Tobit regression.

Variables	Services			Manufacturing		
	All product innovations	**Novel innovations**	**New-to- world innovations**	**All product innovations**	**Novel innovations**	**New-to- world innovations**
Shaping	6.015	2.590	3.164	3.919	9.639*	3.241
	(0.232)	(0.639)	(0.682)	(0.397)	(0.068)	(0.562)
Integrated design	12.75**	3.229	3.125	6.467	16.64***	11.12*
	(0.035)	(0.623)	(0.727)	(0.207)	(0.004)	(0.064)
y2010	–6.645*	–0.697	2.160	–5.532	3.467	6.043
	(0.089)	(0.870)	(0.711)	(0.111)	(0.371)	(0.135)
Design policy	–1.790	–2.941	3.738	–5.066	–15.51**	–8.869
	(0.787)	(0.682)	(0.695)	(0.365)	(0.013)	(0.163)
New business approach	14.40*	12.37	7.276	3.872	4.974	11.76
	(0.054)	(0.122)	(0.491)	(0.615)	(0.558)	(0.156)
Advanced user method	3.614	10.68**	13.30**	1.605	8.580**	-0.735
	(0.385)	(0.018)	(0.032)	(0.660)	(0.0341)	(0.860)
Log(emp)	–3.464**	–3.476**	–6.108***	–0.884	–0.707	0.374
	(0.019)	(0.030)	(0.008)	(0.570)	(0.683)	(0.837)
Share masters	7.576	47.92***	75.26***	4.243	42.56**	72.69***
	(0.491)	(0.000)	(0.000)	(0.805)	(0.023)	(0.000)
R&D dummy	16.58***	17.54***	29.78***	12.29***	12.06**	12.54**
	(0.000)	(0.000)	(0.000)	(0.004)	(0.012)	(0.017)
Age	–0.024	–0.063	0.032	–0.149	0.022	0.009
	(0.877)	(0.712)	(0.895)	(0.144)	(0.842)	(0.943)

Log (export/sales)	0.430	1.833	7.770*	–3.476	6.068	14.68**
	(0.855)	(0.476)	(0.064)	(0.446)	(0.240)	(0.014)
Constant	40.86***	–9.751	–71.01***	32.42***	–31.65***	–65.21***
	(0.000)	(0.220)	(0.000)	(0.000)	(0.003)	(0.000)
Observations	900	900	900	853	853	853
LR chi2	–3517	–2738	–1337	–3548	–2681	–1503
Prob > chi2	0.000	0.000	0.000	0.000	0.000	0.000
Pseudo R2	0.0087	0.0150	0.0527	0.0035	0.0089	0.0244

Robust p-values in parentheses. *** $p < 0.01$, ** $p < 0.05$, * $p < 0.1$ Novel innovations includes new to market and new to world innovative sales.

shown for shares of sales due to: all product innovations, novel product innovations (new to market and new to world), and new-to-the world product innovations.

The models are estimated with Tobit regressions and coefficient estimates for the design roles can be seen as expected changes in the percentage share of innovative sales (where shares are defined from 0 to 100) for the respective design role, in comparison with no systematic design. For example, results for all product innovations show that firms with integrated design have an expected 9.1 percentage points higher share of innovative sales.

In contrast to results for the propensity to innovate, shaping is not found to have a significant impact on innovation performance. For new-to-firm and novel innovation performance, integrated design is significant, implying an average increase of around 9 percentage points. The integrated design coefficient for new to the world innovations is around the same size, though insignificant, indicating greater variance in the relation to sales due to the most novel innovations.

These results thus indicate that the integration of design across different product development activities is related to greater innovation performance. The coefficient for design policy is negative and insignificant or weakly significant across degrees of product novelty. New business areas are, in contrast, positive and significant for new-to-firm innovation and weakly for novel innovations. In both cases, coefficient estimates are large, indicating a 10–11 percentage point increase in the share of innovative sales. Advanced user methods are positive and significant for novel innovations, and in fact, are the only design aspect that is significant for new to the world innovations.

The R&D dummy is significant in all cases, indicating that innovation performance is much higher for R&D firms. Coefficients are also highest for new-to-the-world innovations. The share of highly educated employees is not important for innovation performance overall, but has very strong importance for innovation performance concerning novel innovations. Finally, export intensity is only significant for new-to-the-world innovation performance.

In order to examine differences in innovation performance across different sectors, Table 6 estimates the model for services and for manufacturing separately. The results in Table 6 indicate a number of differences in the role of design for services and manufacturing. We first examine services, where there appears to be a difference in the role of design for innovation performance in general compared with that for novel innovation. First, the integration of design across activities is important for overall innovation performance but not for novel innovation. The same pattern holds for new business approaches. Both of these point to the same indication that for services, the use of design across functions and development activities is important for overall innovation performance. Integrated design is associated with a 13-percentage point increase in the share of innovative sales and new business approaches with a 14-percentage point increase.

However, these factors are not associated with enhanced innovation performance for novel innovation. Instead, advanced user methods are the only design related role that is linked to improved novel innovation performance among service firms. This suggests that deep understanding of customers and their involvement in development processes is particularly important for service firms to successfully develop something novel.

Within manufacturing, integrated design is particularly important for the success of novel innovations, leading to increases of innovation shares of close to 17 percentage points for novel innovations and 11 percentage points for new to the world innovations. A number of design roles are significant when considering all novel innovation but not significant when narrowing the focus to new to the world innovations.

6. Discussion

This research examines how design roles are related to firms' innovation performance. The main contribution of this research is in examining the relation of different design roles to the propensity to innovate and to innovation performance. The analysis is conducted using unique data on the role of design from the Danish Community Innovation Survey (CIS) for 2010 and 2012.

We define shaping as a firm having design focused on aesthetic design or marketing innovation. A firm utilizes integrated design if they have integrated design activities in the innovation process and they use design in one of the fours functions: problem solving, the initiation of innovation development, designers as a part of diverse teams, or branding. Two measures of the strategic use of design are design policy or that they use design to help define new business approaches. The final design-based role is the use of advanced methods to understand user needs.

Firms that use design systematically are much more likely to develop and implement a product innovation than firms that do not use design, both for innovations that are new to the firm but adopted from others and for innovations that are new to their market or new to the world. However, there is little difference between firms that primarily use design for shaping and appearance and firms that also integrate design across development processes. Hence, going from a more standard focus of design on product appearance to the integration of design principles in processes is not associated with greater likelihood to be innovative.

However, when we look instead at the success of these product innovations, measured in terms of shares of innovative sales, the difference is very pronounced. We do not find that shaping is positively associated with innovation performance, while integrated design is positive and significant for both product innovations overall and for novel innovations (which include both new-to-market and new-to-world product innovations), associated with around a nine-percentage point increase in the shares of innovative sales. Similar results are found for the use of design to develop new business approaches. However, for both integrated design and new business approaches, the results are weaker for novel innovation and are insignificant when focusing only on new to the world innovations. The opposite is the case for the use of advanced user methods such as observation and co-creative development together with users. Advanced user methods are not significant for innovations overall, but are highly significant when focusing only on all novel innovations or new to the world innovations. These results thus suggest that an important element of design for novel innovation is gaining a deep understanding of user needs.

We also find that the role of design differs for services and manufacturing. Integrated design is most important for new-to-firm innovation within services while it is most important for novel innovation in manufacturing. In addition, both new business approaches and advanced user methods appear to have a more important role for innovation performance within services.

Roper et al. (2016) look at the benefits from design to innovation for Irish manufacturing plants, where they investigate how the degree of involvement of designers in new product development relates to innovation performance and the propensity for novel product innovation. Overall, they find that the more design is integrated across product development processes, the greater the benefits to innovation performance and the innovative novelty of innovation. While our data on innovations and innovation performance is very much similar to that in Roper et al. (2016), the data used to measure design roles is somewhat different, though there is arguably comparability between our shaping variable and functional specialism (use of designers in only one stage of the product development process), and our integrated design can essentially be seen as a combination of bridging role and continuous involvement (respectively, designer involvement in multiple or all stages of product development).

Our results for integrated design are comparable to results found for the bridging role and continuous involvement in Roper et al. (2016). Our results are less similar to theirs when comparing functional specialization with shaping. While these two measures are not fully comparable, they both represent a systematic yet diminished use of design in product development. And, in contrast to Roper et al. (2016), we do not find significant results for shaping in relation to innovation performance.

Our results thus indicate that the integration of design across the different stages of innovation processes is both associated with greater propensity to innovate and higher performance based on those innovations. The results have a number of implications, both conceptual and managerial.

First, the results provide strong arguments for a broader conceptualization of design that goes beyond aesthetic shaping to include the integration of design activities across the innovation process. The data here, as for example with the dataset used in Roper et al. (2016), allows us to unfold and analyze different aspects of design and its relation to firm performance. These results provide a strong impetus to continue work in attempting to measure different design activities and their impacts in a systematic way.

Second, both the results in general and in particular the strong positive relation found between design and novel innovations suggest a complementary relationship between technical R&D activities and design activities. There is a tendency to view R&D and design functions in an isolation from each other, and sequentially at different stages of the value chain. The results here suggest that design may play an important role in the creation of new knowledge, and enhances the impact of R&D activities on firm performance. Given that there appears to be an important interplay between R&D and design activities over the course of innovation processes, more work should be done to explore how these two activities are connected.

In relation to management, the results suggest that efforts to integrate design activities with other functions throughout the innovation process leads both to greater innovation and increased benefits of innovations for performance. However, strategic policies to incorporate design principles in development activities are not linked to increased performance, and in some cases are found to have a significant, negative relation to performance.

A strength of this study is the fairly detailed data on different forms of design activities, which allows for analysis of the integration of design across innovation processes. As we have already mentioned above, this is an area that could clearly benefit from further work, which would also require greater availability of data on design activities and innovation. Among these areas are the interplay of technical R&D activities and design and more knowledge on how design can be integrated across the innovation value chain.

There are also a number of limitations of our study. Design activities and their integration with other business functions represent very complex, and in some cases abstract, processes that are very difficult to measure. It is important to keep in mind with systematic data of this type that there are limits to how well we can measure the complex role of design in innovation. An additional limitation is the lack of data on design investments. Our data covers a number of aspects concerning design-related activities, but we do not have data on the intensity of these activities, which could be measured through design investments.

Acknowledgements

The authors wish to thank participants at DRUID18, Design scapes Policy Forum, and DRUID Academy Conference for constructive comments. A previous version of the manuscript has been published in a doctoral dissertation: Eklund 2019 *Innovation Capabilities, Design and Cutting Edge: Innovative Growth in the 21st Century*.

References

Berkowitz, M. (1987). Product shape as a design innovation strategy. *Journal of Product Innovation Management*, 4(4): 274–283.

Brown, T. (2008). Design thinking. *Harvard Business Review*, 6.

Creusen, M. and Schoormans, J. (2005). The different roles of product appearance in consumer choice. *Journal of Product Innovation Management*, 22(1): 63–81.

Czarnitzki, D. and Thorwarth, S. (2012). The contribution of in-house and external design activities to product market performance. *Journal of Product Innovation Management*, 29(5): 878–895. https://doi.org/10.1111/j.1540–5885.2012.00935.x.

D'Ippolito, B. (2014). The importance of design for firms' competitiveness: A review of the literature. *Technovation*.

Dell'Era, C., Magistretti, S., Cautela, C., Verganti, R. and Zurlo, F. (2020). Four kinds of design thinking: From ideating to making, engaging, and criticizing. *Creativity and Innovation Management*, 29(2): 324–344. https://doi.org/10.1111/caim.12353.

Eisenman, M. (2013). Understanding aesthetic innovation in the context of technological evolution. *Academy of Management Review*, 38(3): 332–351.

Filippetti, A. (2011). Innovation modes and design as a source of innovation: A firm-level analysis. *European Journal of Innovation Management*, 14(1): 5–26.

Galindo-Rueda, F. and Millot, V. (2015). *Measuring Design and its Role in Innovation*. OECD Science, Technology and Industry Working Papers, 2015/01, OECD Publishing.
Hernández, R. J., Cooper, R., Tether, B. and Murphy, E. (2018). Design, the language of innovation: A review of the design studies literature. *She Ji: The Journal of Design, Economics, and Innovation*, 4(3): 249–274.
Korper, A. K., Patrício, L., Holmlid, S. and Witell, L. (2020). Service design as an innovation approach in technology startups: A longitudinal multiple case study. *Creativity and Innovation Management*, 29(2): 303–323. https://doi.org/10.1111/caim.12383.
Landoni, P., Dell'Era, C., Ferraloro, G., Peradotto, M., Karlsson, H. and Verganti, R. (2016). Design contribution to the competitive performance of SMEs: The role of design innovation capabilities. *Creativity and Innovation Management*, 25(4): 484–499. https://doi.org/10.1111/caim.12165.
Liedtka, J. (2015). Perspective: Linking design thinking with innovation outcomes through cognitive bias reduction. *Journal of Product Innovation Management*, 32(6): 925–938.
Marsili, O. and Salter, A. (2006). The dark matter of innovation: Design and innovative performance in dutch manufacturing. *Technology Analysis and Strategic Management*, 18(5): 515–534.
Mortati, M. (2015). A framework for design innovation: Present and future discussions. *Design Issues*, 31(4): 4–16.
Moultrie, J. and Livesey, F. (2014). Measuring design investment in firms: Conceptual foundations and exploratory UK survey. *Research Policy*, 43(3): 570–587.
Mugge, R. and Dahl, D. W. (2013). Seeking the ideal level of design newness: Consumer response to radical and incremental product design. *Journal of Product Innovation Management*, 30: 34–47.
Na, J. H., Choi, Y. and Harrison, D. (2017). The design innovation spectrum: An overview of design influences on innovation for manufacturing companies. *International Journal of Design*, 11(2).
Norman, D. A. and Verganti, R. (2014). Incremental and radical innovation: Design research vs. technology and meaning change. *Design Issues*, 30(1): 78–96.
Radford, S. K. and Bloch, P. H. (2011). Linking innovation to design: Consumer responses to visual product newness. *Journal of Product Innovation Management*, 28(s1): 208–220.
Rindova, V. P. and Petkova, A. P. (2007). When is a new thing a good thing? Technological change, product form design, and perceptions of value for product innovations. *Organization Science*, 18(2): 217–232.
Roper, S., Micheli, P., Love, J. H. and Vahter, P. (2016). The roles and effectiveness of design in new product development: A study of Irish manufacturers. *Research Policy*, 45(1): 319–329.
Roy, R. and Riedel, J. C. (1997). Design and innovation in successful product competition. *Technovation*, 17(10): 537–594.
Sanderson, S. and Uzumeri, M. 1995. Managing product families: The case of the Sony Walkman. *Research Policy*, 24(5): 761–782.
Santamaría, L., Nieto, M. J. and Barge-Gil, A. (2009). Beyond formal R&D: Taking advantage of other sources of innovation in low-and medium-technology industries. *Research Policy*, 38(3), 507–517.
Utterback, J., Vedin, B.-A., Alvarez, E., Ekman, S., Walsh Sanderson, S., Tether, B. and Verganti, R. (2006). Design-inspired innovation and the design discourse. *Design-Inspired Innovation*, 154–186.
Verganti, R. (2008). Design, meanings, and radical innovation: A metamodel and a research agenda. *Journal of Product Innovation Management*, 25(5): 436–456.
Walsh, V. (1996). Design, innovation and the boundaries of the firm. *Research Policy*, 25(4): 509–529.

Chapter 3

Accelerating Farmers Move on the Irrigation Ladder

Understanding the Potential of Solar-Powered Irrigation in Nigeria

Olufolahan O. Osunmuyiwa,[1,*] *Olalekan S. Sakariyawo*[2] and *Andrew D. Peacock*[3]

1. Introduction

Imagine you are a smallholder vegetable or rice farmer living in a rural area in a low-middle-income country. It has been three months since the government's restrictions on movement. You are unable to visit the farm for work and you do not have an irrigation system installed. Instead, you rely on rainwater to provide irrigation to your crops. However, rainfall has been inconsistent. The lockdown is over, and you have little to nothing to harvest. Like the above scenario, the COVID-19 pandemic has created a huge debate on the vulnerabilities and uncertainties around global and local food systems.

Added to COVID-19, globally farmers are experiencing profound climate pressures prompting the need for farm innovations and transitions that ensure food security, reduce carbon emissions, and promote sustainable food production. In sub-Saharan Africa (SSA), where most farmers are small scale and dependent on rain, increasing climate stress such as droughts and/or late-onset of rain adversely affects crop yield and farm productivity (Lipper et al. 2018). Yet, according to the FAO, annual food production must increase by 80% to meet SSA's food demand by 2040 (FAO 2018).

[1] The Institute for Management of Innovation and Technology (IMIT) Sweden.
[2] COPLANT, Federal University of Agriculture Abeokuta.
[3] EGIS, Heriot Watt University, Edinburgh, Scotland.
Email: a.d.peacock@hw.ac.uk
* Corresponding author: olufolahan.osunmuyiwa@Chalmers.se

For larger economies like Nigeria, where the agricultural sector plays a pivotal role in the economy (accounts for more than 60% of employment and crop production represents 21.2% of the GDP), 80% of food production is primarily done by small-scale rainfed farmers who have poor access to irrigation (Netherlands Enterprise Agency 2019). Nigeria's irrigation challenges are inextricably linked to the challenges in the electricity system. Nigeria's power sector is unable to meet electricity demand with on-grid electricity consumption pegged at 126 kWh per capita (Oyewo et al. 2018). More than 55% of the population lack access to reliable grid electricity with unmet power demand addressed via crude methods such as power outages, load shedding, and the deployment of expensive backup generators. With a low electrification access rate, mostly in rural agricultural communities, mechanisation technologies required to enable crop productivity are missing (Osunmuyiwa and Kalfagianni 2017). To fill the supply gap, Nigeria spends an estimated $22 billion importing food annually (Okojie 2018). With Nigeria's projected population boom, its largely rain-fed farming system and emerging climate threats; the likelihood of food insecurity being amplified is significantly high and can derail Nigeria's targets to ending poverty, hunger, and inequality (UN Sustainable Development Goals 1, 2, and 10)—hence, the urgent need to shift to affordable irrigation systems.

The deployment of renewable technologies such as solar-powered irrigation systems (SPIS) (discussed in Section 2) has been noted as a panacea to address farmers irrigation needs as it enables timely delivery of water required for crop yield, reduces farmers dependence on rain, addresses the energy needs of the farm and mitigates farm production risks (Ali et al. 2016, Hiller et al. 2016, Niajalili et al. 2017, Rizi et al. 2019). SPIS are low carbon, modular, cost-effective, can replace fuel costs of diesel generators, and addresses the non-availability or non-reliability of grid supply in rural communities.

Based on its developmental appeal, there is a significant body of literature exploring the techno-feasibility, cost optimisation or efficiency effects of SPIS. Looking at efficiency and optimisation, Sontake and Kalamkar's (2016) review showed that in Nigeria and Algeria SPIS provided efficiency in water pumping, SPIS performance was seen as satisfactory requiring low operational costs and providing 20 m^3 water daily. Analysing the techno-economic suitability of SPIS for the cultivation of rice, wheat, and vegetables in Bangladesh, Hossain et al. (2015) observed 50% water savings for cultivated eggplants and tomatoes. However, economically, SPIS was not seen as suitable for rice due to a combination of high-water requirements and low recorded yields. Chandel et al. (2015) compared diesel/grid supply with SPIS and found that SPIS were more economically viable with a payback period of 4–6 years. Evaluating the impact of SPIS on cost savings, Treephak et al. (2015) compared the cost implications of SPIS and gasoline pumps on a 1.6 acre of rice land irrigated for twenty-five years. From their analysis, the transition from a gasoline-powered system to SPIS would provide a fuel cost savings of $280 annually.

Despite these extensive body of literature on the techno-economic operationalisations of SPIS (Closas and Rap 2017, Sarkar and Ghosh 2017, Wazed et al. 2018, Xu et al. 2013), previous research offers no empirical insights on how

SPIS become accepted or fail to embed into clear alternatives for fossil-based pumps. To date, sparse attention has been paid towards understanding factors shaping farmers' decision to adopt SPIS. Most research found have either focused on self-reported motivations for domestic use of solar or the effects of learning about solar through informational sources. Furthermore, within the Nigeria context, empirical findings on the influence of SPIS on crop production in Nigeria remains rather sparse. **To address these observed gaps, the objective of this research is two-fold: first, to demonstrate through quantitative modelling, the potential impact of SPIS on crop production in Nigeria, and second, using qualitative case study approaches, to identify factors that define farmers willingness to adopt SPIS and propel them on the irrigation ladder.** To address our research objectives, we look to Rogers' Diffusion of Innovation Theory—which has extensively dealt with how new technologies become adopted, spread in markets, or fail to embed.

Our study makes two main contributions. First, it moves beyond the current narrative of techno-feasibility of SPIS from a power system (that is, capacity of renewables) perspective that is pervasive in the literature and provides comprehensive data on the impact of SPIS on crop production. While the valuation of irrigation impacts can be done using field trials, however, this approach is time consuming, can be disrupted by environmental factors, does not provide enough longitudinal data in a short time frame and most importantly, is expensive. As such, using the Aquacrop modelling, we not only hope to scale the above problems, but also provide an adequate description of crop growth dynamics in relation to energy and water use and long-term effects and impact scenarios of emerging climate threats on crop production in our case study site. Second, using our qualitative approach, we provide insights on the wider economic and socio-cultural dimensions that drive the utilitarian value attached to SPIS. Our research captures the heterogeneous needs of farmers and how the context in which they operate shapes and underpin their willingness to pay for SPIS or capacity to shift existing farming practices to align with SPIS. Finally, from a policy and business perspective, a clear understanding of farmers' motivations to adopt SPIS enables the development of sound farmer-led policies and business models that will accelerate farmers' transition to SPIS. Section 2 explains our conceptual starting point and framework. Section 3 provides an overview of Nigeria's irrigation challenges and sets the context for our choice of methodology. Section 4 explains our methodological approach. In Section 5, we our quantitative and modelling findings. In Section 6, we present our conclusion.

2. Conceptualising the Technology Adoption and Diffusion of Solar-Powered Irrigation Systems

Within the innovation diffusion literature, considerable research has been devoted to analyzing the drivers and barriers to the adoption of new technological systems. This body of work has broadly covered studies on the adoption of new decentralised energy systems, energy efficiency actions, and urban water infrastructure (Geels et al. 2018, Hoffmann et al. 2020, Markard et al. 2020). Diffusion of innovation theory is a well-established framework by (Rogers 2003) which articulates how technologies spread, the timing of adopters, and provides a five-part taxonomy for segmenting of how quickly adoption of new technologies occur. Policy-wise, it is useful for identifying factors that shape and influence levels of innovation adoption, aid in understanding the decision-making process of adopters, and assess whether an innovation aligns and coheres with socio-cultural values of adopters. Consistent with Rogers' model, we further situate users (farmers) as central to SPIS adoption in our analysis. Numerous research within the innovation field has shown that market formation and diffusion of technologies like SPIS is impacted by social construction processes among users (Dewald and Truffer 2012). The adoption of new technologies like SPIS necessitates significant behavioural change on the part of the user, and these changes are largely determined by user expectations and experience, social contexts, and knowledge flows (Blimpo and Cosgrove-Davies 2019, Malen and Marcus 2017). Given the complexities that underpin the adoption of technologies like SPIS, adoption proponents are often advised to have a good grasp of user needs and understand the context and factors that will shape users' decision making about accessing and utilizing proposed technologies (Frederiks et al. 2015, Hirmer and Guthrie 2016). Thus, this research investigates and adapts three of the five identified elements by Rogers, 1996, 2003) that shape the diffusion and spread of new innovations: **(1) knowledge of and function of SPIS innovation's advantage over existing practices; (2) compatibility of SPIS innovation with prior economic valuations, experiences, and routinisation; and (3) complexity, perceived usefulness, and ease-of-use of SPIS innovation.** We decided to exclude the fourth and fifth factors (observability and trialability) as the project was conducted during the 2021 COVID-19 lockdown and only had one day of in-farm demonstrations of SPIS. Below we expand on the three elements.

Knowledge of and function of SPIS innovation's advantage over existing practices: Dealing with new technology requires adaptability and as such, how users accumulate knowledge about new technologies and cognitively identify characteristics that they find desirable affects which technologies are privileged. The role of pre-existing preferences, plans, and needs further alter how the information received is visualised and interpreted (Redmond 2003). This new information stimulates the visualisation of an "end in view," that is fuelled by a pragmatic economic rationalisation and utility maximisation. Some users may further rely on experiences gained by pioneering users or industry actors they trust as their observation of this new technology reduces uncertainty around the technical, economic, and social compatibility of the technology. Such referential knowledge may result in patterns of neighbourhood or cluster diffusion (Dewald and Truffer 2012). However, in cases where there are information asymmetries between actors involved in the technology markets (financiers, suppliers, or agri-field agents) about the economic benefits of the technology to the users or the risks involved; this increases the users' risk aversion towards the technology (Blasch et al. 2017).

Compatibility of SPIS innovation with prior economic valuations, experiences, and routinisation: Studies evaluating the adoption of socially desirable technologies show that cost, income, and economic valuations play a huge role in users' willingness to pay for technologies (Frederiks et al. 2015). Such economic valuations are typically two-pronged, the cost of technology adoption, and the cost of operations (van der Kroon et al. 2014). While the immediate cost of purchase, might not deter the potential user, yet income flow and predictability might pose a threat to the adoption of the technology especially when such uptake requires a recurrent payment from an individual or household with an irregular income (Blimpo and Cosgrove-Davies 2019). For instance, Lee et al. (2016) assessment of users' willingness to pay for grid access across villages in Kenya shows that high poverty prevalence and liquidity constraints posed significant barriers to householders' willingness to connect to electricity. Based on these liquidity constraints, grid developers often design solutions that mimic current cost patterns for candles, for example, pay-as-you-go or as the case of irrigation those that mimic diesel/petrol costs (Grimm et al. 2019). New technologies like SPIS become embedded when users consider them crucial to provide functional and financial advantages over dominant technologies and services (Urmee and Md 2016).

Complexity, perceived usefulness, and ease-of-use of SPIS innovation: How users interpret the use conditions of new technologies affects the shift between technology adoption and behavioural adjustments (Dewald and Truffer 2012). This relates to the user's perceived long-term value of the technology and how they weigh the suitability of such technologies with existing irrigation artefacts. Suitability can also be viewed in terms of the instrumental value of the technology. In the case of SPIS, this can be the ease of integration with other farm technologies, forms of economic savings from the replacement of fossil pumps with SPIS, the time benefits and convenience offered by a transition to SPIS (ability to remotely automate SPIS), a surge in income from an increased acreage, or increased irrigation processes due to higher power capacity (Hirmer and Guthrie 2016). Users' interpretation of technology usefulness can also be influenced by their experience of and concern about emerging trends like climate change. For example (Waibel et al. 2018), found that high concern about climate change and the perceived impact of climate change by farmers on crop production led to higher investment in irrigation by over one-fifth of farming households in the Dak Lak province of Vietnam. However, as noted by Zilberman et al. (2012) the adoption of SPIS technology as an adaptation to climate change will be largely dependent on farmers' access to credit, the underlying agro-ecology and supporting institutions that can promote or hinder adoption.

3. Case Context-Nigeria

Nigeria has one of the largest potentials for irrigation development in Africa at an estimated 3.2 million ha. It has an estimated 220,000 ha under irrigation utilising boreholes and petrol diesel pumps (Vermillion 2004). Statistics suggest that of the c.156 km^3 of water available on an annual basis from surface and groundwater resources, only 8 km^3 is currently exploited (FAO 2016). Therefore, only 1% of the cultivated area is currently irrigated (FAO 2016), leaves farmers vulnerable to weather variability resulting in regular shortfalls in production. The absence of irrigation not only increases production risks but also hinders farmers' capacity to transition from subsistence to commercial farming. Furthermore, recent changes in climatic conditions such as the late onset of rainfall and droughts have begun to adversely affect crop production with severe implications for food security (FEWS NET-USAID 2018). If such climate vulnerability is not addressed and farmers continue to rely on rainfed production, Nigeria will not be able to meet its food needs.

Different developmental actors have begun to support the wider adoption of irrigation in Nigeria. For instance, in 2014, the World Bank under the Transformation Irrigation Management in Nigeria (TRIMING) project, provided USD 495.3 million to support 140,000 farmers uptake of irrigation, enhance the efficiency of existing irrigation infrastructure, rehabilitate, and modernize irrigation schemes while mobilising private sector investment (World Bank 2014). Similarly, the government has begun to pursue several policies such as the Master Plan for Irrigation and Dam Development, a draft Of National Irrigation Policy and Water Resources Infrastructure Operation and Maintenance Policy—all to encourage the development of small- to large-scale public and private irrigation systems (Adelodun and Choi 2018, FAO 2016). These policies are anchored on the use of fossil fuel-based infrastructure (grid, diesel, or petrol) that are both environmentally unsustainable and economically illiterate, resulting in farmers having to bear large capital, operational, and maintenance costs. Besides the capital and operational costs, maintenance, and replacement costs of petrol pumps are generally 2–4 times more than SPIS. Thus, rather than lifting farmers out of poverty, such systems might further entrench poverty. This reinforces the urgency of investigating how affordable, scalable, climate-smart irrigation systems required to address farmers and agri-ecosystem needs and meet increasing food demand can be deployed.

3.1 Case study communities-Ogun

Ogun State was chosen as a case study site based on discussions with institutional collaborators working on climate impacts on agriculture in Nigeria. Ogun State is situated in South-West Nigeria with an estimated population of 5.2 million people. The state has a tropical climate of two wet and dry seasons. The dry season starts from November to March. Annual rainfall value ranges between 1,400 and 1,500 mm with a relatively high temperature of an average of 32°C.

Ogun state is split into four agricultural zones (Abeokuta, Ijebu, Ikenne, and Ilaro) by the Ogun State Agricultural Development Programme (OGADEP) for ease of implementing agricultural programmes and administrative purposes. For our study, we focused on two (Abeokuta and Ilaro) out of the four agricultural zones. Specifically, our quantitative and qualitative research study was conducted across five local government areas, namely Odeda, Ayetoro, Ifo, Egbado, and Yewa North. Numerous studies have shown that climatic parameters such as rainfall patters, and minimum and maximum temperatures are the vital climatic elements required for agricultural production. In recent years, Ogun State has begun to experience the late onset of rainfall, early cessation of annual rainfall, and frequent flooding. This has led to alterations in growing seasons with significant impacts on yield outputs. Maize is one of the main crops produced in Ogun State. It is typically prevalent among small-scale farmers as this is a household staple and farmers are guaranteed a stable market to sell their produce. Policy-wise, maize has become one of the major crops and value chains supported under the Ogun State Agricultural Linkage Programme. Yet, maize cultivation remains largely rain-fed in Ogun State. Emerging climate conditions highlighted above can potentially worsen the supply and demand of maize products. In the next section, we explore the potential impact of solar-powered irrigation systems in addressing these climate impacts.

3.2 Understanding the potential impact of solar-powered irrigation systems in Nigeria

The deployment of renewable technologies such as solar-powered irrigation systems (SPIS) has been noted as a panacea to address farmers' irrigation needs. SPIS has the potential to eliminate rain dependency in the timely delivery of water required for crop growth whilst addressing the energy needs of the farm, thus mitigating farm production risks (Ali et al 2016, Hiller et al. 2016, Niajalili et al. 2017, Rizi et al. 2019). SPIS are low carbon, modular, cost-effective, can replace fuel costs of diesel generators and address the non-availability or non-reliability of grid supply in rural communities.

The potential for SPIS in Nigeria is very high due to the country's rich solar irradiation potentials. Estimates indicate that the deployment of solar photovoltaic (PV) modules to cover 1% of Nigerian land area would generate 207 TWh annually, seven times the amount of electricity generated in 2014 (Oyewo et al. 2018). Operationally, and per the International Renewable Energy Agency (IRENA) estimates, Nigeria cumulatively installed 28 MW of solar capacity in 2019. Policy-wise, the government considers solar PV as one of the pathways to meet its increasing energy demands and electrification goals—as such, government policies and institutional financing are currently designed to boost domestic utilisation and growth of solar. For instance, Nigeria's Renewable Energy Master plan targets 500 MW of installed solar PV capacity by 2025. Furthermore, there is a proliferation of solar micro/mini grids at the household and community levels to power small appliances, markets, rural communities, and companies. According to the Efficiency For Access Coalition (2019), Nigeria has become one of the fastest growing markets for total volume for solar products sold globally. In 2019 alone, off-grid solar systems (50–100 W output) sales in Nigeria doubled in value from \$23 million in 2018 to \$63 million in 2019 and related solar products had an import value estimated to be \$150 m. Annually, Nigeria's off-grid market is estimated to yield \$10 billion in revenue and \$6 billion in savings for Nigerian homes and businesses (Netherlands Enterprise Agency 2021).

Based on Nigeria's solar growth trajectory, there is a huge opportunity for SPIS to develop alongside mini and micro grid development in the Nigerian energy market and drive Nigeria's irrigation revolution. The recent decline in the cost of PV and batteries has made PV more competitive with diesel. Yet for small-scale farmers, the high upfront installation cost continues to deter potential adoption of solar. For context, the average cost of installing a 4-kW solar PV system including costs for storage for domestic use is estimated at $9,090 (Netherlands Enterprise Agency 2021). Strikingly, a look at the solar/SPIS supply chain in Nigeria shows why cost remains a barrier for the adoption of solar/SPIS among low-income earners (see Figure 1). There is an absence of commercial-scale solar manufacturing, and as such, most solar products found in the domestic market are imported. While there is a big distribution market for solar panels for domestic energy use, yet SPIS-related sales and distribution remain sparse except for one or two solar companies.

We argue that outside of cost implications, knowledge about SPIS in general and its potential impact plays a huge role in farmers' willingness to adopt SPIS. Most small-scale farmers still have limited knowledge of and access to solar pumps. A review of irrigation equipment used among small-scale farmers in Nigeria shows that farmers who practice irrigation often rely on diesel-powered water pumps. More importantly, there remains a lack of quantifiable evidence on the potential impact of SPIS on farm yield, water use, energy costs, and farm productivity in Nigeria. We argue that the availability of such data may enhance farmers willingness to adopt SPIS, and enable the development of services, market structures, and policies relevant for the farming policy ecosystem. Thus, using a quantitative modelling tool (Aquacrop) and case study approach, we explore the potential impact of SPIS on crop production in Ogun State, identify and evaluate the potential factors that may influence farmers move to SPIS.

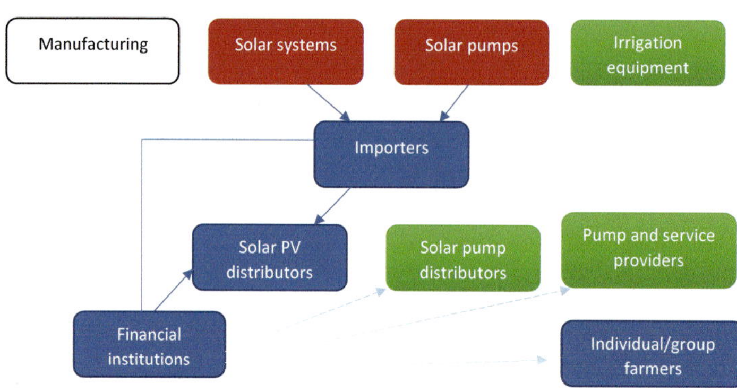

The red boxes represent the non-existent parts of the value chain; the green boxes represent the sparsely existent part of the value chain; the blue boxes represent the well-established part of the value chain.

Figure 1: Solar-powered irrigation supply-chain in Nigeria.

4. Methodological Approach

Typically, the analysis of the factors underpinning users' adoption of SPIS has been largely based on qualitative research. However, quantitative models hold a dominant place in policymaking around sustainability issues and shape decisions on a range of vital issues such as electrification, renewable generation, and long-term agri-energy planning. Therefore, this research combines both qualitative and quantitative modelling approaches to understand the impact of SPIS on crop production and identify factors driving farmers' willingness to adopt SPIS. An in-depth needs analysis of this nature is best done at the local level, as findings from such exercise can help policymakers design case-specific financial and policy interventions that can be targeted to boost cropping outputs.

4.1 Modelling

The technical potential of SPIS on the yield of Maize crops being grown in the Ogun State province in South West Nigeria was evaluated using the FAO Aquacrop model. Aquacrop is a crop growth model that is used to simulate yield response to water of herbaceous crops (see Raes et al. 2009, Steduto et al. 2009 for a comprehensive overview of the model approach and computational structure). The core of the growth engine is described in Equation 1 where B is aboveground biomass production, WP^* is crop water productivity normalized for evapotranspiration, and T_r is crop transpiration (mm). The yield is subsequently derived from aboveground biomass production using a crop-specific harvest index.

$$B = WP^* \sum T_r \qquad (1)$$

When well-calibrated for a crop, the model allows simulation of a wide range of field management practices, including the irrigation approach. Crop input parameters were procured from a calibrated Aquacrop model for a maize crop growing in Zaria, in the Guinea Savannah agro-ecological zone in Nigeria (Oiganji et al. 2016). The soil texture was assumed to be sandy loam (Epuh et al. 2019) and the soil fertility was assumed to be non-limiting with optimal weeding practices. Daily weather data for the period 1991 to 2020 was obtained from the NASA Langley Research Centre (LaRC) POWER Project funded through the NASA Earth Science/Applied Science Program (NASA 2021) using the latitude 7.606 and longitude 2.996 as a proxy location for Ogun State. The daily weather data was used to compute daily reference evapotranspiration (ET_0) using the ASCE Penman-Monteith equation (Walter et al. 2000). The required daily irrigation volume was derived from ET_0 using a single crop coefficient approach with FAO crop coefficients for maize (Allen et al. 1998). To assess the technical potential of SPIS, it was assumed that daily irrigation requirement was always met. A maize cropping period of 90 days was assumed with sowing date triggered by the date after March 1st in the candidate year when 5% of cumulative rainfall for the period March to June had occurred. This process is akin to methods described by Adefisan and Abatan (2015) for estimating a rainfall onset date for specific seasons in different agro-climatic zones of Nigeria. For each year rainfed yield and irrigated yield was calculated as dry weight (tonnes per hectare).

4.2 Qualitative data/capacity-building workshop

The capacity-building workshop served several purposes. First, it was deployed as a tool to educate farmers about irrigation in general and demonstrate the efficacy of

SPIS more specifically. Second, it was designed to support farmers' knowledge of the market structures and supporting instruments around SPIS. Third, to generate and gain knowledge on farmers willingness to pay for SPIS and finally to co-design potential ownership, management, and governance structures of SPIS that are acceptable to farmers in Nigeria. As with the methodology of workshops, we began with discussions around preliminary findings from the surveys as well as our climate modelling with the hope that participants might collaborate in the co-generation of knowledge around barriers to the adoption of SPIS and co-produce potential ownership and service structures around SPIS. The two-day event was held at the Federal University of Agriculture Abeokuta in February 2021 with n = 135 people from different aspects of the SPIS and farming value-chain: (i) farmers (n = 123 from four different local government areas of Ogun State) (ii) irrigation suppliers (n = 4), Agric extension workers (n = 2), and academia (n = 6). Participants were invited based on their prior participation in the survey conducted at the start of the study. The reason for selecting the same set of farmers twice was to gain indepth qualitative information about their long-term experiences with changing climate and how they have adapted to this new reality. The exercise also served as an opportunity to continue the knowledge dissemination around SPIS which began during the survey collection. In the second part of the workshop, participants were asked to focus on current and future challenges around the adoption of SPIS. Two questions guided this phase of the workshop: What is the impact of climate change on small-scale farmers and what scale of knowledge do farmers have about this challenge? Where do farmers belong on the irrigation transition ladder and what kind of irrigation technologies, knowledge, and system specifications do farmers require to accelerate their growth on the transition ladder? The question on climate change was included to understand whether participants had observed changes in climate in the last ten years and identify the impact of this observation on their views around the compatibility of SPIS with this new reality. Participants were asked to discuss the impact of the observed climate changes on farming in their region (e.g., crop failure or lower yield) and provide measures taken to adapt to these new climate realities (e.g., invest more in irrigation). The climate question served as a basis to subsequently discuss farmers position on the irrigation ladder and the potential for SPIS. The exercise combined plenary, participatory, and on-site knowledge demonstrations in which farmers were invited and prompted to share thoughts and experiences around solar irrigation pumps and associated smart technologies that can enable farm efficiency. The workshop lasted for seven hours each day. It was conducted in the Yoruba language. Discussions were recorded with consent from all participants. Key themes from the discussion are analysed in the results section. In sum, the questionnaire, modelling, and the indepth qualitative workshop methods complemented each other. The questionnaires provided a macro-overview of farmers' irrigation challenges and knowledge of SPIS, the qualitative workshop made it possible to investigate indepth the context and factors serving as barriers to SPIS as shown by the questionnaires, while the modelling addresses farmers key concerns about the usability and impact of SPIS thus providing a clear picture of the potential pathway of the transition to SPIS.

5. Results and Discussions

The results of the modelling and qualitative findings are presented separately below. In the first part, we engage with findings from our crop modelling. In the second, we present findings from the workshop and discuss issues raised by farmers as critical barriers and enablers to the adoption of SPIS.

5.1 The modelled potential impact of spis on yield output in Ogun-state Nigeria

We used a crop water-production model (Aquacrop), calibrated for maize cropping in Nigeria and applied it to a daily weather dataset that encompasses the period 1990–2020 (see Figure 2). This allowed the impact of weather variability (e.g., warming periods, late onset of rainfall, early cessation of annual rainfall, and frequent flooding) during a specific cropping season (March–July) on crop production to be estimated. This longitudinal assessment permitted a more robust estimate to be made of the potential uplift associated with the transition from rainfed to irrigated production, underpinned by the adoption of SPIS. From 1990 to 2020, the Aquacrop model determined that there were five years where rainfall levels were insufficient to support a crop (Figure 2). In each case, this occurred where a short period of rain in early March, sufficient to trigger the start of the cropping period was followed by a period of drought. If the crop had been sown at the end of March, sufficient rainfall occurred to support the crop over the next ninety days. Of the twenty-five years where a crop was produced, the yield ranged from 0.26 T/Ha (2000) to 3.45 T/Ha (1997). The average across all 30 years was 1.84 T/Ha, this in comparison to reported average maize yields for the period 1991–2020 of 1.52 T/Ha for Nigeria (FAO 2021) and 1.40 T/Ha for the period 1996–2017 for Ogun State (Oluwatimilehin and Ayanlade 2021). The provision of SPIS creates yield potential for the region of 3.79 T/Ha (average value for period 19902020), a twofold increase in output.

The modelled yield was found to be a non-linear function of irrigation applied. When expressed as a percentage of crop evapotranspiration (ET_c) for the candidate

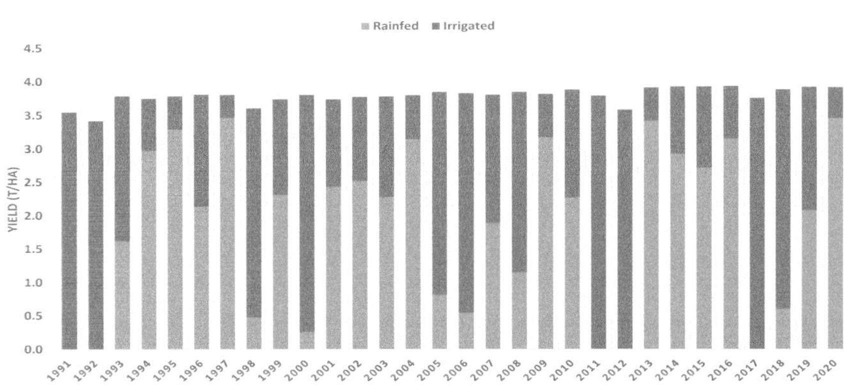

Figure 2: Rainfed yield versus irrigated yield modelled by Aquacrop for the period 1991–2020.

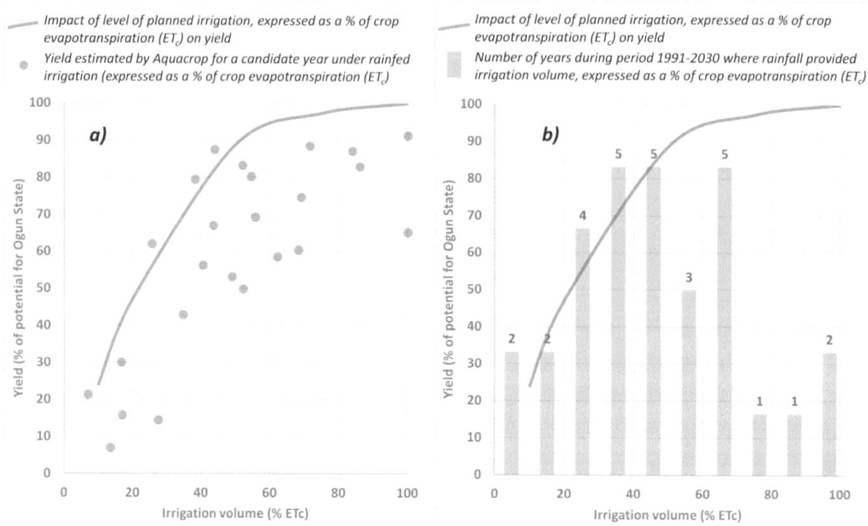

Figure 3: Yield as a percentage of potential versus irrigation as a percentage of crop evapotranspiration and the percentage of ETc achieved by rainfall during the period 1991 to 2020; a) each year and b) years grouped by percentage of ETc.

year, a clear relationship can be found, with decreasing impact on yield achieved when irrigation exceeded 50% of ET_c (Figure 3). Considering only rainfall, the output from the modelling study can be seen to approximate this relationship, with both lower yield values and scatter function of the timing of the rain events (Figure 3a). When the complete thirty-year period is considered, there were eighteen years when the volume of rainfall was below 50% ETc, leaving twelve years where it exceeded 50% (Figure 3b).

The increased yield attributable to the application of irrigation to supplement rainfall impairs the water use efficiency (WUE) of maize production. With rainfed maize production, the average WUE of resultant production was 903 m^3/T to produce the average yield of 1.84 T/Ha compared to 1248 m^3/T to produce 3.79 T/Ha under irrigated production. The impact of any transition toward irrigated agriculture in support of increasing food production and farmers' incomes must be cognisant of the potential impact on water resources to ensure that the solution is sustainable. Rather than optimizing irrigation to achieve maximum yield, improved WUE can be achieved by applying only a proportion of the calculated evapotranspiration with only a small impact on maize production. This would also have an impact on SPIS system costs, with lower irrigation volumes resulting in lower pump run times and subsequent PV size. The improved performance indicted with the single crop cycle modelled here emphasises the impact of the SPIS technology in increasing the resilience of rainy season cropping, effectively eliminating the influence of the sowing date on production. Performance uplift will also be derived from extending crop production into dry seasons.

5.2 Qualitative results

This section presents the profile of farmers in our capacity workshop and findings from the workshop on the factors shaping farmers preferences and willingness to adopt solar-powered irrigation systems.

5.2.1 Farmer/user profile

Participant farmers varied in age, with 60% of farmers between the ages of 33–58. A larger proportion of the participants (74.1%) were male while 25.9% were female. The average plot of land owned and farmed by participants was 2.58 plots. Income wise, 62.4% of farmers earned between USD 10 and USD 600 per annum while only 8.3% of farmers earned over USD2000 in the previous year. Averagely, 50.2% of farmers had 20 years experience in farming and indicated that they engaged in other-agricultural activities that include direct selling in the market (60%), livestock (44.9%), farm labour (12.2%), and artisan work (11.2%). Only a small percentage were solely focused on farming. Crops planted by the sampled farmers were maize (93.7%), cassava (91.7%), and vegetables (59.00%). The majority of the farmers (94.1%) signified that farming was both for household consumption and sales purposes while only 4.9% planted for sales only.

To characterize the scale of irrigation access, the types of technologies used as well as the barriers to the adoption of SPIS, farmers were asked to provide data on the current irrigation technologies used. Most farmers (99%) relied on rain as the primary source of irrigation. Of these 99%, 41% combined rain with watering cans to support irrigation while 22% relied on a combination of surface irrigation and rain. Only 1.95% of farmers relied on sprinklers for irrigation. The overt reliance on rain mirrors the national average of less than 2% irrigation (described in Section 1). Gauging farmers perception and observations around climate change in the last ten years, 82.4% of all sampled farmers reportedly experienced droughts or absence of rain. Among the farmers that experienced droughts, 47.3% described their experience in the previous year as severe while 28.8% viewed their experience of droughts as mildly severe. When asked about access to innovative weather tools, 74.1% of farmers indicated that access to real-time data on weather changes was difficult to get (see Figure 4).

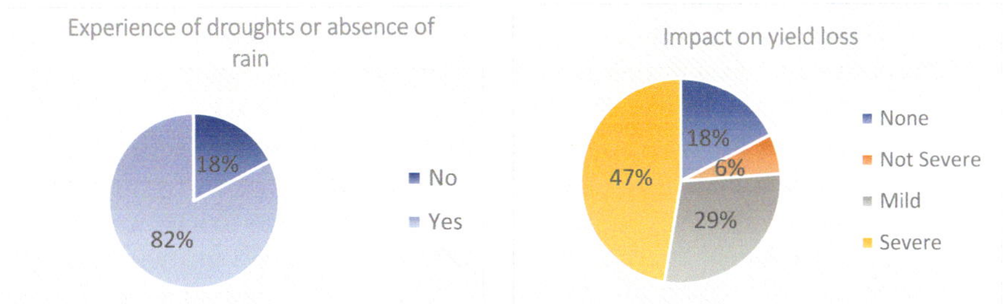

Figure 4: Farmers' experience and analysis of the severity of droughts.

The frequency of irrigation can be a big indicator of farmers willingness to transition to the irrigation ladder. 40.5% of participant farmers irrigated at an average of twice daily, 2% irrigated three times a day, 14.1% irrigated once daily, while 39% of farmers indicated that irrigation was not a daily practice. For most farmers, the mean cost of irrigation was USD100 per season. When asked about irrigation preferences, 99% of farmers preferred to move to some form of automated or mechanised type of irrigation. Farmers were asked about access to irrigation technologies and 84.4% of the respondents found it difficult to access available irrigation options or markets. This mirrors the (see Figure 1) characterization of the irrigation market in Nigeria, where the irrigation manufacturing supply chain is virtually nonexistent. Acquiring and using irrigation pumps (grid-tied) requires access to electricity and enables the meaningful application or use of innovations like precision irrigation. When asked about access to electricity, 91.2% of farmers found it difficult to access electricity even though most farmers indicated that they could afford to pay for this service.

5.2.2 Evaluating factors shaping farmers preferences and willingness to pay for SPIS

Based on our conceptual framework and interactions with farmers, these factors were found to have influenced farmers willingness to adopt SPIS: (1) knowledge of and function of SPIS innovations gained by interacting with other farmers; (2) compatibility of SPIS innovation with prior economic valuations, experiences, and routinisation; and (3) Perceived usefulness and compatibility of SPIS due to observed climate change by farmers.

Knowledge of and function of SPIS innovations gained by interacting with other farmers: Knowledge plays a crucial role in how farmers identify the desirable characteristics of SPIS and how they become embedded into farming practices. When asked about prior knowledge of SPIS, only 24% of farmers knew about solar irrigation pumps or precision-based irrigation systems. Most farmers acknowledged that despite extension service outreaches around diesel-based pumps, there has been no knowledge dissemination or access to regular training on the adoption of solar-powered irrigation technologies to amplify farm productivity. This knowledge gap can perhaps be traced to the fact that key solar irrigation supply chain market actors: manufacturers, wholesalers, retailers, consultants, and irrigators that can facilitate the uptake of SPIS (Figure 1) are currently missing. The absence of in-country solar pump manufacturers removes the possibility of a local champion. Also, while the potential market is large, SPIS-related sales and distribution networks are sparse with only one or two participating companies. As such, despite the benefits of SPIS, most small-scale farmers still have limited knowledge of and access to solar pumps.

For participants who were aware of SPIS, this came because of their exposure to and interactions with other farmers under the National Fadama irrigation scheme (farmers supported by the government). These interactions led to a referential knowledge of irrigation, its potential impacts on crop yield and willingness to adopt SPIS. As described by one participant: "*Our counterparts under the Federal Fadama irrigation scheme got higher yield in comparison to what we had in the previous season. As for us, unless we go to the river which can sometimes be miles away, there is no way to irrigate our farms in the dry season.*" In this quote, the participant explicitly stated their observation of the yield growth among the Fadama irrigated farms. The discussion also signalled the absence of water infrastructure and suggest an implicit awareness of the impact this has on farmers' capacity to integrate irrigation technologies on their farms. Except for a few rice farmers who were close to rivers, most participants did not have access to open water sources or tube wells in their immediate surroundings. As described by one participant, "*We need water for everything. There is nothing a farmer wants to do that he does not require water, which is why we hope this capacity building can help us address our water issues. We also hope the government can take note of the current changes in the climate and support us with water infrastructure*." Participants in this case attached production function and yield valuation to the availability of water supply. Such high valuations lend themselves to farmers' willingness to adopt SPIS as they viewed the use of SPIS as an opportunity to focus on a diverse range of high-value crops. The above quotes ties with Hirmer and Cruickshank (2014) arguments that user investment in new technologies can be facilitated by their exposure to synthesised information or demonstrations that help the user shift from abstraction to familiarisation. In the case of SPIS for our participants, such information entailed, for instance, the impact of SPIS on local water impacts, compatibility with existing irrigation infrastructure, or farming practices.

Compatibility of SPIS innovation with prior economic valuations, experiences, and routinisation: As noted by Blasch et al. (2017), users often decide on technology adoption by evaluating the lifetime cost, accounting for the initial purchase price, net present value (economic, social, or environmental) of the technology, and future spending on operations. As observed in our study, critical to participants' willingness to transition to SPIS was the cost of purchase and the economic value of SPIS. For most participants, the seasonality of harvests means that a one-off payment for irrigation technologies is highly unfeasible. As described by one participant: "*Our income is not stable because we do not farm year-round. It will be difficult to pay for a solar pump without some form of instalment system*." Furthermore, challenges around income flow and predictability were highlighted by most participants. The recent farmer/herder crisis in the area was identified as a threat to income flow, liquidity and, more importantly, a core barrier to the installation of irrigation systems. As described by one participant and corroborated by others: "*Herders are migrating into our farmlands and destroying crops and water sources. If we invest in an expensive irrigation system, how would we protect it from these invasions*?" For these farmers, if unable to farm due to fears of herder invasion, then the adoption of irrigation technologies to enhance productivity becomes redundant. Yet, most participants concluded that the security barriers can be scaled once irrigation technologies are modular enough to be moved around. Furthermore, farmers were willing to embrace pay-as-you-go business/cost models that mimicked diesel cost options.

Perceived usefulness and compatibility of SPIS due to observed climate change by farmers: Users' interpretation of SPIS compatibility was influenced by their experience of and concern about emerging climate trends. In their description of the observed temperature changes during cropping seasons, two temperature observations were highlighted: (i) late onset of rain at the beginning of the planting season and (ii) intense rainfall at the end of the cropping season. When asked about the reference point for both observations, participants cited two main knowledge sources. First, farmers compared their yield outputs in the last twenty years and acknowledged that their crops performed better in the first ten years than the last ten years due to timely rainfall. Second, some farmers had been given mini-weather stations by the state's Ministry of Agriculture. These farmers were trained on how to use the equipment to read temperature and share the information with their counterparts. As explained by one of the participants: "*We were given mini-weather stations because we had a micro-climate in our area and could not rely on the weather station in the bigger town. Yet, I do not think the weather station works as there were so many instances where we were predicted to have rain and it did not happen. Essentially all the weather data provided failed us.*" Those who got weather information from the extension workers also complained that the information provided was typically inaccurate as this was usually generated from local weather stations that were far from farmers. For most farmers, the possibility to achieve two annual cropping cycles and enhance cropping intensity has diminished. When presented with the opportunity to shift to SPIS, farmers who had experienced severe damages due to the late onset of rain in the previous cropping season showed more willingness to sign-up for SPIS. Their prior experience of climate-related shocks made the adoption of irrigation more urgent for them than their counterparts. As described by one of the participants: "*Everyone here can testify to it that we lost a lot of money last season due to late rains. We have begun to observe the same pattern this year. Any technology that would help us address these potential losses would be a huge blessing.*"

6. Conclusion

The Nigerian farming sector is subject to increasing disruption in production due to climate uncertainty. The sector is dominated by smallholders whose economic profile often means that the impact of crop failure can be catastrophic. There is an urgent need to transition farming production towards climate-smart and sustainable innovations to mitigate these production risks, whose instance is forecasted to increase. A fundamental element of this strategy is the adoption of irrigation technologies such as SPIS, allowing farmers to move towards irrigated methods of practice. For this study, we first established the local performance benefit made possible by SPIS adoption. Modelling work undertaken in this study indicates that the technology can potentially double yields by increasing the resilience of rainy season cropping. Additional production increases are also possible by allowing cropping to be extended into the dry seasons. Whilst essential, this quantified potential can only be one input to a transition narrative that places the new technology within existing and discrete farming practices. Additional inputs that focus on user (i.e., farmer) interactions with SPIS and how these might shape adoptions patterns are also necessary. As such using participatory workshop, we were able to generate user preferences. Workshop outputs indicated that farmers expressed a lack of knowledge about where to purchase or how

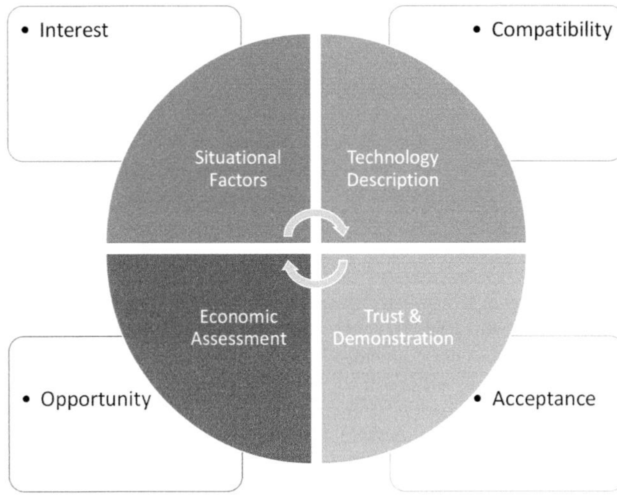

Figure 5: Workshop outputs.

to deploy and operate SPIS with this compounded by opacity in supporting business models. These factors created key barriers that affected technology acceptance and need to be addressed to allow acceleration of uptake. The contours of an engagement process can perhaps be viewed when the output of the workshops is studied (Figure 5). Situational factors can create initial interest in the technology, for instance where farmers had direct experience of crop loss because of changes in rainfall patterns. Technology compatibility requires appropriate technological descriptions to allow the farmers to place the technology within their existing artefacts and practices. Whether the technology represents a real opportunity for the farmer can only be addressed by economic assessment, revealing the upside, costs, and acquisition model. This information then has to be provided and demonstrated through trusted organisations, reflecting that existing farming practices may be discrete and have been embedded over more than one generation.

References

Adefisan, E. A. and Abatan, A. A. (2015). Agroclimatic zonning of Nigeria based on rainfall characteristics and index of drought proneness. *Journal of Environment and Earth Science*, 5(12): 115–128.

Adelodun, B. and Choi, K. -S. (2018). A review of the evaluation of irrigation practice in Nigeria: Past, present and future prospects. *African Journal of Agricultural Research*, 13(40): 2087–2097. https://doi.org/10.5897/AJAR2018.13403.

Ali, A., Bahadur D. Rahut and Behera, B. (2016). Factors influencing farmers' adoption of energy-based water pumps and impacts on crop productivity and household income in Pakistan. *Renewable and Sustainable Energy Reviews*, 54: 48–57. https://doi.org/10.1016/j.rser.2015.09.073.

Allen, R. G., Pereira, L. S., Raes, D. and Smith, M. (1998). *Crop Evapotranspiration-Guidelines for Computing Crop Water Requirements* (No. 56).

Blasch, J., Boogen, N., Filippini, M. and Kumar, N. (2017). Explaining electricity demand and the role of energy and investment literacy on end-use efficiency of Swiss households. *Energy Economics*, 68: 89–102. https://doi.org/10.1016/j.eneco.2017.12.004.

Blimpo, M. P. and Cosgrove-Davies, M. (2019). *Electricity Access in Sub-Saharan Africa Uptake, Reliability, and Complementary Factors for Economic Impact* (Africa Dev). World Bank. https://doi.org/10.1596/978–1–4648–1361–0.

Chandel, S. S., Nagaraju Naik, M. and Chandel, R. (2015). Review of solar photovoltaic water pumping system technology for irrigation and community drinking water supplies. *Renewable and Sustainable Energy Reviews*, 49: 1084–1099. https://doi.org/10.1016/j.rser.2015.04.083.

Closas, A. and Rap, E. (2017). Solar-based groundwater pumping for irrigation: Sustainability, policies, and limitations. *Energy Policy*, 104: 33–37. https://doi.org/10.1016/J.ENPOL.2017.01.035.

Dewald, U. and Truffer, B. (2012). The local sources of market formation: Explaining regional growth differentials in german photovoltaic markets. *European Planning Studies*, 20(3): 397–420. https://doi.org/10.1080/09654313.2012.651803.

Efficiency For Access Coalition. (2019). *The State of the Off-Grid Appliance Market.*

Epuh, E., Jimoh, N. O., Orji, M. J. and Daramola, O. E. (2019). Application of remote sensing, GIS and hydrogeophysics to groundwater exploration in Ogun State: A case study of OGD-sparklight estate. *Nigerian Journal of Environmental Sciences and Technology*, 3(2): 370–385.

Fao. (2016). *National Investment Profile Water for Agriculture and Energy-nigeria.*

Fao. (2018). *The Future of Food and Agriculture: Alternative Pathways to 2050* (vol. 148).

Fews net-usaid. (2018). *Revised Livelihoods Zone Map and Descriptions for Nigeria a Report of the Famine Early Warning Systems Network (fews net).*

Frederiks, E. R., Stenner, K. and Hobman, E. V. (2015). Household energy use: Applying behavioural economics to understand consumer decision-making and behaviour. *Renewable and Sustainable Energy Reviews*, 41: 1385–1394. https://doi.org/10.1016/j.rser.2014.09.026.

Geels, F. W., Schwanen, T., Sorrell, S., Jenkins, K. and Sovacool, B. K. (2018). Reducing energy demand through low carbon innovation: A sociotechnical transitions perspective and thirteen research debates. *Energy Research and Social Science*, 40(June 2017): 23–35. https://doi.org/10.1016/j.erss.2017.11.003.

Grimm, M., Lenz, L., Peters, J. and Sievert, M. (2019). Demand for off-grid solar electricity—Experimental evidence from rwanda. *Ssrn*, 1–36. https://doi.org/10.2139/ssrn.2921081.

Hiller, M., Zahner, A., Priesemann, C., Magrath, J., Freling, R., Vinci, S., Parajuli, B., Komor, P. and Lavergne, M. (2016). Solar pumping for irrigation: Improving livelihoods and sustainability. In *Oxfam Arunabha Ghosh and Shalu Agrawal CEEWWorld Bank* (Issue June).

Hirmer, S. and Cruickshank, H. (2014). The user-value of rural electrification: An analysis and adoption of existing models and theories. *Renewable and Sustainable Energy Reviews*, 34: 145–154. https://doi.org/10.1016/j.rser.2014.03.005.

Hirmer, S. and Guthrie, P. (2016). Identifying the needs of communities in rural Uganda: A method for determining the 'user-perceived value' of rural electrification initiatives. *Renewable and Sustainable Energy Reviews*, 66: 476–486. https://doi.org/10.1016/j.rser.2016.08.037.

Hoffmann, S., Feldmann, U., Bach, P. M., Binz, C., Farrelly, M., Frantzeskaki, N., Hiessl, H., Inauen, J., Larsen, T. A., Lienert, J., Rg Londong, J., Lü, C., Maurer, M., Mitchell, C., Morgenroth, E., Nelson, K. L., Scholten, L., Truffer, B. and Udert, K. M. (2020). A research agenda for the future of urban water management: Exploring the potential of nongrid, small-grid, and hybrid solutions. *Environ. Sci. Technol.*, 54: 5312–5322. https://doi.org/10.1021/acs.est.9b05222.

Hossain, M. A., Hassan, M. S., Mottalib, M. A. and Hossain, M. (2015). Feasibility of solar pump for sustainable irrigation in Bangladesh. *International Journal of Energy and Environmental Engineering*. https://doi.org/10.1007/s40095–015–0162–4.

Lee, K., Brewer, E., Christiano, C., Meyo, F., Miguel, E., Podolsky, M., Rosa, J. and Wolfram, C. (2016). Electrification for "under Grid" households in Rural Kenya. *Development Engineering*, 1. https://doi.org/10.1016/j.deveng.2015.12.001.

Lipper, Leslie, McCarthy, N., Zilberman, D., Asfaw, S. and Branca, G. (2018). *Climate Smart Agriculture Building Resilience to Climate Change* (Natural Re). Springer Nature. http://www.springer.com/series/6360.

Malen, J. and Marcus, A. A. (2017). Promoting clean energy technology entrepreneurship: The role of external context. *Energy Policy*, 102: 7–15. https://doi.org/10.1016/j.enpol.2016.11.045.

Markard, J., Geels, F. W. and Raven, R. (2020). Challenges in the acceleration of sustainability transition. *Challenges in the Acceleration of Sustainability Transition*. https://doi.org/10.1088/1748–9326/ab9468.
NASA. (2021). *Nasa Power | Prediction Of Worldwide Energy Resources*. https://power.larc.nasa.gov/.
Netherlands Enterprise Agency. (2019). *The Vegetable and Potato Sector Nigeria*.
Netherlands Enterprise Agency. (2021). *Solar Report Nigeria*.
Niajalili, M., Mayeli, P., Naghashzadegan, M. and Poshtiri, A. H. (2017). Techno-economic feasibility of off-grid solar irrigation for a rice paddy in Guilan province in Iran: A case study. *Solar Energy*, 150: 546–557. https://doi.org/10.1016/j.solener.2017.05.012.
Oiganji, E., Igbadun, H. E., Mudiare, O. J. and Oyebode, M. A. (2016). Calibrating and validating AquaCrop model for maize crop in northern zone of Nigeria. *Agricultural Engineering International: CIGR Journal*, 18(3): 1–13.
Okojie, J. (2018). *N10bn Lost Over Non-implementation of Tomato Policy - Businessday NG*. Business Day. https://businessday.ng/agriculture/article/n10bn-lost-non-implementation-tomato-policy/.
Oluwatimilehin, I. A. and Ayanlade, A. (2021). Agricultural community-based impact assessment and farmers' perception of climate change in selected Ecological Zones in Nigeria. *Agriculture & Food Security*, 10(1): 1–17.
Osunmuyiwa, O. and Kalfagianni, A. (2017). The Oil Climax: Can Nigeria's fuel subsidy reforms propel energy transitions? *Energy Research and Social Science*, 27: 96–105. https://doi.org/10.1016/j.erss.2017.03.003.
Oyewo, A. S., Aghahosseini, A., Bogdanov, D. and Breyer, C. (2018). Pathways to a fully sustainable electricity supply for Nigeria in the mid-term future. *Energy Conversion and Management*, 178: 44–64. https://doi.org/10.1016/J.ENCONMAN.2018.10.036.
Parvaresh Rizi, A., Ashrafzadeh, A. and Ramezani, A. (2019). A financial comparative study of solar and regular irrigation pumps: Case studies in eastern and southern Iran. *Renewable Energy*, 138: 1096–1103. https://doi.org/10.1016/j.renene.2019.02.026.
Raes, D., Steduto, P., Hsiao, T. C. and Fereres, E. (2009). Aquacrop-The FAO crop model to simulate yield response to water: II. main algorithms and software description. *Agronomy Journal*, 101(3): 438–447. https://doi.org/10.2134/AGRONJ2008.0140S.
Redmond, W. H. (2003). Innovation, diffusion, and institutional change. *Journal of Economic Issues*, 37(3): 665–679. https://doi.org/10.1080/00213624.2003.11506608.
Rogers, E. (2003). *Diffusion of Innovations*. Free Press.
Sarkar, M. N. I. and Ghosh, H. R. (2017). Techno-economic analysis and challenges of solar powered pumps dissemination in Bangladesh. *Sustainable Energy Technologies and Assessments*, 20: 33–46. https://doi.org/10.1016/j.seta.2017.02.013.
Sontake, V. C. and Kalamkar, V. R. (2016). Solar photovoltaic water pumping system—A comprehensive review. *Renewable and Sustainable Energy Reviews*, 1038–1067. https://doi.org/10.1016/j.rser.2016.01.021.
Steduto, P., Hsiao, T. C., Raes, D. and Fereres, E. (2009). Aquacrop-the FAO crop model to simulate yield response to water: I. Concepts and underlying principles. *Agronomy Journal*, 101(3): 426–437. https://doi.org/10.2134/AGRONJ2008.0139S.
Treephak, K., Thongpron, J., Somsak, D., Saelao, J. and Patcharaprakiti, N. (2015). An economic evaluation comparison of solar water pumping system with engine pumping system for rice cultivation. *Japanese Journal of Applied Physics*, 54(8): 08KH01. https://doi.org/10.7567/JJAP.54.08KH01/XML.
Urmee, T. and Md, A. (2016). Social, cultural and political dimensions of off-grid renewable energy programs in developing countries. *Renewable Energy*, 93: 159–167. https://doi.org/10.1016/j.renene.2016.02.040.
Van Der Kroon, B., Brouwer, R. and Van Beukering, P. J. H. (2014). The impact of the household decision environment on fuel choice behavior. *Energy Economics*, 44: 236–247. https://doi.org/10.1016/J.ENECO.2014.04.008.
Vermillion, D. (2004). *Developing Water Users Associations for Small Dams and Fadama Areas in Nigeria. Report to the Nigeria Country Programme of the FAO*.
Waibel, H., Pahlisch, T. H. and Völker, M. (2018). Farmers' perceptions of and adaptations to climate change in southeast asia: The case study from Thailand and Vietnam. *In*: Lipper, L., McCarthy, N., Zilberman, D., Asfaw, S. and Branca, G. (eds.). Climate Sm. Natural Resource Management and Policy, vol 52. Springer.

Walter, I. A., Allen, R. G., Elliott, R., Jensen, M. E., Itenfisu, D., Mecham, B., Howell, T. A., Snyder, R., Brown, P., Echings, S. and Spofford, T. (2000). ASCE's standardized reference evapotranspiration equation. *Watershed Management and Operations Management*, 1–11.

Wazed, S. M., Hughes, B. R., O'Connor, D. and Kaiser Calautit, J. (2018). A review of sustainable solar irrigation systems for Sub-Saharan Africa. *Renewable and Sustainable Energy Reviews*, 81(June 2017): 1206–1225. https://doi.org/10.1016/j.rser.2017.08.039.

World Bank. (2014). *Nigeria*. http://data.worldbank.org/country/nigeria#cp_fin.

Xu, H., Liu, J., Qin, D., Gao, X. and Yan, J. (2013). Feasibility analysis of solar irrigation system for pastures conservation in a demonstration area in Inner Mongolia. *Applied Energy*, 112: 697–702. https://doi.org/10.1016/J.Apenergy.2013.01.011.

Zilberman, D., Zhao, J. and Heiman, A. (2012). *Adoption Versus Adaptation, with Emphasis on Climate Change*. https://doi.org/10.1146/annurev-resource-083110–115954.

Energy

Chapter 4

Energy Utilization and Output Dynamics in the Middle East and North Africa Countries

Is the Export- and Globalization-Led Hypothesis Valid?

Stephen Taiwo Onifade,[1] *Festus Victor Bekun*[2,4] and *Andrew Adewale Alola*[3,4,*]

1. Introduction

The Middle East and North Africa (MENA) region consist of several countries in a vast area of land spreading across the northern part of the continent of Africa up to the Middle East region in Asia. Over the years, trading activities have notably grown in this region, especially the trade of raw primary commodities such as crude petroleum as many of the countries in this region are vastly endowed with oil and gas reserves. As of 2019, countries in the Middle East region alone accounts for about 48.08% of the global proved crude oil reserves, and the MENA region altogether account for about 45.39% of total global crude oil export (BP 2020). This abundant raw commodity endowment partially explains the trend of the growing volume of trade that was witnessed in the region as production in oil and gas industry does not only stimulate larger exports but also helps in footing large import bills through

[1] School of Finance and Accounting, Department of Economics, University of Vaasa, FI-65200 Vaasa, Finland.
[2] International Logistics and Transportation, Istanbul Gelisim University, Istanbul, Turkey.
[3] CREDS-Centre for Research on Digitalization and Sustainability, Inland Norway University of Applied Sciences, 2418, Elverum, Norway.
[4] Adnan Kassar School of Business, Lebanese American University, Beirut, Lebanon.
Emails: stephentaiwo.onifade@gmail.com; fbekun@gelisim.edu.tr
* Corresponding author: andrew.alola@hotmail.com

abundant foreign reserves from oil revenues. As such, trade dynamics are expected to have created attendant effects on the general economic performances of the MENA countries over the years.

In the era that is characterized with the global drive for sustainable economic growth amidst the global goals vis-à-vis the Sustainable Development Goals (SDGs) of the United Nations Development Programme (UNDP 2015), growth dimensions have consistently been amplified. While the account of export in trade especially for energy commodities might be significant for the case of the MENA countries, generally, other factors are potential source of economic expansion. For instance, the early work of Kraft and Kraft (1978) has been largely complimented, thus reaffirming the aspects of energy development (Khalid et al. 2018, Alola et al. 2020, Doğan et al. 2021, Nwani 2021). In addition, the new global order of economic competition amidst the SDGs (SDG-1, SDG-8, and SDG-10) for economic prosperity and reduction of inequality, is recalibrating the economic aspects of trade globalization (Brown and Lauder 1996, Joshua et al. 2020). With a broader economic intuition, the augmentation of the growth model (Solow 1956) has provided more relevant aspects of economic development.

While the global concerns for a more sustainable environment continues to rise, a review of the export-led growth hypothesis (ELGH) vis-à-vis the growing energy demand and rapid urbanization would be beneficial for both the economic and environmental sustainability of the countries in this part of the world. Firstly, because of the roles that hydrocarbon-based activities play in the overall economic performance of the region, and secondly, to address the growing demands for environmental sustainability even in the era of trade globalization. Hence, this study focuses on empirically exploring the subject matter of the ELGH while considering the possible implications of energy dynamics for both the economic and environmental sustainability of selected MENA countries including Algeria, Egypt, Iran, Qatar, Saudi Arabia, Kuwait, Oman, and Morocco. The study addresses the salient issues of sustainable development from the perspectives of trade globalization and energy consumption amidst rapid urbanization in the region which has often been overlooked in the extant studies on the ELGH for the area. The novelty of the study is both from the variable augmentation (by employing export in trade and trade globalization) and model specification (the utilization of both cointegration and quantile regression).

The entire study has been structured into four sections. Section one (1) contains the introduction, Section two (2) discusses the extant studies, Section three (3) provides detailed information on the empirical methodology, data sources, and the discussions of results, while Section four (4) concludes the study with policy recommendations.

2. Literature Review

The debate on economic growth catalyst has been a topical issue in the growth literature. This has led to lots of growth models like the classical growth model built on the Cob-Douglas production function (Douglas 1976), Solow growth model (Solow 1956), and more. Additional growth models have been recorded in the growth literature like the export-led growth hypothesis (ELGH); this hypothesis highlights the nexus between export and economic growth. The export-induced economic growth is well documented in development and international economic literature. The baseline assumption of neoclassical growth theory advances that export is integral and pertinent to economic productivity which by extension increases global market more competitive. However, the export and economic growth in inconclusive in the extant literature.

The flourishing literature on ELGH varies based on the sample period, data, and econometrics procedure used. Empirically in Bangladesh, Love and Chandra (2005) explored the validity of the ELGH over the period 1972–2000. The study showed a uni-directional causality running from economic growth to export. Thus, it failed to give credence to the ELGH in Bangladesh over the sample period. More recently, Marwan et al. (2013) did not find support for a causality connection between exports and economic growth for Sudan's economy. Similarly, Gokmenoglu et al. (2015) explored the case of Costa Rica using Johansen co-integration and Granger Causality test as estimation techniques to investigate the long-run relationship and causality between exports and economic growth for Costa Rica. Their empirical results fail to find support for ELGH. Instead, the study found a economic growth-export relationship as reported by the Granger causality analysis over annual frequency period from 1980–2013. The study of Quaicoe et al. (2017) using Vector error correction model (VECM) join the strands of literature that fail to find validity for the ELGH for the Ghanaian experience. On the other hand, there exist studies that find support for the validity of the ELGH. Balassa (1978) investigated the nexus between economic growth and exports for eleven selected developing economies. The study findings submit that exports drive economic growth thereby affirming the validity of the ELGH for the sampled blocs. Tyler (1981) examined ELGH for fifty-five countries. The study found a positive relationship between export and economic growth over the sampled period. More studies have found validation for the ELGH (see Wacziarg (2000), Mamun and Nath (2005), Herrerias and Orts (2010)).

There is a flourishing study on the nexus between impact of export diversification, energy consumption on economic growth for different regions, and different techniques to explore the effect of other macroeconomic indicators on economic growth. For instance, Herzer and Nowak-Lehmann (2006) investigated the impact of export diversification in a production function framework for an annual frequency period from 1962–2001 in Chile. The study shows that export is pertinent to the economic growth of the Chilean economy. More studies in the energy literature have augmented the production function with foreign trade volume to explore the effect of foreign trade on economic growth. Examples of such studies are Shahbaz et al. (2013) using autoregressive distributed lag methodology investigated the long run and causality relationship between financial development, foreign trade, and capita in China's economy over 1971–2011. The study demonstrated a long-run relationship between the outlined variables. While the long-run regression shows a positive correlation between foreign trade and energy consumption on economic growth, in the case of China. Another study for Australia conducted by Rahman and Mamun (2016) investigated the long-run and causality relationship between energy use and international trade on GDP growth. The study submits that there exists a long-run bound between the outlined variable. On the other hand, the causality analysis reveals a feedback casualty relationship between trade and economic growth. In a panel of fifteen countries, Usman et al. (2020) examined the impacts of financial development, disaggregated energy consumption (renewable and non-renewable energy usage), and trade openness on economic growth. The study found that both renewable energy utilization and trade openness significantly influences GDP growth. The outcome of Le (2020) for forty-six Emerging Market and Developing Economies shows that financial development energy use and trade openness positively and significantly affect the GDP growth while the study also validates the energy induces growth hypothesis where energy is a key indicator for GDP growth over investigated period. The study of Zafar et al. (2019) is in line with Le (2020) however for the case of Asia-Pacific Economic Cooperation (APEC) countries.

Given the trajectory of literature highlighted, looking from another perspective, studies of the nexus between export and economic growth while accounting for trade and financial development on economic growth is well established in the literature. However, an investigation into ELGH while addressing the role of urbanization, trade, globalization, and primary energy consumption on economic growth for selected Middle East and North Africa (MENA) countries is limited. To this end, the present study distinct from the previous literature on the following fronts: (i) This study incorporates primary energy consumption, globalization, urbanization, and trade openness in the ELGH framework for selected Middle East and North Africa (MENA) countries which have received less documentation in the extant literature. The variables highlighted in this study aligns with the United Nations Sustainable Development Goals (SDGs) target to be achieved by 2030 which has received fewer entries in the literature. Thus, this study seeks to fill this void. (ii) The scope of this study is timely for selected Middle East and North Africa (MENA) countries, which will serve as policy documented for other emerging economies. (iii) This study provides a battery of econometrics contribution by the adoption of quantile regression that accounts for different tails of the data in conjunction with Augmented Mean Group (AMG), Common Correlated Effects Mean Group (CCEMG), and Dumitrescu and Hurlin for direction of causality analysis among the outlined variables. Our study circumvents cross-sectional dependency analysis by utilizing second-generational estimators, thus providing robust coefficients for ample policy construction.

3. Data and Methodology

The data set consists of a panel of observations on all variables of interest for the included countries in the study from 1990 spanning to 2018. The main explanation and justification for the sample range and the number of countries included in the study is based on data availability. The level of availability of reliable data points is still very low in many of the MENA countries as majority of these countries are still developing. Besides, the developmental process that could pave way for reliable and continuous data collection in some countries in the region are often characterized by various setbacks including political instability, corruption, various humanitarian challenges, conflicts, and displacements of the people among other issues. As such, the study is limited to eight selected countries including, Algeria, Egypt, Iran, Qatar, Saudi Arabia, Kuwait, Oman, and Morocco. Data points were sourced from the World Bank Development Indicator WDI (2020), the Globalization Index of the KOF Swiss Economic Institute (Gygli et al. 2019), and British Petroleum Statistical Review of World Energy (BP 2020). Equation (1) was set up to explore the nexus between globalization in trade, energy consumption, and the economic performance of the selected countries:

$$LnY_{it} = \alpha + \beta_1 LnREXP_{it} + \beta_2 LnTiG_{it} + \beta_3 LnECN_{it} + \beta_4 LnUrB_{it} + \mu_{it} \quad (1)$$

From Equation (1), the variable LnY represents the real gross domestic product (GDP) per capita (current \$) for country i in time t while the variable $LnREXP$ denotes the real export of goods and services. The ratio of the sum of import and export of a country measured as a percentage of the country's GDP has been used in extant studies to proxy for trade openness and in some cases, it is simultaneously used to measure the impacts of globalization (Bal et al. 2016, Joshua et al. 2020). However, rather than using this variable to capture the effect of globalization in trade, the KOF index for trade globalization ($LnTiG$) was adopted for the study. The index provides a broader approach to measuring globalization in trade by accounting for not just the export components of trade as commonly used in the ELGH, or the traditional trade openness variable alone, but the index also accounts for other crucial factors that impact international trade dynamics such as trade regulations, taxes, tariffs, and agreements. As such, the $LnTiG$ variable provide necessary control for important trade impacting factors within the context of trade globalization. The KOF index has been applied in empirical study especially for the OECD countries and the BRICS economies (Lee et al. 2017, Haseeb et al. 2019, Saint Akadiri et al. 2020). The variable $LnECN$ denotes primary energy consumption per capita (Gigajoule per capita) and it was used to capture the impact of energy consumption in each country while μ_{it} is the error term for the model given estimations from the observations from each country. Also, we incorporated the variable $LnUrB$ (urban population measures as percentage of total population) into the model to capture the possible effects of urbanization among the countries as this variable has been strongly debated and associated with energy dynamics in contemporary studies (Shahbaz et al. 2019, Hassan et al. 2019, Onifade et al. 2021, Shahbaz et al. 2021). Lastly, the model was defined in terms of per capita because the countries under investigations vary in different regards when considering their population size, the size of individual country's labor force, and the GDP size among other issues (Yavari and Mohseni 2012, Luo 2020).

3.1 Summary statistics and correlation analysis

The correlation analysis amongst the four variables was performed and the obtained correlation coefficients are provided in Table 1.

The simple correlation results from Table 1 shows that there is a weak correlation between per capital income and the trade components while both energy consumption

Table 1: Summary statistics and correlation matrix.

	LnY	LnREXP	LnTiG	LnECN	LnUrB
Mean	2.083001	8.758959	1.728305	2.065120	1.843739
Maximum	3.240322	10.11450	1.899490	3.057744	2.000000
Minimum	1.164281	7.635893	1.343358	1.075986	1.630000
Std. Dev.	0.474586	0.450564	0.142830	0.554114	0.118202
Skewness	–0.140240	–0.021450	–0.882963	–0.001618	–0.333959
Observations	232	232	232	232	232
Correlation Matrix					
	LnY	**LnREXP**	**LnTiG**	**LnECN**	**LnUrB**
LnY	1				
p-value	–				
LnREXP	0.2790***	1			
p-value	(0.0000)	–			
LnTiG	0.4558***	–0.1652**	1		
p-value	(0.0000)	(0.0117)	–		
LnLnECN	0.8508***	0.0610	0.5457***	1	
p-value	(0.0000)	(0.3543)	(0.0000)	–	
LnUrB	0.7205***	-0.0612	0.5319***	0.8899***	1
p-value	(0.0000)	(0.3529)	(0.0000)	(0.0000)	–

Source: Author's computation. Note: ***, ** and * are 1%, 5% and 10% significant level, respectively.

and urbanization have strong correlation with income level. Given that there is substantial level of economic proximity among the countries in the panel study considering that economic activities are predominantly driven by oil and gas sector in the region, it is highly imperative to account for possible cross-section dependency among variables of interest. This would help in providing a better insight into the choice of appropriate unit-root and cointegration tests accordingly. Besides, the significance of the CD test has been emphasized in the literature (De Hoyos and Sarafidis 2006, Eberhardt and Bond 2009, Dong et al. 2018). Hence, the results of three cross-sectional dependency tests were reported for the study. The outputs of Pesaran (2015) LM Test, Pesaran (2007) CD Test, and the estimate from Breusch and Pagan (1980) LM Test, have been reported in Table 2.

Table 2: Cross-sectional dependency (CD) test.

Model	Pesaran (2007) CD Test	Pesaran (2015) LM Test	Breusch and Pagan (1980) LM Test
LnY – f (LnREXP, LnTiG, LnECN, LnUrB)	10.40***	–2.73***	70.25***
p-value	(0.0000)	(0.0060)	(0.0000)

Source: Author's computation. Where ***, ** and * are 1%, 5%, and 10% significant level, respectively.

From the results in Table 2, given the statistical significance of the various test statistics, we reject the null hypothesis of no cross-sectional dependence and therefore conclude that there is cross-sectional dependence in the panel study. As a result, the study adopted panel unit root tests that can conveniently accommodate the challenges of presence of CD in the panel observations.

3.2 Panel unit root test

The need to conduct a unit root test while analyzing economic relationships with time series data has been highlighted in the empirical literature (Katsiampa 2017, Onifade et al. 2020a, Taiwo et al. 2020). Having confirmed cross-section dependency in Table 2, applying the traditional panel unit root tests like the Fisher Augmented Dickey-Fuller (Dickey and Fuller 1979), Philips and Perron (1988), and Im et al. (2003) could create spurious results as regarding the actual unit-root properties of variables and ultimately lead to a flaw in the corresponding cointegration relationship among the variables in the panel study. Hence, the Pesaran (2007) panel IPS and CIPS tests that can accommodate unit root test in the presence of cross-sectional dependency were applied for the unit root analysis. The tests were carried out on all the variables at the level and the first difference, using the model that allows for individual intercepts and trend and the results are provided in Table 3.

The unit root test results show that all variables are non-stationary at the level. However, all variables were stationary at the first difference and therefore integrated order of one I(1). This result justifies the adoption of the cointegration techniques for the study in line with empirical studies (Adedoyin et al. 2020, Shahbaz et al. 2018a, Shahbaz et al. 2018b).

Table 3: Panel unit root test.

VARIABLES	CIPS		IPS	
	Level	First Difference	Level	First Difference
	C&T	C&T	C&T	C&T
LnY	–2.620	–4.802***	–2.0713	–5.0224***
LnREXP	–2.725	–4.292***	–1.9812	–5.0725***
LnTiG	–2.467	–5.104***	–1.9206	–5.1054***
LnECN	–2.168	–5.061***	–2.0016*	–5.8335***
LnUrB	–0.750	–3.316***	–1.8605	–3.2968***

Source: Author's computation. Variables are in natural logarithms and the unit root tests were reported alongside the subscripts *, **, *** representing the statistical significance of the outputs at 10%, 5%, and 1% significance levels respectively. C&T represent model with both constant and trend. We report the t-bar stat for the IPS.

3.3 Panel cointegration test

When variables are non-stationary in the level and stationary in the first difference, we can apply the Johansen co-integration test to ascertain the long-run relationship between the variables (Shrestha and Bhatta 2018, Çoban et al. 2020). However, caution must be exercise in choice of approach since not all panel cointegration techniques can conveniently accommodate the presence of cross-sectional dependency. Hence, the study utilizes Westerlund (2007) technique for examining the cointegration relationship among the variables. Given panel variables Y and X following the expression in Equation 2, the Westerlund (2007) cointegration technique is based on the error correction mechanism (ECM).

$$\Delta Y_{it} = \pi_i d_i + \theta_i (Y_{it-1} + Y_i^* X_{it-1}) + \sum_{j=1}^{m} \theta_{ij} \Delta Y_{it-j} + \sum_{j=0}^{m} \delta_{ij} \Delta X_{it-j} + \varepsilon_{it} \quad (2)$$

where, $\pi_i^* = (\pi_{1i}, \pi_{2i})^*$, denotes the vector of parameters in the model. The $d_t = (1 - t)^*$, denotes the deterministic mechanisms, θ_i while represents the error correction parameter. The technique produces four major statistics based on the least square estimation and corresponding significance of the adjustment term θ_i of the ECM model in Equation 2 to identify the existence of cointegration. The Westerlund (2007) cointegration technique produces both group statistics (Gτ and Gα) and the panel statistics (Pτ and Pα) as shown in Equations 3 and 4, respectively.

$$G\tau = \frac{1}{N} \sum_{i=1}^{N} \frac{{}^\wedge \alpha i}{SE({}^\wedge \alpha i)} \qquad G\alpha = \frac{1}{N} \sum_{i=1}^{N} \frac{T {}^\wedge \alpha i}{{}^\wedge \alpha i(1)} \quad (3)$$

From Equation 3, $^\wedge \alpha\, i$ is represented by $SE(^\wedge \alpha\, i)$ as the standard error while the semiparametric kernel approach of $\alpha\, i\,(1)$ follows $^\wedge\, \alpha\, i\,(1)$.

$$P\tau = \frac{{}^\wedge \alpha i}{SE(\alpha i)} \qquad P\alpha = T\alpha^\wedge \quad (4)$$

From Equation 4, the panel mean estimations show that the entire panel is co-integrated. Following contemporary studies (Alola et al. 2020, Shahbaz et al. 2018a, b), we reported the estimates for the Westerlund (2007) cointegration technique as shown in Table 4.

From Table 4, the Westerlund (2007) panel cointegration techniques report a combination of four test statistics. According to the results, both group statistics (Gτ and Gα) and the panel statistics (Pτ and Pα) have P-values that are less than 5% to reject the null hypothesis that there is no co-integration relationship among the variables. Hence, the confirmation of a long-run relationship between the variables.

Table 4: Panel cointegration test (Westerlund 2007).

Statistics	Value	Robust P-value
Gτ	–2.521***	(0.000)
Gα	–7.330***	(0.000)
Pτ	–5.413***	(0.000)
Pα	–8.921***	(0.000)

Source: Author's computation. The subscripts *, **, *** represent the statistical significance of the outputs at 10%, 5%, and 1% significance levels, respectively.

3.4 Panel regressions approach

Having established the long-run relationship between the variables, we proceed to panel regression estimations. The Common Correlated Effects Mean Group (CCEMG) estimator, the Augmented Mean Group (AMG), and the Quantile regression (QR) techniques were adopted for the study as these techniques are among the second-generation regression approaches that have been noted to be capable of producing robust estimate for models with cross-sectional dependence (Mensi et al. 2014, Halliru et al. 2020, Moreno-Izquierdo et al. 2020). The CCEMG panel estimator of Kapetanios et al. (2011) based on the original approach of Pesaran (2006) was applied with regards to Equation 5 while the AMG estimator of Eberhardt and Bond (2009) and Eberhardt and Teal (2010) follows the Equation 6.

$$Y_{it} = \alpha_i + \beta_i X_{it} + \gamma_i Y^*_{it} + \delta_i X^*_{it} + \theta_i UCF_t + \mu_{it} \tag{5}$$

$$\Delta Y_{it} = \alpha_i + \beta_i \Delta X_{it} + \sum_{t=1}^{T} \pi_t D_t + \varphi_i UCF_t + \mu_{it} \tag{6}$$

From Equation 5, the Y^* and X^* denote the average values of the panel variables Y and X while the *UCF* variable denotes the unobserved common effects. Subsequently, from Equation 6, given that Δ and D are the difference operator and a dummy component, respectively, the corresponding AMG operator given N number of countries follows the expression in Equation 7.

$$\text{AMG} = \frac{1}{N}\sum_{i=1}^{N} \varphi_i \tag{7}$$

The country-specific estimates of coefficients are denoted by φ_i in Equation 7. Having taken into cognizance the outputs from the initial tests, in addition to the CCEMG and the AMG, the quantile regression (QR) as developed by Koenker and Bassett (1978) was also applied to examine the impacts of the explanatory variables in the initial base model of the study following the expression in Equation 8.

$$QLnY_{it}(\tau / \chi_{it}) = \phi_i^{(\tau)} + \phi_1^{(\tau)} LnOPN_{it} + \phi_2^{(\tau)} LnTiG_{it} + \phi_3^{(\tau)} LnLab_{it} + \phi_4^{(\tau)} LnEXR_{it} + \phi_5^{(\tau)} LnECN_{it} + \mu_{it} \tag{8}$$

From Equation 8, the β represents the slope parameter for each regressor at given quantile (τ) while $QLnY_{it}(\tau/\chi_{it})$ represent the τth conditional quantile of the given dependent variable. χ_{it} captures the vector of the regressors for the quantile (τ) with respect to individual country i in the panel study at a given time t while μ_{it} represents the disruption term of the given vector. The estimated slope parameters ϕ_1 and ϕ_2 are expected to come out positive and significant for the affirmation of the trade-led growth hypothesis. The estimations from the QR were used as sensitivity test to complement the findings from the CCEMG and the AMG approaches for the purpose of enhanced empirical analysis since the QR approach has a unique strength of accounting for heterogeneity and other important issues like variations in distribution of data observations for obtained coefficient estimates (Angrist et al. 2006, Alsayed et al. 2020). The results of the combined estimation techniques were presented in Table 5.

Table 5: QR and CCEMG estimates.

Techniques	CCEMG	AMG	Quantile Regression (QR)									
Dependent (var): LnY			$\tau = 0.05$	$\tau = 0.10$	$\tau = 0.20$	$\tau = 0.30$	$\tau = 0.40$	$\tau = 0.50$	$\tau = 0.60$	$\tau = 0.70$	$\tau = 0.80$	$\tau = 0.90$
LnREXP	0.41200 **	0.4430 ***	0.0532 *	0.0706 **	0.1033 ***	0.1050 ***	0.1197 ***	0.1666 ***	0.1494 **	0.1937 ***	0.1658 *	0.2196 ***
P-value	(0.0230)	(0.0000)	(0.0600)	(0.0235)	(0.0001)	(0.0007)	(0.0022)	(0.0010)	(0.0161)	(0.0080)	(0.0542)	(0.0002)
LnTiG	–1.0492 ***	–0.6887 **	1.1401 ***	1.0319 ***	0.4127 ***	0.3773 ***	0.3230 *	0.0891	–0.1277	–0.5036 *	–0.9284 **	–1.8769 ***
P-value	(0.000)	(0.0450)	(0.0000)	(0.0000)	(0.0008)	(0.0044)	(0.0520)	(0.8050)	(0.6944)	(0.0530)	(0.0248)	(0.0000)
LnECN	0.5551	0.3453 *	0.6221 ***	0.7460 ***	0.7605 ***	0.7686 ***	0.7779 ***	0.8047 ***	0.8744 ***	0.9795 ***	0.9608 ***	0.7955 ***
P-value	(0.1220)	(0.086)	(0.0000)	(0.0000)	(0.0000)	(0.0000)	(0.0000)	(0.0000)	(0.0000)	(0.0000)	(0.0000)	(0.0000)
LnUrB	1.8595	0.1004	0.2995	–0.7834 ***	–0.5935 ***	–0.4169 *	–0.2827	–0.1036	–0.1841	–0.6933 **	–0.4261	0.7912 **
P-value	(0.7350)	(0.9710)	(0.6182)	(0.0013)	(0.0053)	(0.0776)	(0.1896)	(0.5994)	(0.4091)	(0.0482)	(0.3343)	(0.0242)
Observation	232	232	232	232	232	232	232	232	232	232	232	232
Wald test	23.08 ***	29.19 ***										
p-value	(0.0001)	(0.0000)										
No. Regressors	4	4	4	4	4	4	4	4	4	4	4	4
No. Group	8	8	8	8	8	8	8	8	8	8	8	8

Source: Author's computation. Variables are in natural logarithms and the estimates were reported alongside the subscripts *, **, *** representing the statistical significance of the outputs at 10%, 5%, and 1% significance levels, respectively.

4. Results and Discussion

From Table 5, estimates from both CCEMG and AMG estimators are similar in terms of the obtained coefficient signs. The AMG outcomes appear to be slightly better compared to the CCEMG nevertheless both reflect relatively close interpretations of relationships among variables in the model. From the outcome, real export, energy consumption, and urbanization have positive impacts on income levels among the countries. The positive impacts of real export and energy consumption on economic growth were significant while that of urbanization was not. A percentage increase in real export and energy consumption is expected to stimulate income level by 0.44% and 0.34%, respectively. This result strongly upholds the export-led growth hypothesis (ELGH) for the MENA region. The findings further support the claims from several studies regarding the stimulating impacts of energy consumption on economic growth in different economies (Dogan et al. 2020, Onifade et al. 2021, Ivanovski et al. 2021).

On the other hand, the impact of trade globalization variable was negative and significant as a percentage rise in trade globalization reduces growth by 0.68%. This is an indicator that the gains from trade among MENA countries in the panel study are still being inhibited by several trade-impacting factors including tariffs, regulation, and agreements. Findings from the QR technique corroborated the other two approaches and the QR estimates were much more insightful as the technique provided the estimates based on the observations across different quantiles.

Estimates from the QR are significant for both real export and energy consumption across all quantiles while those of trade globalization and urbanization vary significantly across quantiles. The QR approach clearly show that both real export and energy consumption were important growth stimulant as the positive impacts of these variables were significant across all quantiles. On the other hand, trade globalization demonstrated significant positive impact on growth only at the initial levels as shown by the output for the lower quantile, however, the impacts turn out to be negative at higher quantiles thus corroborating the output from the other approaches. Finally, urbanization exerts negative impacts on income levels among the country, however, the impacts were generally unstable across all quantiles.

In a nutshell, the results regarding both energy consumption and urbanization have significant implications for the region at large. A look at the trend of all variables in Figure 1 show that both variables have witnessed significant upward trend in the region, especially in the decades after the advent of oil and gas production in commercial quantity in the region. In fact, as of 2014, majority of the MENA countries in the study including Qatar, Oman, Saudi Arabia, Algeria, and Iran have fossil fuel energy consumption accounting for over 99% of the total energy consumption, while urban population could account for as high as 99.18% and 100% of the total population in Qatar and Kuwait (WDI 2020). While the present study has revealed the significant role of energy consumption in economic growth, several studies have elaborated on the dangers of fossil energy consumption vis-à-vis the exacerbation of growing carbon emission levels (Hanif 2018, Hanif et al. 2019, Newell and Simms 2020, Shahbaz et al. 2021). As such, current findings are indicators for policymakers to strategize towards investment in clean energy to help maintain sustainable environment for the region's growing urbanization level.

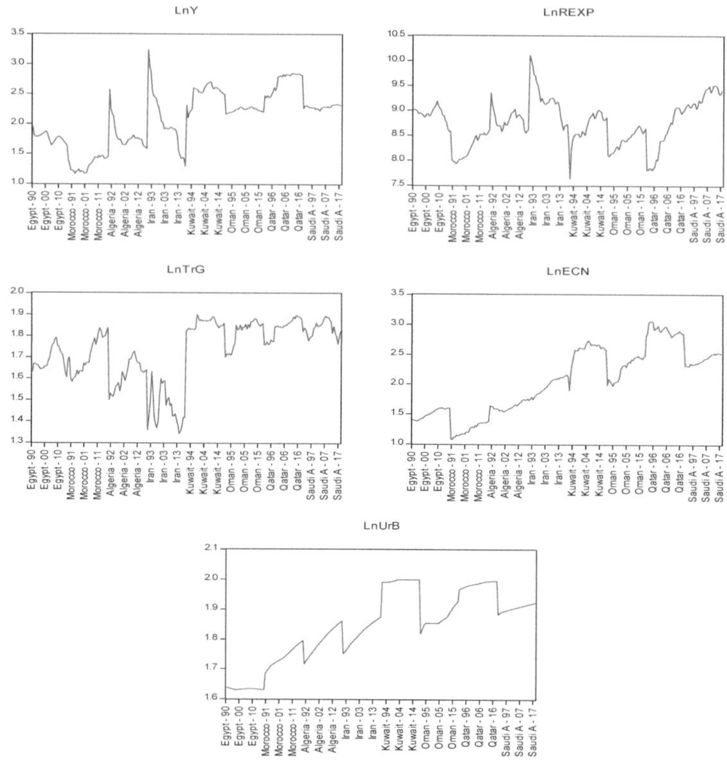

Figure 1: Graphical representation of the variables.

Lastly, we have provided relevant diagnostic tests of slope equality for the QR estimates, and a variance inflation factor (VIF) test. The quantile slope equality test results based on the significance of the Wald test outputs show that the obtained slope coefficients are significantly different across the quantile levels while the VIF test also confirmed that the model is not suffering from multicollinearity issues as the VIFs were below the threshold of 10. The quantile slope equality test and the VIF test were presented in Table 7 and Table 8. The null hypothesis of slope equality is accepted for all regressors between the 20th and 40th quantile but the hypothesis is rejected at different quantiles for each of the variable for instance, while the null can be accepted for energy consumption at the 30th quantile, it can be rejected at the 50th, 70th, and 90th quantiles. As such, the test further corroborated the conclusions from the initial estimates that the impacts of the regressors on economic growth varies across quantiles.

4.1 Panel causality tests

To consolidate the findings, we explored the causality nexus among the variables using Dumitrescu and Hurlin (DH 2012) panel causality tests in line with the observed importance of such tests in existing empirical studies (Adedoyin et al. 2020, Dong et al. 2018, Onifade et al. 2020b, Le and Ozturk 2020) and the obtained results are detailed out in Table 6.

Table 6: Dumitrescu and Hurlin (DH) Panel Causality tests.

Variables	LnY	LnREXP	LnTiG	LnECN	LnUrB	Direction of Causality
LnY	–	1.65884	2.23580 **	3.0843 **	7.8514 ***	LnY → LnTiG, LnECN, LnUrB
p-value		(0.3250)	(0.0480)	(0.0006)	(0.0000)	
LnREXP	4.6282 ***	–	1.5356	2.3816 **	3.6335 ***	LnREXP → LnY, LnECN, LnUrB
p-value	(0.0000)		(0.4400)	(0.0259)	(0.0000)	
LnTiG	4.1965 ***	2.9556 ***	–	4.9951 ***	5.4254 ***	LnTiG → LnY, LnREXP, LnECN, LnUrB
p-value	(0.0000)	(0.0013)		(0.0000)	(0.0000)	
LnECN	4.0405 ***	1.7724	2.7228 ***	–	15.5751 ***	LnECN → LnY, LnTiG, LnUrB
p-value	(0.0000)	(0.2381)	(0.0049)		(0.0000)	
LnUrB	1.9402	2.7197 ***	2.2056 *	5.3503 ***	–	LnUrB → LnREXP, LnTiG, LnECN
p-value	(0.1419)	(0.0050)	(0.0542)	(0.0000)		

Source: Author's computation. Variables are in natural logarithms the test statistics were reported alongside the subscripts *, **, *** representing the statistical significance of the outputs at 10%, 5%, and 1% significance levels, respectively.

Based on the result of the DH panel causality test in Table 6, the trio of real exports, energy consumption, and trade globalization granger cause economic growth among the countries. In addition, there was a bi-directional causality between economic growth, energy consumption, and trade globalization. The causality results also reveal that both urbanization and rising income levels are drivers of energy consumption as these variables granger cause the level of energy consumption among the countries in the panel study.

5. Conclusion and Recommendations

This study has established the existence of a long-run relationship between real export, globalization in trade, energy consumption, urbanization, and the economic growth performances of selected MENA countries between 1990 and 2018. The empirical results through a combination of second-generation techniques including the Augmented Mean Group (AMG), Common Correlated Effects Mean Group (CCEMG), and Quantile Regression (QR) affirm the validity of the export-led growth hypothesis (ELGH) for selected countries in the MENA region including Algeria, Egypt, Iran, Qatar, Saudi Arabia, Kuwait, Oman, and Morocco. However, when other salient trade impacting factors like regulations, taxes, tariffs, and agreements were accounted for, trade globalization only demonstrate significant positive impacts on income levels at lower quantiles while the distribution of the impacts turn out to be negatively significant at higher quantiles.

5.1 Policy and recommendation

The result is an indication that the countries in the MENA region are yet to implement sufficient policy measures and reforms that can position the region towards harnessing the expected gains from trade. On the other hand, while urbanization among the countries demonstrate varying degrees of impacts on growth, energy consumption came out clear as important growth stimulant as its impact was clearly significant across all quantiles. Finally, evidence from the Granger Causality corroborated the estimated coefficients as both the trade and the energy components granger cause the income levels for the panel study.

Hence, we recommend implementations of policies frameworks that can facilitate removal of various trade barriers by ensuring favorable tax regimes and trade regulations. Besides, strengthening of the institutions among the countries in the region to facilitate transparency and removal of bottlenecks in trade related matters would also be beneficial for the entire region to be well positioned in benefiting from trade dynamics in contemporary times.

Furthermore, based on the empirical evidence from the study, although energy consumption is highly significant for the regional growth, we would also recommend that the authorities in the MENA region should ensure more commitments toward investment in clean energy and urban infrastructures. Firstly, doing this would go a great extent in helping the region in maintaining energy driven growth vis-à-vis income sustainability as such investments would be critical to the positioning of the region for imminent energy transition which may create possible distortions on the growth trajectory for the region. Secondly, the clean energy and urban investments would also help in ensuring environmental sustainability of the region while trade policies are constantly reviewed to harness the benefits of trade among these countries in an increasingly globalized world.

References

Adedoyin, F. F., Bekun, F. V. and Alola, A. A. (2020). Growth impact of transition from non-renewable to renewable energy in the EU: The role of research and development expenditure. *Renewable Energy*, 159: 1139–1145.

Ahmed, M. (2010). Trade competitiveness and growth in the Mena region. *The Arab World Competitiveness Review*, 2010: 23–26.

Alola, A. A., Akadiri, S. S. and Usman, O. (2020). Domestic material consumption and greenhouse gas emissions in the EU-28 countries: Implications for environmental sustainability targets. *Sustainable Development*.

Alsayed, A. R., Isa, Z., Kun, S. S. and Manzi, G. (2020). Quantile regression to tackle the heterogeneity on the relationship between economic growth, energy consumption, and CO_2 emissions. *Environmental Modeling and Assessment*, 25(2): 251–258.

Angrist, J., Chernozhukov, V. and Fernández-Val, I. (2006). Quantile regression under misspecification, with an application to the US wage structure. *Econometrica*, 74(2): 539–563.

Bal, D. P., Dash, D. P. and Subhasish, B. (2016). The effects of capital formation on economic growth in India: Evidence from ARDL–bound testing approach. *Global Business Review*, 17(6): 1388–1400.

Balassa, B. (1985). Exports, policy choices, and economic growth in developing countries after the 1973 oil shock. *Journal of Development Economics*, 18(1): 23–35.

BP. (2020). Statistical Review of World Energy June 2020. Available at: http://www.bp.com/statisticalreview.

Breusch, T. S. and Pagan, A. R. (1980). The Lagrange multiplier test and its applications to model specification in econometrics. *The Review of Economic Studies*, 47(1): 239–253.

Brown, P. and Lauder, H. (1996). Education, globalization and economic development. *Journal of Education Policy*, 11(1): 1–25.

Çoban, O., Onifade, S. T., Yussif, A. B. and Haouas, I. (2020). Reconsidering trade and investment-led growth hypothesis: New evidence from Nigerian economy. *Journal of International Studies*, 13(3): 98–110. DOI:10.14254/2071–8330.2020/13–3/7.

De Hoyos, R. E. and Sarafidis, V. (2006). Testing for cross-sectional dependence in panel-data models. *The Stata Journal*, 6(4): 482–496.

Dickey, D. A. and Fuller, W. A. (1979). Distribution of the estimators for autoregressive time series with a unit root. *Journal of the American Statistical Association*, 74: 427–31.

Doğan, B., Driha, O. M., Balsalobre Lorente, D. and Shahzad, U. (2021). The mitigating effects of economic complexity and renewable energy on carbon emissions in developed countries. *Sustainable Development*, 29(1): 1–12.

Dogan, E., Altinoz, B., Madaleno, M. and Taskin, D. (2020). The impact of renewable energy consumption to economic growth: A replication and extension of. *Energy Economics*, 90: 104866.

Dong, K., Sun, R., Li, H. and Liao, H. (2018). Does natural gas consumption mitigate CO_2 emissions: Testing the environmental Kuznets curve hypothesis for 14 Asia-Pacific countries. *Renewable and Sustainable Energy Reviews*, 94: 419–429.

Douglas, P. H. (1976). The Cobb-Douglas production function once again: Its history, its testing, and some new empirical values. *Journal of Political Economy*, 84(5): 903–915.

Dumitrescu, E. I. and Hurlin, C. (2012) Testing for Granger non-causality in heterogeneous panels. *Econ. Model.*, 29: 1450–1460.

Eberhardt, M. and Bond, S. (2009). Cross-section dependence in nonstationary panel models: A novel estimator.

Gokmenoglu, K. K., Sehnaz, Z. and Taspinar, N. (2015). The export-led growth: A case study of Costa Rica. *Procedia Economics and Finance*, 25: 471–477.

Gygli, Savina, Florian Haelg, Niklas Potrafke and Jan-Egbert Sturm. (2019). The KOF globalisation index—Revisited. *Review of International Organizations*, 14(3): 543–574. https://doi.org/10.1007/s11558–019–09344–2.

Halliru, A. M., Loganathan, N., Hassan, A. A. G., Mardani, A. and Kamyab, H. (2020). Re-examining the environmental Kuznets curve hypothesis in the Economic Community of West African States: A panel quantile regression approach. *Journal of Cleaner Production*, 276: 124247.

Hanif, I. (2018). Impact of fossil fuels energy consumption, energy policies, and urban sprawl on carbon emissions in East Asia and the Pacific: A panel investigation. *Energy Strategy Reviews*, 21: 16–24.

Hanif, I., Raza, S. M. F., Gago-de-Santos, P. and Abbas, Q. (2019). Fossil fuels, foreign direct investment, and economic growth have triggered CO_2 emissions in emerging Asian economies: Some empirical evidence. *Energy*, 171: 493–501.

Haseeb, A., Xia, E., Saud, S., Ahmad, A. and Khurshid, H. (2019). Does information and communication technologies improve environmental quality in the era of globalization? An empirical analysis. *Environmental Science and Pollution Research*, 26(9): 8594–8608.

Hassan, S. T., Xia, E., Huang, J., Khan, N. H. and Iqbal, K. (2019). Natural resources, globalization, and economic growth: Evidence from Pakistan. *Environmental Science and Pollution Research*, 26(15): 15527–15534.

Herrerias, M. J. and Orts, V. (2010). Is the Export-led growth gypothes is enough to account for China's growth? *China and World Economy*, 18(4): 34–51.

Herzer, D. and Nowak-Lehmann, D. F. (2006). Export diversification, externalities and growth: Evidence for Chile. https://doi.org/10.1016/j.ribaf.2017.07.052.

Hossain, T. and Maitra, B. (2020). Monetary policy, trade openness and economic growth in India under monetary-targeting and multiple-indicator approach regimes. *Arthaniti: Journal of Economic Theory and Practice*, 19(1): 108–124.

Im, K. S., Pesaran, M. H. and Shin, Y. (2003). Testing for unit roots in heterogeneous panels. *Journal of Econometrics*, 115: 53–74.

Ivanovski, K., Hailemariam, A. and Smyth, R. (2021). The effect of renewable and non-renewable energy consumption on economic growth: Non-parametric evidence. *Journal of Cleaner Production*, 286: 124956.

Japan and the Asian tigers. *Japan and the World Economy*, 20(2): 155–174. https://doi.org/10.1016/j.japwor.2006.10.001.

Johansen, S. and Juselius, K. (1990). Maximum likelihood estimation and inference on cointegration with application to the demand for money. *Oxford Bulletin of Economics and Statistics*, 52: 169–209.

Joshua, U., Adedoyin, F. F. and Sarkodie, S. A. (2020). Examining the external-factors-led growth hypothesis for the South African economy. *Heliyon*, 6(5): e04009.

Joshua, U., Salami, O. M. and Alola, A. A. (2020). Toward the path of economic expansion in Nigeria: The role of trade globalization. *Journal of Labor and Society*, 23(2): 205–220.

Katsiampa, P. (2017). Volatility estimation for Bitcoin: A comparison of Garch models. *Economics Letters*, 158: 3–6.

Kapetanios, G., Pesaran, M. H. and Yamagata, T. (2011). Panels with non-stationary multifactor error structures. *Journal of Econometrics*, 160(2): 326–348.

Khalid, A. M., Sharma, S. and Dubey, A. K. (2018). Developing an indicator set for measuring sustainable development in India. In *Natural Resources Forum* (Vol. 42, No. 3, pp. 185–200). Oxford, UK: Blackwell Publishing Ltd.

Koenker, G. Bassett. (1978). Quantile regression. *Econometrica*, 46: 33–50.

Koenker, R. and Hallock, K. F. (2001). Quantile regression. *Journal of Economic Perspectives*, 15(4): 143–156.

Kraft, J. and Kraft, A. (1978). On the relationship between energy and GNP. *The Journal of Energy and Development*, 401–403.

Le, H. P. (2020). The energy-growth nexus revisited: The role of financial development, institutions, government expenditure and trade openness. *Heliyon*, 6(7): e04369.

Le, H. P. and Ozturk, I. (2020). The impacts of globalization, financial development, government expenditures, and institutional quality on CO_2 emissions in the presence of environmental Kuznets curve. *Environmental Science and Pollution Research*, 27(18): 22680–22697.

Lee, C. C., Lee, C. C. and Chiou, Y. Y. (2017). Insurance activities, globalization, and economic growth: New methods, new evidence. *Journal of International Financial Markets, Institutions and Money*, 51: 155–170.

Love, J. and Chandra, R. (2005). Testing export-led growth in Bangladesh in a multivariate VAR framework. *Journal of Asian Economics*, 15(6): 1155–1168. https://doi.org/10.1016/j.asieco.2004.11.009.

Luo, W. (2020). Demography and economic growth: The effect of tax composition. *Applied Economics Letters*, 1–6.

Mahadevan, R. and Suardi, S. (2008). A dynamic analysis of the impact of uncertainty on import-and/or export-led growth: The experience of Japan and the Asian Tigers. *Japan and the World Economy*, 20(2): 155–174.

Mamun, K. A. and Nath, K. (2005). Export-led growth in Bangladesh: A time series analysis. *Applied Economics Letters*, 12: 361–364.

Marwan, N. F., Kadir, N. A. A., Hussin, A., Zaini, A. A., Rashid, M. E. A. and Helmi, Z. A. G. (2013). Export, aid, remittance and growth: Evidence from Sudan. *Procedia Economics and Finance*, 7: 3–10.

Mensi, W., Hammoudeh, S., Reboredo, J. C. and Nguyen, D. K. (2014). Do global factors impact BRICS stock markets? A quantile regression approach. *Emerging Markets Review*, 19: 1–17.

Moreno-Izquierdo, L., Rubia-Serrano, A., Perles-Ribes, J. F., Ramón-Rodríguez, A. B. and Such-Devesa, M. J. (2020). Determining factors in the choice of prices of tourist rental accommodation. New evidence using the quantile regression approach. *Tourism Management Perspectives*, 33: 100632.

Newell, P. and Simms, A. (2020). Towards a fossil fuel non-proliferation treaty. *Climate Policy*, 20(8): 1043–1054.

Nwani, C. (2021). Taking Venezuela back to the sustainability path: The role of financial development and economic integration in low-carbon transition. In *Natural Resources Forum*. Oxford, UK: Blackwell Publishing Ltd.

Onifade, S. T., Alola, A. A., Erdoğan, S. and Acet, H. (2021). Environmental aspect of energy transition and urbanization in the OPEC member states. *Environmental Science and Pollution Research*, 1–12.

Onifade, S. T., Ay, A., Asongu, S. and Bekun, F. V. (2020a). Revisiting the trade and unemployment nexus: Empirical evidence from the Nigerian economy. *Journal of Public Affairs*, e2053.

Onifade, S. T., Çevik, S., Erdoğan, S., Asongu, S. and Bekun, F. V. (2020b). An empirical retrospect of the impacts of government expenditures on economic growth: New evidence from the Nigerian economy. *Journal of Economic Structures*, 9(1): 6.

Pesaran, M. H. (2007). A simple panel unit root test in the presence of cross section dependence. *Journal of Applied Econometrics*, 22(2): 265–312.

Pesaran, M. H. (2015). Testing weak cross-sectional dependence in large panels. *Econom. Rev.*, 34(6-10): 1089–1117.

Phillips, P. C. and Perron, P. (1988). Testing for a unit root in time series regression. *Biometrika*, 75(2): 335–346. https://doi.org/10.1093/biomet/75.2.335.

Quaicoe, A., Aboagye, A. Q. Q. and Bokpin, G. A. (2017). Assessing the impact of export processing zones on economic growth in Ghana.

Rahman, M. M. and Mamun, S. A. K. (2016). Energy use, international trade and economic growth nexus in Australia: New evidence from an extended growth model. *Renewable and Sustainable Energy Reviews*, 64: 806–816. https://doi.org/https://doi.org/10.1016/j.rser.2016.06.039.

Saint Akadiri, S., Alola, A. A., Olasehinde-Williams, G. and Etokakpan, M. U. (2020). The role of electricity consumption, globalization and economic growth in carbon dioxide emissions and its implications for environmental sustainability targets. *Science of the Total Environment*, 708: 134653.

Shahbaz, M., Haouas, I. and Van Hoang, T. H. (2019). Economic growth and environmental degradation in Vietnam: Is the environmental Kuznets curve a complete picture? *Emerging Markets Review*, 38: 197–218.

Shahbaz, M., Khan, S. and Tahir, M. I. (2013). The dynamic links between energy consumption, economic growth, financial development and trade in China: Fresh evidence from multivariate framework analysis. *Energy Economics*, 40: 8–21. https://doi.org/https://doi.org/10.1016/j.eneco.2013.06.006.

Shahbaz, M., Shahzad, S. J. H., Mahalik, M. K. and Hammoudeh, S. (2018a). Does globalisation worsen environmental quality in developed economies? *Environmental Modeling and Assessment*, 23(2): 141–156.

Shahbaz, M., Shahzad, S. J. H., Mahalik, M. K. and Sadorsky, P. (2018b). How strong is the causal relationship between globalization and energy consumption in developed economies? A country-specific time-series and panel analysis. *Applied Economics*, 50(13): 1479–1494.

Shahbaz, M., Sharma, R., Sinha, A. and Jiao, Z. (2021). Analyzing nonlinear impact of economic growth drivers on CO_2 emissions: Designing an SDG framework for India. *Energy Policy*, 148: 111965.

Shrestha, M. B. and Bhatta, G. R. (2018). Selecting appropriate methodological framework for time series data analysis. *The Journal of Finance and Data Science*, 4(2): 71–89.

Solow, R. M. (1956). A contribution to the theory of economic growth. *The Quarterly Journal of Economics*, 70(1): 65–94.

Taiwo, S., Alagöz, M. and Erdoğan, S. (2020). Inflation, oil revenue, and monetary policy mix in an oil-dependent economy: Empirical insights from the case of Nigeria. *International Journal of Business*, 7(2): 96–109.

Tyler, W. G. (1981). Growth and export expansion of developing countries: Some empirical evidence. *Journal of Development Economics*, 9(1): 121–30.

Ulaşan, B. (2015). Trade openness and economic growth: Panel evidence. *Applied Economics Letters*, 22(2): 163–167.

Usman, M., Makhdum, M. S. A. and Kousar, R. (2020). Does financial inclusion, renewable and non-renewable energy utilization accelerate ecological footprints and economic growth? Fresh evidence from 15 highest emitting countries. *Sustainable Cities and Society*, 102590. https://doi.org/https://doi.org/10.1016/j.scs.2020.102590.

UNDP. (2015). United Nations Sustainable Development Goals. Available at: https://unstats.un.org/sdgs Accessed June 5, 2023.

Wacziarg, R. (2000). Measuring the dynamic gains from trade. Research Paper Series, Research Paper No. 1654.

Westerlund, J. 2007. Testing for error correction in panel data. *Oxford Bulletin of Economics and Statistics*, 69: 709–748.

World Development Indictors, WDI. (2020). https://data.worldbank.org/ Accessed Sept 2020.

Yavari, K. and Mohseni, R. (2012). Trade liberalization and economic growth: A case study of Iran. *Journal of Economic Policy Reform*, 15(1): 13–23.

Zafar, M. W., Shahbaz, M., Hou, F. and Sinha, A. (2019). From nonrenewable to renewable energy and its impact on economic growth: The role of research and development expenditures in Asia-Pacific economic cooperation countries. *Journal of Cleaner Production*, 212: 1166–1178. https://doi.org/https://doi.org/10.1016/j.jclepro.2018.12.081.

Appendix

Table 7: Wald tests for slopes equality.

	0.05 vs. 0.1	0.1 vs. 0.2	0.2 vs. 0.3	0.3 vs. 0.4	0.4 vs. 0.5	0.5 vs. 0.6	0.6 vs. 0.7	0.7 vs. 0.8	0.8 vs. 0.9
LnREXP	–0.0174	–0.0327	–0.0016	–0.0147	–0.0468	0.0171	–0.0442	0.0278	–0.0537
LnTiG	0.1081	0.6192***	0.0353	0.0543	0.2339	0.2169	0.3758*	0.4248	0.9484 ***
LnECN	–0.1239 **	–0.0144	–0.0080	–0.0092	–0.0267	–0.0696*	–0.1050 **	0.0186	0.1652 ***
LnUrB	1.0829 **	–0.1898	–0.1766	–0.1342	–0.1791	0.0805	0.5092 **	–0.2671	–1.2173 ***
Wald Test (Chi-Sq. Stat)		224.3977							
P-value		0.0000							

Table 8: Variance inflation factor table.

Variables	VIF
LnREXP	1.13
LnTiG	1.52
LnECN	5.45
LnUrB	5.16
Mean VIF	3.32

Chapter 5

Evaluating the Significance and Trends in Energy Management Practices

The Case of the Organized Industrial and Service Sectors of Central Anatolia Province, Konya-Turkey

Vehbi Meşin[1] and *Stephen Taiwo Onifade*[2,*]

1. Introduction

The relationship between humanity and energy is almost as old as the existence of man. Energy consumption, which has been developing in line with the needs of humanity over the decades has become an indispensable element of peoples' lives till date (Dernbach 2007, Mattick et al. 2010, Duran et al. 2022, Gyamfi et al. 2022). In this context, as a phenomenon that directly affects the welfare levels of societies today, energy is more or less the basic inputs of all economic sectors, including agriculture, industry, construction, trade, transportation, communication, financial institutions, housing, and services among other sectors. Thus, energy and economy are directly interconnected, and several studies have even linked economic growth to energy use (Balsalobre-Lorente and Álvarez-Herranz 2016, Bekun et al. 2021, Apergis and Payne 2010, Gyamfi et al. 2021, Erdoğan et al. 2022, Onifade et al. 2021, Alola et al. 2021, Dingru et al. 2023). As such, countries need to constantly strategies on energy policies to ensure access to energy resources, stable use of energy, and promote overall energy efficiencies for societal benefits.

[1] Graduate School of Natural and Applied Sciences, Department of Energy Systems Engineering, Kocaeli University, Turkey.
[2] School of Finance and Accounting, Department of Economics, University of Vaasa, FI-65200 Vaasa, Finland.
Email: vehbi.mesin@boun.edu.tr
* Corresponding author: stephentaiwo.onifade@gmail.com

The need for energy and the demand for energy continue to increase with each passing day, with the effect of urbanization, economic growth, and various demographic developments across the globe. In 2017, for instance, primary energy consumption was observed to have increased by 2.2% compared to previous year (2016), with the effect of strong economic growth in the world economy in the same year (BP 2020). The observed rise in energy consumption has been noted to be a very significant growth if the figure was juxtaposed with the 1.7% average energy consumption growth in the preceding ten years (KPMG 2019). At the same time, it is expected that global energy demand would increase substantially by 2050 even as the global Gross Domestic Product (GDP) rises, thus creating more concern about environmental pollution as carbon intensive energy consumption are expected to rise (World Bank 2006). Historically, GDP growth is one of the major stimulating factors for energy demand historically, and it has been noted that carbon dioxide (CO_2) emissions rise by about 1% for every percentage growth in GDP (World Bank 2006, Gazbir 2018). Therefore, to meet the anticipated increases in energy demand vis-à-vis the global GDP growth, necessary steps must be taken in terms of ensuring energy efficiency on a global scale.

The Turkish economy's energy demands are growing and major factors such as economic growth, industrialization, population expansion, and urbanization are the main parameters that fuel the energy demand in Turkey as well as in many other emerging countries as evident in empirical studies (Gazbir 2018, Aydin 2014, Onifade et al. 2021a, Ağbulut 2019, Onifade et al. 2021b, Alola and Onifade 2022, Bekun et al. 2022, Onifade and Alola 2022). Turkey's energy policy has been shaped around issues such as giving priority to energy supply security and realizing sustainable energy supply by taking environmental factors into account. Its energy policy is basically aimed to diversify resources, increase resource exploration activities, strengthen infrastructure, consume energy in a highly efficient manner, use renewable energy resources and domestic coal more efficiently, and increase initiatives in energy technologies. The main developmental projects that are aimed at increasing energy efficiency are usually implemented in Turkey by the Energy Savings Coordination Board, which is supervised by the Ministry of Energy and Natural Resources. On the other hand, the institution responsible for minimizing environmental pollution via better efficiency in the use of energy resources is the National Energy Savings Center (UETM) and this institution is also affiliated to the General Directorate of Electrical Power Resources Survey and Development Administration (EIE). Since 1990, following the targeted goal of maximizing the energy saving capacity in Turkey with operational measures that will reduce investment costs by 40% as outlined by the Ministry of Energy and Natural Resources (UETM 2011), the EIE and the Energy Efficiency Test Tool Program (UETM) have been carrying out studies relating to energy savings in the device and equipment industries sector.

In the current study, we set out to assess the level of importance attached to energy efficiency and the tendency of companies towards integrating efficiency measures based on energy efficiency policies in Turkey. In this context, we applied a quantitative research technique by conducting a survey on companies in the Anatolia province of Konya. The research provides important benefits to stakeholders as it

Figure 1: A map of Turkey showing the major organized industrial hubs.
Source: Authors' compilation using data from Turkish organized industrial regions guide.

provides first hand information and directions to policymakers on handling the matters of energy efficiency vis-à-vis the obtained results of the field work. Furthermore, the surveyed province (Konya) offers additional benefits for studies like this as it is one of the largest provinces with the major organized industrial hubs in Turkey as seen in Figure 1. As a matter of fact, Konya is one of the important provinces that bears the employment and industrial burdens of the Turkish economy, with a total of seventy industrial sites, seventeen with ministry support and fifty-three without ministry support, all within ten organized industrial zones. Furthermore, since a large part of the global energy demand is met with exhaustible resources, such as fossil fuels which are gradually decreasing in terms of reserves (Shafiee and Topal 2009, Abas et al. 2015, Bozkaya et al. 2022, Onifade 2022), the pressing need for more efficiency in terms of energy usage becomes more pertinent. As such, considering that energy demands are also fast growing in Turkey, the present study further contributes to the growing literature on the Turkish energy sector from the experience of the Anatolian province of Konya.

The remaining aspect of the study features the literature review in Section 2, the research design and methods in Section 3, and the findings in Section 4. The last section summarizes the study and encompasses policy recommendations to stakeholders.

2. Literature Review

In recent years, there has been an increase in energy demand due to various developments in the industrial sector throughout the world among other factors. As such, many studies have been conducted by researchers and various projects have been carried out by international organizations and countries to ensure improvements and more efficient ways of utilizing energy resources. Many of the studies have focused on different sectors and various industries have been scrutinized as case studies including manufacturing industries, textiles industries, automobiles, and some other heavy metal industries. García-Quevedo and Jové-Llopis (2021) noted that the empirical findings from the strategies to boost energy efficiency in industry via the instrumentality of energy-saving approaches are generally inconclusive.

Thollander and Ottosson (2010) examined the energy-intensive sub-sectors in Sweden and discussed the practices related to energy management. They focused on the cases of the pulp and paper industry alongside the foundry industry. Their study revealed that almost one out of every three paper and pulp producers and two out of every five examined foundries did not take steps to ensure that there are proper measurements of the energy intensities in their processes. Moreover, they also discovered that, around one out of every five paper andpulp producers and around 50% of the foundries did not have any specific energy management strategy on a long-term basis. It was concluded that only 40% of the paper and pulp enterprises and only 25% of the foundry enterprises that were operating in Sweden as evaluated in the study have expected practices for a successful energy management system. In another study by Gordić et al. (2010), the developments in energy efficiency management in industrial enterprises in Serbia were investigated using the case of car manufacturers in the country. They emphasized the aspect of energy management systems that reduce the production costs of the factories and increase the profitability rates. In addition, auxiliary enterprises and consumptions that are the basis of energy consumption were evaluated during the energy audit process. Their study showed that several energy saving measures can be implemented for various energy means including steam, hot water, compressed air, electricity, and water. Furthermore, it was pointed out that the applications in these areas could easily be applied in other metal-intensive enterprises.

In another study by Hasanbeigi and Price (2012), energy efficiency technologies and energy consumption were examined for the case of the textile industry around the world. A detailed review of the energy intensities in existing sub-processes and energy savings opportunities based on energy measurements and their return on investment was provided in the study. They noted that there is insufficient information on strategies or measures for energy efficiency in the global textile industry. However, following the identification of about 184 energy-efficiency measures relating to the textile industry, their study revealed that there are several possible energy efficiency strategies that the textile industry can harness, even at a lower payback period.

The study of Kanchiralla et al. (2021) also examined the energy end-use dynamics for the case of the Swedish food industry. Their findings showed that about 66.6% of the energy used is attributed to the production process in this industry while just around 33.3% is attributed to the inherent support processes. They further noted that the heat processing is the most intensive aspect of the unit processes as far as energy use and CO_2 emissions are concerned for production while the space heating is the most intensive aspects of energy use and CO_2 emissions in terms of support processes.

As for Turkey, some studies have been conducted with regards to the issues of energy efficiency in various industrial branches of the country following the nation's rapidly developing economy and increasing energy demand in the past couple of decades (Ervural et al. 2016, Ates and Durakbasa 2012, Önüt and Soner 2006, Cin et al. 2021). According to the 2017 data of the Turkish Statistical Institute (TUIK 2020), electricity consumption of 138 million, 804 thousand, 506-megawatt hour (MWh) was recorded for the industry and service sectors. The manufacturing industry was the sector that consumed the most electricity with about 91 million, 870 thousand, 286 MWh of electricity consumption. Thus, the manufacturing industry, with about 66.2% of total amount make up the sector with the largest share in total electricity consumption. Within the industry and service sectors, about 78.6% of electricity was consumed in the production of goods and services, and 10.8% in lighting and other electrical office equipment.

Hence, following these developments, energy management practices started gaining some attention to boost energy efficiency in companies and also bearing in mind that Turkey is mainly dependent on energy imports (Ediger and Berk 2011, Bulut and Muratoglu 2018). It was noted in the study of Ates and Durakbasa (2012) that there aren't sufficient energy management standards in Turkey as their study revealed that just about 22% of the understudied firms have some standard corporate energy management practices. In another study, Önüt and Soner (2006) examined the energy management practices of hotels in the southern province of Antalya in Turkey. The study also showed that just about 25% of the understudied hotels can be said to be efficient in their energy management practices. The study of Ervural et al. (2016) on the other hand only consider the examination of the efficiency of the renewable energy prospects of the provinces in Turkey. Their findings also show that some regions are inefficient. They noted that the efficiency is high in Marmara region as well as the Black Sea region. The province (Konya) examined in this study is however not within these two regions.

Overall, there are very few similar studies that reveal the importance of energy efficiency for companies and especially for those in the central Anatolian region. However, it is very imperative to determine the level of importance attached to energy efficiency and tendency of companies operating in any field to adhere to energy management practices. This would help to reveal the approaches adopted for efficiency in general, and to also make sense of the reasons for the basic motivations or reluctance in this sense. Furthermore, determining the level of importance and trends will guide the planning of legal regulations and incentives related to energy management practices in a more target-oriented manner. The focal point of energy management in factories is to execute certain energy programs to ensure coordination among all business units to save energy. Hence, this study explores the significance of energy efficiency and the observable trends among companies in terms of efficiency policies in Turkey.

3. Research Design

In this part of the study, we provide explanations on the research design and procedure including the population used and sample selection, how the data were collected, and how the data were analyzed within the framework of the examined hypotheses. To determine the significance and trends in energy efficiency across industrial and service sectors, we tried to reach a synthesis through the findings obtained by conducting a survey on the opinions of the experts on the subject matter of efficiency in the areas of interest. A cross-sectional analysis approach and general scanning model were then used for the analysis. General scanning models are scanning methods that are applied over a whole population or a sample determined from the population to reach a general opinion about such population usually consisting of more than one element (Karasar 2007).

3.1 Research population and sample selection

The sample size for the study is limited to the industrial firms and companies in the service sector that could provide trusted data as fully registered member of the Konya Chamber of Commerce. Thirty Konya manufacturing industry firms and twenty service-providing organizations were randomly selected from the population of available registered organizations. The designed questionnaires were voluntarily administered and twenty-nine Konya manufacturing industry firms and seventeen educational institutions under the service sector category eventually agreed to fill out the questionnaires. During the preliminary assessments of the collected data, three questionnaires (one from manufacturing and two others from the service sector) would be deemed invalid were encountered. As such, a total of forty-three samples representatives, twenty-eight of which are manufacturing industry firms and fifteen of which are service-providing institutions, formed the sample of the research.

3.2 Data collection and procedure

In this study, the questionnaire form was used as a data collection tool. The questionnaire form consists of three parts: it consists of general information about companies, the importance of energy use in terms of energy efficiency, and the measuring of energy efficiency levels of companies in terms of energy use.

In the first part of the survey, general information about the companies can be found, including the sector in which the company belongs, the number of years the company has been operating, the number of people employed in the business, the percentage (%) share of the energy costs in the company's facilities in the total operating costs on an annual basis, the measurement of the amount of energy spent for business activities in the company's facilities. The form consists of nine-item evaluations about whether an energy manager is employed in the company, whether service is received from energy-efficiency assessment companies, whether the company employees are trained on energy use, and the measurement of the use of renewable energy. In the second part of the questionnaire, the "Importance of Energy Use in terms of Energy Efficiency Scale" was selected by taking the opinions of experts. Within this scale, a 5-point Likert type (1=strongly disagree, 2=disagree, 3=undecided, 4=agree, 5=strongly agree or) was used, which explores the opinions of companies about the importance of energy use in terms of energy efficiency based on a total of twenty-two evaluations. In the third and last part of the questionnaire, the evaluations about the energy efficiency tendency levels of the companies in terms of energy use were asked. This is also done using twenty-two questions within the 5-point Likert type form containing percentage values (1=0–20%, 2=21%–40%, 3=41%–60%, 4=61%–80%, 5=81%–100%).

Moving on, descriptive statistics (mean, standard deviation, median value, minimum, maximum, number, and percentile) were given for categorical and continuous variables. In addition, the homogeneity of the variances, which is one of the prerequisites of the parametric tests, was checked with the "Levene" test. Normality assumption was checked with the "Shapiro-Wilk" test. When evaluating the differences between the two groups, the "Student's Test" was used if the parametric test prerequisites are met; if not, the "Mann Whitney-U test" was used. For comparison of three or more groups, a One-Way Analysis of Variance and Tukey HSD test from multiple comparison tests were used, whereas Kruskal Wallis test and Bonferroni-Dunn test was utilized for multiple comparison tests. We proceed to examine the following hypotheses to achieve the research objective:

H_1 : Considering the energy use by firms, the energy efficiency trend differs significantly according to the characteristics of the firm.

H_{1a}: Considering the energy use by firms, the energy efficiency trend varies according to sectors.

H_{1b}: Considering the energy use by firms, energy efficiency varies according to the year of activity.

H_{1c}: Considering the energy use by firms, the energy efficiency trend differs according to the number of people employed in the business.

H_{1d}: Considering the energy use by firms, the energy efficiency trend differs according to the annual total cost/energy cost ratio.

H_{1e}: Considering the energy use by firms, the energy efficiency trend differs according to the measurement of the amount of energy consumed.

H_{1f}: Considering the energy use by firms, the energy efficiency trend differs according to the employment of energy managers.

H_{1g}: Considering the energy use by firms, the energy efficiency trend differs according to the status of firms receiving service in from efficiency companies.

H_{1h}: Considering the energy use by firms, the energy efficiency trend differs according to the energy training status of the employees.

H_1 : Considering the energy use by firms, the energy efficiency trend differs according to the status of the existing studies on alternative energy use.

4. Results and Analysis

In this section, we present and analyze the findings obtained from the survey exercises. The results are presented in tables followed by the corresponding explanations of each tabular presentation.

4.1 Frequency analysis of participating firms

As seen in Table 1, under Panel 1, 65.1% of the companies participating in the research are in the industry sector and 34.9% are in the service sector. In the distribution of the sample according to the sector in which the companies are found, there is a 30.2%

Table 1: Frequency Distribution.

Panel 1: Frequency Distribution by Sector of the Firm		
53.5	**Frequency**	**Percent**
Industry	28	65.1
Service	15	34.9
Total	43	one hundred
Panel 2: Frequency Distribution of Firms by Activity Period		
	Frequency	**Percent**
1–4 Years	3	7.0
5–9 Years	6	14.0
10–14 Years	11	25.6
15 Years and Over	23	53.5
Total	43	one hundred
Panel 3: Frequency Distribution by the Number of Persons Employed by the Firms		
	Frequency	**Percent**
10–50 People	23	53.5
51–100 People	6	14.0
101–150 People	3	7.0
151 and Over Persons	11th	25.6
Total	43	one hundred
Panel 4: Frequency Distribution According to the Annual Energy Cost of the Firms' Facilities		
	Frequency	**Percent**
less than 25%	32	74.4
26–35%	10	23.3
36–49%	1	2.3
Total	43	one hundred
Panel 5: Frequency Distribution According to How Firms Measure the Amount of Energy Spent		
	Frequency	**Percent**
Bills	35	81.4
Single Counter	4	9.3
does not measure	4	9.3
Total	43	one hundred
Panel 6: Frequency Distribution by Companies' Employment of Energy Managers		
	Frequency	**Percent**
Yes	3	7.0
No	40	93.0
Total	43	one hundred

Table 1 contd. ...

...Table 1 contd.

Panel 7: Frequency Distribution by Firms Receiving Services from Energy Efficiency Firms		
	Frequency	**Percent**
Yes	8	18.6
No	35	81.4
Total	43	one hundred
Panel 8: Frequency Distribution According to the Energy Usage Training of the Firms to their Employees		
	Frequency	**Percent**
Once a Month	one	2.3
Every 6 Months	6	14.0
Once a Year	9	20.9
no	27	62.8
Total	43	one hundred
Panel 9: Frequency Distribution by Companies' Studies Related to Alternative Energy Use		
	Frequency	**Percent**
Yes	14	32.6
No	29	67.4
Total	43	one hundred

difference between the industry sector and the service sector. In Panel 2, 7% of the companies participating in the research have been operating for 1–4 years, 14% for 5–9 years, 25.6% for 10–14 years, 53.5% for 15 years or more. In the sample, 53.5% of the companies have been operating for at least 15 years. According to Panel 3, 53.5% of the companies participating in the research also employ 10–50 people, 14.0% 51–100 people, 7.0% 101–150 people, 25.6% employ 151 and more people. Hence, the firms employing 10–50 people are the majority in the sample accounting for about 53.5% of the population.

According to Panel 4, 74.4% of the companies participating in the research spend less than 25% on energy costs based on their annual total cost, 23.3% spend 26–30% on energy costs based on their annual total cost, while only 2.3% of them spend 36–49% on energy costs based on their annual total cost. Among the companies participating in the research in the sample, with a rate of 74.4%, the companies that spend less than 25% of their annual total costs on energy are in the majority. According to Panel 5, 81.4% of the companies participating in the research measure the amount of energy they consume with invoices, and 9.3% measure the amount of energy they consume with a single meter. It was discovered that about 9.3% of them did not measure the amount they spend on energy at all.

Moving on to Panel 6, only 7.0% of the companies participating in the research employ energy managers while 93.0% do not employ energy managers. According to Panel 7, 18.6% of the companies received service from energy efficiency companies, while more than 80% did not receive service from energy efficiency companies. Moving on to Panel 8, only 2.3% of the companies participating in the research provide energy use training to their employees once a month, 14.0% every six months, and 20.9% once a year while about 62.8% do not provide energy use training to their employees. Lastly, looking at the Panel 9, it can be seen that only 32.6% of the companies participating in the research have studies on the use of alternative energy, while 67.4% do not have any study on the use of alternative energy.

4.2 Factor and reliability analysis findings regarding scales

Within the scope of the study, reliability and validity analyzes for the factors that determine the energy use of companies in terms of energy efficiency and the energy efficiency tendency levels of companies in terms of energy consumption were meticulously carried out. The Cronbach's alpha (α) coefficient was used for reliability analysis while factor analysis was applied for the validity analysis. The level of suitability with factor analysis was evaluated with Bartlett's test of sphericity and adequacy of sample size (KMO) with sample adequacy statistics. The results are shown in Table 2.

In Table 2, factor analysis for the evaluations of the factors determining energy use in companies in terms of energy efficiency is shown. Accordingly, there is no item in the questionnaire with an alpha value of less than 0.20. And since the reliability of twenty-two items was high, we continued with all items without removing them. In Table 3, the Cronbach's alpha (α) reliability coefficient calculated for the twenty-two items used in the application is detailed out. The questionnaire can be said to be reliable since the estimated Cronbach's alpha (α) reliability coefficient was 0.873 which is reasonably above 0.80 level. Barlett test can be used to explained whether the data obtained from the study group is suitable for explanatory factor analysis with respect to the Kaiser–Meyer–Olkin test (KMO test) (Pituch and Stevens 2015). A high KMO value means that each variable in the scale can be predicted perfectly by other variables. If the values are zero or close to zero, no interpretation can be made based on these values, since there is a dispersion in the correlation distribution. Following the KMO test, if the value is less than 0.50, the factor analysis cannot be continued. In the validity study of the questionnaire, factor analysis was performed with the Varimax method on the data collected on the items in the questionnaire and the findings are also included in the Table 3. When the total variance explained in this context is examined, it is seen that there is 1 factor among twenty-two items based on the application data, and this factor measures 69.775% of the examined feature based on the measurement tool. Within the scope of studies in social sciences, the total explained variance value should be around 55% or more. Hence, 69.775% is sufficient for the total explained variance value within the scope of our study. In this context, the validity and reliability analyze of the evaluations that determine the energy efficiency trend levels of the participating firms/organizations in terms of energy use are shown in the Table 4.

Table 4 shows the factor analysis for the evaluations of the factors that determine the energy efficiency trend in energy use by companies. Accordingly, there is no item in the questionnaire with an alpha value of less than 0.20. The Cronbach's alpha (α) coefficient of the questionnaire used in the research is shown in the Table 5.

Table 2: Factors determining energy use in firms in terms of energy efficiency.

	Average that will apply if the item is removed from the scale	The variance that will apply if the item is removed from the scale	Item total correlations	Reliability coefficient applicable if the item is removed from the scale
Energy-saving power tools should be used.- (Class A)	88.33	103.42	0.21	0.92
Electric motor efficiency is important in energy use	88.07	103.07	0.45	0.91
Energy-saving lighting should be used	87.93	101.88	0.71	0.90
Existing dimmer system should be used in lighting	88.84	98.23	0.58	0.90
An automatic control system sensitive to daylight should be used in lighting	88.30	100.88	0.55	0.91
Daylight sensors should be used in outdoor lighting	88.07	100.73	0.78	0.90
Pressurized sodium lamps should be used in outdoor lighting	89.60	105.48	0.70	0.92
Motion sensors should be used in public areas	88.47	101.45	0.42	0.91
Leak detection and monitoring records should be made in terms of energy efficiency in electric and lighting systems	88.09	101.71	0.64	0.90
In your opinion, monitoring systems are important in terms of energy efficiency in electricity and lighting systems	88.35	99.95	0.60	0.90
Boiler compressor room design should be done	88.33	98.84	0.74	0.90
To ensure heating efficiency, a heating pump should be used	88.74	101.91	0.37	0.91
An automatic control system should be used for heating, cooling, and ventilation	88.26	99.48	0.64	0.90
In terms of energy efficiency, waste heat should be used	88.53	97.87	0.71	0.90
Ceiling fans should be used to cool the interior, exterior, and common areas	89.07	100.83	0.38	0.91
Indoor entrances should be designed to prevent heat loss	88.09	100.75	0.78	0.90

Table 2 contd. ...

...Table 2 contd.

	Average that will apply if the item is removed from the scale	The variance that will apply if the item is removed from the scale	Item total correlations	Reliability coefficient applicable if the item is removed from the scale
Heat insulation system should be used according to the conditions of usage areas	88.14	98.36	0.74	0.90
Thermostats that can be locked at room temperature should be used	88.63	96.24	0.57	0.91
Sunscreen, film glass, etc., to reduce the cooling load materials should be used	88.30	101.07	0.52	0.91
Air-conditioning systems should be designed to benefit from energy savings and be highly efficient	88.12	97.49	0.70	0.90
In terms of energy efficiency in heating, cooling and ventilation systems, leak detection and monitoring should be recorded	88.21	100.12	0.74	0.90
In your opinion, there should be monitoring systems in terms of energy efficiency in heating, cooling, and ventilation systems	88.23	98.14	0.65	0.90

Table 3: Cronbach alpha reliability and validity coefficient of the questionnaire.

Cronbach Alpha Reliability			
	Number of Items		Reliability Coefficient
Survey Used in the Application	22		0.873
Validity Coefficient of the Questionnaire			
Factor	**Square Sum of Varimax Rotating Factor Loads**		
	Total	**Disclosed variance %**	**Cumulative Variance %**
1	8,455	69,775	69,775

Kaiser-meyer-olkin sampling adequacy: 0.745 Bartlett's sphericity test chi-square value 756.25 degrees of freedom 231 p = 0.001.

Table 5 also shows the Cronbach's alpha (α) reliability coefficient calculated for twenty-two items used in SPSS. Accordingly, the Cronbach's alpha (α) reliability coefficient calculated was 0.910. It can be said that the questionnaire is quite suitable since this coefficient is above 0.80. In the validity study of the questionnaire, factor analysis was performed on the data collected on the items in the questionnaire with

Table 4: Factors determining the energy efficiency tendency in energy consumption by firms.

	Average that will apply if the item is removed from the scale	The variance that will apply if the item is removed from the scale	Item total correlations	Confidence coefficient to apply if the item is removed from the scale
Energy-saving electrical appliances are used (Class A)	55.77	373.90	0.72	0.93
In terms of efficiency in energy use, importance is given to the efficiency of electric motors	56.00	385.71	0.60	0.93
Energy-saving lighting is used	55.40	381.01	0.69	0.93
Existing dimmer system is used in lighting	56.95	382.33	0.62	0.93
In lighting, an automatic control system sensitive to daylight is used	57.14	389.98	0.44	0.94
Daylight sensors are used in outdoor lighting	56.72	394.02	0.34	0.94
Pressurized sodium lamps are used in outdoor lighting	57.40	385.58	0.62	0.93
Motion sensors are used in public areas	56.05	391.52	0.35	0.94
Leakage detection and monitoring are recorded in terms of energy efficiency in electricity and lighting systems	56.67	373.42	0.67	0.93
Monitoring systems are considered important in terms of energy efficiency in electricity and lighting system	56.30	372.31	0.70	0.93
The design of the boiler compressor room has been made	56.30	385.36	0.51	0.94
A heating pump is used to ensure heating efficiency	56.91	387.18	0.54	0.94
An automatic control system is used for heating, cooling, and ventilation	56.16	373.57	0.68	0.93
In terms of energy efficiency, waste heat should be used	56.98	371.55	0.79	0.93
Ceiling fans are used to cool the interior, exterior, and common areas	56.93	384.78	0.54	0.94
Indoor entrances are designed to prevent heat loss	56.35	371.85	0.73	0.93

Table 4 contd. ...

...Table 4 contd.

	Average that will apply if the item is removed from the scale	The variance that will apply if the item is removed from the scale	Item total correlations	Confidence coefficient to apply if the item is removed from the scale
Heat insulation system is used according to the conditions of usage areas	55.98	382.88	0.62	0.93
Thermostats with lockable set value at room temperature are used	56.53	373.02	0.68	0.93
Sunscreen, film glass, etc., to reduce the cooling load materials are used	56.16	380.85	0.60	0.93
Air-conditioning systems are designed to benefit from energy savings and be highly efficient	56.05	367.19	0.81	0.93
Leak detection and monitoring are recorded in terms of energy efficiency in heating, cooling, and ventilation systems	56.77	372.28	0.75	0.93
In your opinion, monitoring systems are considered important in terms of energy efficiency in heating, cooling, and ventilation systems	55.93	380.07	0.56	0.94

Table 5: Cronbach alpha reliability and validity coefficient of the questionnaire.

Cronbach Alpha Reliability			
	Number of Items		Confidence Coefficient
Survey Used in the Application	22		0.910
Validity Coefficient of the Questionnaire			
Factor	Square Sum of Varimax Rotating Factor Loads		
	Total	Disclosed Variance %	Cumulative Variance %
1	9.359	68.143	68.143

KMO sample adequacy: 0.755 Bartlett's sphericity test chi-square value 865.220 Degree of freedom 231 $p = 0.001$.

the Varimax method and the findings are shown in the Table 5. When the total variance explained in this context is examined, it is seen that there is 1 factor among twenty-two items based on the application data, and this factor measures 68.143% of the feature accounted for by the measurement tool and this amount is considered sufficient for the total explained variance within the scope of our study.

4.3 Difference descriptive statistics

The difference between the average and standard deviations of the answers given to each item by the factors relating to the evaluation scales of the firms and organizations participating in the research in terms of their energy efficiency and the energy efficiency tendency levels are shown in the Table 6. In Table 6, the average of each item varies between 0–4. A high mean indicates high impact. The significance and trend assessments will be discussed in detail from Table 6 in the concluding section.

4.4 Hypotheses testing and findings

In the study, descriptive statistics (mean, standard deviation, median value, minimum, maximum values, and percentile) were given for categorical and continuous variables. In addition, the homogeneity of the variances, which is one of the prerequisites of the parametric tests, was checked with the "Levene" test while the normality assumption was checked with the "Shapiro-Wilk" test. When it is desired to evaluate the differences between the two groups, "Student's t Test" if it meets the parametric test prerequisites; if not, the "Mann Whitney-U test" was used. One-Way Analysis of Variance and Tukey HSD test from multiple comparison tests were used for comparison of three or more groups, and Kruskal Wallis test and Bonferroni-Dunn test from multiple comparison tests were used. The data evaluation was carried out with IBM SPSS Statistical package for Windows, Version 25.0. Armonk, NY: IBM Corp. The $p < 0.05$ and $p < 0.01$ levels were considered statistically significant.

According to Table 7, when energy use is taken into consideration by companies, the differences in energy efficiency level according to the sectorial trends are shown with the "Student's t test". According to this, when energy use is taken into consideration by the firms, the energy efficiency trend shows a statistically significant difference in terms of the sector ($p < 0.05$). This situation shows that industrial firms have a higher tendency towards efficient use of energy compared to the firms in the service sector.

According to Table 8, when energy use is taken into consideration by companies, the differences in energy efficiency trend according to the operating period are shown with the "One-Way Analysis of Variance". Accordingly, when energy use is put into consideration by companies, the tendency for energy efficiency does not differ statistically in terms of operating time ($p > 0.05$).

According to Table 9, when energy use is taken into consideration by companies, the differences in energy efficiency trend according to the number of people employed are shown with the "Kruskal Wallis Test". Accordingly, when energy use is considered by companies, the energy efficiency trend does not statistically significantly differ in terms of the number of people employed ($p > 0.05$).

From Table 10, the differences in the energy efficiency trend according to the annual total cost/energy cost ratio are shown with the "One-Way Analysis of Variance" when energy use is considered by companies. Based on the findings, when energy use is taken into account by companies, the energy efficiency trend does not

Table 6: Statistics describing differences in evaluations determining the importance of energy use and the tendency of energy efficiency for firms.

	N	Minimum	Maximum	Mean	Std. Deviation
The importance and trend difference of using electrical appliances in energy saving	43	–4.00	4.00	0.930	1,907
The difference in importance and trend of electric motor efficiency in energy use	43	–2.00	4.00	1,419	1,401
The importance and trend difference of using energy-saving lighting	43	–1.00	4.00	0.953	1,214
The importance and trend difference of the use of dimmer systems in lighting systems	43	–2.00	4.00	1,605	1,237
The importance and trend difference of using daylight sensitive automatic control systems in lighting	43	–2.00	4.00	2,326	1,584
The importance and trend difference of using daylight sensors in outdoor lighting	43	–1.00	4.00	2,140	1,684
The importance and trend difference of using pressurized sodium lamps in outdoor lighting	43	–2.00	3.00	1,279	1,403
The importance and trend difference of using motion sensors in public areas	43	–2.00	4.00	1,070	1,316
The difference in importance and tendency of making leak detection and monitoring records in terms of energy efficiency in electrical and lighting systems	43	–2.00	4.00	2,070	1,624
The importance and trend difference of giving importance to monitoring systems in terms of energy efficiency in electrical and lighting systems	43	–1.00	4.00	1,442	1,637
The importance and trend difference of the design of the boiler and compressor room	43	–1.00	4.00	1,465	1,316
The importance and trend difference of using a heating pump in terms of ensuring heating efficiency	43	–1.00	4.00	1,651	1,307
The importance and trend difference of using automatic control system for heating, cooling, and ventilation	43	0.00	4.00	1,395	1,365
The importance and trend difference of using waste heat in terms of energy efficiency	43	–1.00	4.00	1,930	1,470
The difference in importance and tendency of using ceiling fans for cooling indoor, outdoor, and common areas	43	–1.00	4.00	1,349	1,289
The difference in importance and tendency of designing indoor entrances to prevent heat loss	43	0.00	4.00	1,744	1,382

Table 6 contd. ...

...Table 6 contd.

	N	Minimum	Maximum	Mean	Std. Deviation
The difference in importance and tendency of using a thermal insulation system according to the situation in the usage areas	43	–1.00	4.00	1,326	1,169
The importance and trend difference of using thermostats with lockable set value at room temperature	43	–2.00	4.00	1,395	1,678
Sunscreen, film glass, etc., to reduce the cooling load. difference in importance, and tendency of using materials	43	–1.00	4.00	1,349	1,494
The difference in importance and trend of designing air-conditioning systems in a way that will benefit from energy savings and be highly efficient	43	–1.00	4.00	1,419	1,435
In terms of energy efficiency in heating, cooling and ventilation systems, the importance and trend difference of leak detection and monitoring recording	43	0.00	4.00	2,047	1,396
The difference in importance and trend of monitoring energy efficiency in heating, cooling, and ventilation systems	43	0.00	4.00	1,395	1,365
Total Difference of Factors	43	–6	67	32,30233	31,673

Table 7. Sectorial differences in energy efficiency tendency based on energy use.

	n	Mean	Std. Deviation	Std. Error	Test Statistics	p-value
Industry	28	36.11	17.58	3.32	1,963	0.04
Service	15	25.20	19.46	5.03		

show a statistically significant difference in terms of annual total cost to energy cost ratio at ($p > 0.05$).

According to Table 11, when energy use is taken into consideration by companies, the differences in energy efficiency trend according to the measurement of the amount of energy consumed are shown with the "Kruskal Wallis Test". Accordingly, when energy use is considered by companies, the energy efficiency trend shows a statistically significant difference in terms of the measurement of the amount of energy consumed ($p < 0.05$).

According to Table 12, when energy use is taken into consideration by firms, the differences in energy efficiency trend according to energy manager employment are shown with the "Mann Whitney-U test". Accordingly, when energy use is considered by companies, the energy efficiency trend does not show a statistically significant difference in terms of employment of energy managers ($p > 0.05$).

According to Table 13, when energy use is taken into consideration by companies, the energy efficiency trend, and the differences according to the status of receiving

Table 8: Differences in energy efficiency tendency based on firm operating period.

	n	Mean	Std. Deviation	Std. Error	95% Confidence Limits		Minimum	Maximum	Test Statistics	p-value
					Lower Boundary	Upper Limit				
1–4 years	3	18.00	6.56	3.79	1.71	34.29	12.00	25.00	0.783	0.511
5–9 years	6	37.83	23.04	9.41	13.65	62.02	4.00	67.00		
10–14 years	11th	31.00	21.99	6.63	16.23	45.77	-6.00	67.00		
15 years and above	23	33.35	17.05	3,56	25.97	40.72	9.00	62.00		
Total	43	32.30	18.78	2.86	26.52	38.08	-6.00	67.00		

Table 9: Differences in energy efficiency tendency based on the number of employed person.

	n	Mean	Std. Deviation	Std. Error	95% Confidence limits		Minimum	Maximum	Test Statistics	p-value
					Lower Boundary	Upper Limit				
10–50 people	23	31.87	19.06	3.97	23.63	40.11	-6.00	67.00	0.112	0.952
51–100 people	6	31.33	25.34	10.34	4.75	57.92	4.00	61,00		
101–150 people	3	28.33	4.73	2.73	16.59	40.07	23.00	32.00		
151 and above	11th	34.82	18.54	5.59	22.36	47.27	9.00	62.00		
Total	43	32.30	18.78	2.86	26.52	38.08	-6.00	67.00		

Table 10: Differences in energy efficiency tendency based on annual total cost/energy cost ratio in firms' facilities.

	n	Mean	Std. Deviation	Std. Error	95% Confidence Limits		Minimum	Maximum	Test Statistics	p-value
					Lower Boundary	Upper Limit				
less than 25%	32	30.44	18.38	3.25	23.81	37.07	–6.00	67.00	0.63	0.539
26–35%	10	38.10	20.78	6.57	23.23	52.97	9.00	67.00		
36–49%	one	34.00					34.00	34.00		
Total	43	32.30	18.78	2.86	26.52	38.08	–6.00	67.00		

Table 11: Differences in energy efficiency tendency based on the measurement method of the amount of energy consumed.

	n	Mean	Std. Deviation	Std. Error	95% Confidence Limits		Minimum	Maximum	Test Statistics	p-value
					Lower Limit	Upper Limit				
Bills	35	28.74	17.06	2.88	22.88	34.60	-6.00	62.00	4.04	0.025*
single counter	4	44.25	20.65	10.32	11.40	77.10	21.00	67.00		
Not measured	4	51.50	19.62	9.81	20.28	82.72	23.00	67.00		
Total	43	32.30	18.78	2.86	26.52	38.08	-6.00	67.00		

Table 12: Differences in efficiency based on energy manager employment.

	n	Mean	Std. Deviation	Std. Error	95% Confidence Limits		Minimum	Maximum	Test Statistics	p-value
					Lower Limit	Upper Limit				
Yes	3	38.00	13.75	7.94	3.85	72.15	23.00	50.00	0.29	0.592
No	40	31.88	19.17	3.03	25.74	38.01	-6.00	67.00		
Total	43	32.30	18.78	2.86	26.52	38.08	-6.00	67.00		

services from energy efficiency assessment companies are shown with the "Mann Whitney-U test". Accordingly, when energy use is considered by firms, the tendency for energy efficiency does not differ significantly in terms of receiving services from evaluation firms ($p > 0.05$).

According to Table 14, when energy use is taken into consideration by companies, the differences in energy efficiency trends according to the energy training status of the employees are shown with the "One-Way Analysis of Variance". Accordingly, when energy use is considered by companies, the tendency of energy efficiency does not significantly differ in terms of the energy training status of the employees ($p > 0.05$).

According to Table 15, when energy use is considered by companies, the differences in energy efficiency trends compared to the status of the companies' renewable energy use studies are shown with the "Student's t test". Accordingly, when energy use is taken into consideration by firms, the tendency of energy efficiency does not significantly differ in terms of the prevailing status of the studies on renewable energy use of firms ($p > 0.05$).

5. General Evaluation and Conclusions

We utilized a cross-sectional analysis approach while incorporating general scanning approach for assessing the significance of energy efficiency and energy management practices using the case of the organized industrial and service sectors of Central Anatolia Province, Konya-Turkey. Within the framework of the information obtained through the literature review, a questionnaire consisting of carefully designed forty-four questions was administered among participating provincial firms in the industrial sector and organizations in service sector.

Based on the findings, there is a statistically significant difference in energy efficiency and trends in management practices in terms of the sector in which a participating firm or organization is located. The companies in the industry sector have a higher energy efficiency tendency than the companies in the service sector. Also, about 53.5% of the firms and organization in the sample have been operating for fifteen years or more and it was discovered that the energy efficiency trend does not show a statistically significant difference in terms of operating time. This situation shows that the operating time of the companies is not important when determining efficiency levels and prevailing energy management policies. Conversely, it has been observed that firms with a shorter operating period compared to others have a high awareness of the importance and a better tendency for energy efficiency. More than 50% of the participating firms or organization in the research employ 10–50 people, 14.0% 51–100 people, 7.0% 101–150 people, and 25.6% employ 151 and more people. From the result, energy efficiency trend does not show a statistically significant difference in terms of the number of people employed. Hence, the fact that the number of people employed is high does not necessarily mean that the importance and trend of energy efficiency will exist. In this regard, the firm size or the firm's operating time experience show that there should be no straight logic in determining efficiency policy proposals.

Table 13: Differences according to the status of getting service from energy efficiency evaluation firms.

	n	Mean	Std. Deviation	Std. Error	95% Confidence Limits		Minimum	Maximum	Test Statistics	p-value
					Lower Limit	Upper Limit				
Yes	3	38.00	13.75	7.94	3.85	72.15	23.00	50.00	0.35	0.56
No	40	31.88	19.17	3.03	25.74	38.01	-6.00	67.00		
Total	43	32.30	18.78	2.86	26.52	38.08	-6.00	67.00		

Table 14: Differences according to employees' energy training status.

	n	Mean	Std. Deviation	Std. Error	95% Confidence Limits		Minimum	Maximum	Test Statistics	p-value
					Lower Limit	Upper Limit				
once a month	one	41.00					41.00	41.00	0.346	0.559
every 6 months	6	21.50	17.84	7.28	2.78	40.22	4.00	53.00		
once a year	9	35.00	17.27	5.76	21.73	48.27	9.00	57.00		
no	27	33.48	19.57	3.77	25.74	41.22	-6.00	67.00		
Total	43	32.30	18.78	2.86	26.52	38.08	-6.00	67.00		

Table 15. Differences according to the current status of studies related to renewable energy use.

	n	Mean	Std. Deviation	Std. Error	95% Confidence Limits		Minimum	Maximum	Test Statistics	p-value
					Lower Limit	Upper Limit				
Yes	14	29.86	16.66	4.45	20.24	39.48	4.00	57.00	0.35	0.559
No	29	33.48	19.89	3.69	25.92	41.05	-6.00	67.00		
total	43	32.30	18.78	2.86	26.52	38.08	-6.00	67.00		

Annual energy expenditure of 2.3% of firms corresponds to 36–49% of their annual total expenditures. This rate is 26–30% for 23.3% of the firms and less than 25% for about 74.4% of the firms. The energy efficiency trend does not show a statistically significant difference in terms of annual total cost to the energy cost ratio. However, as a result, it would not be wrong to say that energy expenditures constitute 1/4 of the annual expenditures of approximately 2/3 of the companies participating in the research. Furthermore, about 81.4% of the companies participating in the research measure the amount of energy they consume with utility bills and 9.3% measure the amount of energy they consume with a single meter, while 9.3% do not measure the amount of energy they consume. It was discovered that the energy efficiency trend differed statistically significantly in terms of measuring the amount of energy consumed. The energy efficiency trend does not show a statistically significant difference in terms of employment of energy managers or in terms of receiving services from evaluation companies. It is seen that the energy use of companies employing energy managers is also high. This situation may arise from a legal requirement perspective, as well as it requires considering the possibility that the company may have a tendency on the subject.

Moving on, only 2.3% of the companies participating in the research provide energy use training to their employees once a month and 14.0% every six months, while 20.9% provide trainings once a year. On the other hand, about 62.8% of them give no energy use training to their employees. Also, the energy efficiency tendency does not show a statistically significant difference in terms of the energy training status of the employees. Furthermore, more than 67% do not have a study on the use of renewable energy. In addition, considering the energy use by the companies, the energy efficiency trend shows a statistically significant difference on sectorial basis but does not differ statistically significantly in terms of the current status of the companies' renewable energy use studies. For applying solar energy system as one of the most common renewable energies sources to be implemented with rooftop projects in industrial facilities would require the original structure of the building to be ready for this. Hence, renewable energy project installation becomes incompatible or inappropriate in buildings that are built without making provision for such a project in their blueprint structure. In this case, it is essential that new industrial facilities and buildings be licensed to withstand this static load while conducting energy efficiency policy studies.

Moving on, the difference in the importance and tendency of using automatic control systems that are sensitive to daylight in lighting was calculated as (2,326 ± 1,584). According to this result, although daylight sensitive control systems are found useful, they are not preferred. Reasons for this could include the inadequacy of the infrastructure of the building installations, the culture of managing the lighting control panels from a single point, and the lack of necessary technical guidance. The difference in importance and trend of using daylight sensors in outdoor lighting was calculated as (2.140 ± 1.684). It is estimated that this difference is due to some of the aforementioned reasons. in terms of energy efficiency in heating, cooling and ventilation systems, the difference in importance and trend of leak detection and monitoring is calculated as (2,047 ± 1,396). In addition to the fact that this issue is of great importance, the inadequacy or low awareness of companies providing leak detection and monitoring services is estimated to be the reason for the low trend. It has been calculated that the use of electrical appliances in energy saving is the lowest with the difference in energy efficiency importance level (0.930 ± 1.907). The efficiency of energy saving devices in the use of electrical appliances has been sufficiently known and accepted by the society. The difference in the importance and trend of energy-efficient lighting evaluation was calculated as (0.953 ± 1.214). The main reason for this is that the widely produced energy-saving lamps are long-lasting, healthy, and accessible. Also, it was observed that the difference between the importance and tendency of using motion sensors in public areas is (1.070 ± 1.316). If we examine the result of this evaluation, we can say that the hygiene and usefulness of sensor systems increases this rate of preference. Finally, based on the analysis and on-site evaluations, some aspects in the survey questions that are thought not to be adequately understood by the participants include (a) the design of the boiler compressor room should be done, (b), leakages detection and monitoring records should be made for efficiency in heating, cooling, and ventilation systems, and (c), leakages detection and monitoring records should be made for efficiency in electrical and lighting systems.

5.1 Policy recommendation

To ensure energy efficiency for companies, it is necessary to address energy-saving investments and energy efficiency policies at a macro level. Especially large-scale companies in the manufacturing industry sector, whose energy costs are over 20% of their annual costs, need to make provisions for energy efficiency and savings in the general budget by evaluating energy projects vis-à-vis their investment costs before turning to energy savings. At the same time, it is necessary to analyze how soon the energy saving investments will return as profitability. In addition, the measures to be taken to use energy resources efficiently and economically in the facilities of the companies; automation-controlled monitoring of all systems, replacement of lighting systems with energy-saving PLC (programmable logical control unit) lamps, installation of motion systems sensors, installation of dimmer systems, adequate use of daylight, and selection of UPS systems with appropriate power are considered important in terms of energy efficiency. Also, considering that some participants lack information on some scales in terms of energy efficiency, institutions and organizations working on energy efficiency and those developing policies should give more attention to the evaluations of these salient areas by organizing informative, training, and seminar programs cutting across both the manufacturing industry and service sector. In addition, it is equally important to carry out awareness-raising activities to sensitize the public.

In addition, the energy efficiency trend does not show a statistically significant difference in terms of employment of energy managers or in terms of receiving services from evaluation companies and the number of companies that receive energy efficiency services among the companies participating in the survey is quite low. In this context, considering this situation while determining the fund and support mechanisms in the field of energy efficiency, deducting the amount to be spent on efficiency studies for companies from taxes or to encourage companies with loans with favorable interest rates should be considered to curb productivity losses, leakages, and inefficient practices. Also, quite a low proportion of the companies participating in the survey provided energy use training to their employees in some way. Again, while establishing a fund and support mechanism in the field of energy efficiency, it would be appropriate to direct companies to support in-house energy efficiency trainings and to carry them out in a professional line. In addition, since renewable energy project installations can become incompatible or inappropriate in firms' buildings that are built without making provision for such a project in the blueprint structural design. It is further recommended that new industrial facilities and buildings be supervised and licensed to withstand this static load while conducting energy efficiency policy studies.

References

Abas, N., Kalair, A. and Khan, N. (2015). Review of fossil fuels and future energy technologies. *Futures*, 69: 31–49. https://doi.org/10.1016/j.futures.2015.03.003.

Ağbulut, Ü. (2019). Turkey's electricity generation problem and nuclear energy policy. *Energy Sources, Part A: Recovery, Utilization, and Environmental Effects*, 41(18): 2281–2298. https://doi.org/10.1080/15567036.2019.1587107.

Alola, A. A., Adebayo, T. S. and Onifade, S. T. (2021). Examining the dynamics of ecological footprint in China with spectral Granger causality and quantile-on-quantile approaches. *International Journal of Sustainable Development and World Ecology*, 1–14. https://doi.org/10.1080/13504509.2021.1990158.

Alola, A. A. and Onifade, S. T. (2022). Energy innovations and pathway to carbon neutrality in Finland. *Sustainable Energy Technologies and Assessments*, 52: 102272. https://doi.org/10.1016/j.seta.2022.102272.

Apergis, N. and Payne, J. E. (2010). Renewable energy consumption and economic growth: Evidence from a panel of OECD countries. *Energy Policy*, 38(1): 656–660. https://doi.org/10.1016/j.enpol.2009.09.002.

Ates, S. A. and Durakbasa, N. M. (2012). Evaluation of corporate energy management practices of energy intensive industries in Turkey. *Energy*, 45(1): 81–91. https://doi.org/10.1016/j.energy.2012.03.032.

Aydin, G. (2014). Modeling of energy consumption based on economic and demographic factors: The case of Turkey with projections. *Renewable and Sustainable Energy Reviews*, 35: 382–389. https://doi.org/10.1016/j.rser.2014.04.004.

Balsalobre-Lorente, D. B. and Álvarez-Herranz, A. (2016). Economic growth and energy regulation in the environmental Kuznets curve. *Environmental Science and Pollution Research*, 23(16): 16478–16494. https://doi.org/10.1007/s11356–016–6773–3.

Bekun, F. V., Gyamfi, B. A., Onifade, S. T. and Agboola, M. O. (2021). Beyond the environmental Kuznets Curve in E7 economies: Accounting for the combined impacts of institutional quality and renewables. *Journal of Cleaner Production*, 127924. https://doi.org/10.1016/j.jclepro.2021.127924.

Bekun, F. V., Onifade, S. T., Agboola, P. and Altuntaş, M. (2022). How do technological innovation and renewables shape environmental quality advancement in emerging economies: An exploration of the E7 bloc?. *Sustainable Development*, 30(4): 1–13. http://dx.doi.org/10.1002/sd.2366.

British Petroleum BP. (2020). Statistical Review of World Energy June 2020. Available at: http://www.bp.com/statisticalreview. Accessed May 2021.

Bulut, U. and Muratoglu, G. (2018). Renewable energy in Turkey: Great potential, low but increasing utilization, and an empirical analysis on renewable energy-growth nexus. *Energy Policy*, 123: 240–250. https://doi.org/10.1016/j.enpol.2018.08.057.

Bozkaya, Ş., Onifade, S. T., Duran, M. S. and Kaya, M. G. (2022). Does environmentally friendly energy consumption spur economic progress: Empirical evidence from the Nordic countries? *Environmental Science and Pollution Research*, 29(46): 1–11. https://doi.org/10.1007/s11356–022–23452–4.

Cin, R., Acuner, E. and Onaygil, S. (2021). Analysis of energy efficiency obligation scheme implementation in Turkey. *Energy Efficiency*, 14(1): 1–21. https://doi.org/10.1007/s12053–020–09914-z.

Dingru, L., Onifade, S. T., Ramzan, M. and AL-Faryan, M. A. S. (2023). Environmental perspectives on the impacts of trade and natural resources on renewable energy utilization in Sub-Sahara Africa: Accounting for FDI, income, and urbanization trends. *Resources Policy*, 80: 103204. https://doi.org/10.1016/j.resourpol.2022.103204.

Dernbach, J. C. (2007). Overcoming the behavioral impetus for greater US energy consumption. *Pac. McGeorge Global Bus. and Dev. LJ*, 20: 15. Available at https://heinonline.org/HOL/Contents?handle=hein.journals/tranl20andid=1andsize=2andindex=andcollection=journals.

Duran, M. S., Bozkaya, Ş., Onifade, S. T. and Kaya, M. G. (2022). Nuclear energy consumption and energy-driven growth nexus: A system GMM analysis of 27 nuclear utilizing countries across the globe. *Environmental Science and Pollution Research*, 29(46): 1–9. https://doi.org/10.1007/s11356–022–22951–8.

Ediger, V. Ş. and Berk, I. (2011). Crude oil import policy of Turkey: Historical analysis of determinants and implications since 1968. *Energy Policy*, 39(4): 2132–2142. https://doi.org/10.1016/j.enpol.2011.01.058.

Ervural, B. C., Ervural, B. and Zaim, S. (2016). Energy efficiency evaluation of provinces in Turkey using data envelopment analysis. *Procedia-Social and Behavioral Sciences*, 235: 139–148. https://doi.org/10.1016/j.sbspro.2016.11.009.

Erdoğan, S., Onifade, S. T., Altuntaş, M. and Bekun, F. V. (2022). Synthesizing urbanization and carbon emissions in Africa: How viable is environmental sustainability amid the quest for economic growth in a globalized world? *Environmental Science and Pollution Research*, 29(45): 1–14. https://doi.org/10.1007/s11356–022–18829–4.

García-Quevedo, J. and Jové-Llopis, E. (2021). Environmental policies and energy efficiency investments. An industry-level analysis. *Energy Policy*, 156: 112461. https://doi.org/10.1016/j.enpol.2021.112461.

Gazbir. (2018). Energy Status in the World and Turkey. Available at gazbir.org.tr/uploads/page/Dunya-ve-Turkiye-Enerji-Gorunumu.pdf?_x_tr_sl=trand_x_tr_tl=enand_x_tr_hl=tr Access January 2022.

Gordić, D., Babić, M., Jovičić, N., Šušteršič, V., Končalović, D. and Jelić, D. (2010). Development of energy management system–Case study of Serbian car manufacturer. *Energy Conversion and Management*, 51(12): 2783–2790. https://doi.org/10.1016/j.enconman.2010.06.014.

Gyamfi, B. A., Onifade, S. T., Nwani, C. and Bekun, F. V. (2021). Accounting for the combined impacts of natural resources rent, income level, and energy consumption on environmental quality of G7 economies: A panel quantile regression approach. *Environmental Science and Pollution Research*, 1–13. https://doi.org/10.1007/s11356–021–15756–8.

Gyamfi, B. A., Bekun, F. V., Onifade, S. T. and Altuntaş, M. (2022). Significance of air transport to tourism-induced growth hypothesis in E7 economies: Exploring the implications for environmental quality. *Tourism: An International Interdisciplinary Journal*, 70(3): 339–353. https://hrcak.srce.hr/279096.

Hasanbeigi, A. and Price, L. (2012). A review of energy use and energy efficiency technologies for the textile industry. *Renewable and Sustainable Energy Reviews*, 16(6): 3648–3665. https://doi.org/10.1016/j.rser.2012.03.029.

https://ekbbelgesi.com.tr/ Access Date: 17 March 2020.

Kanchiralla, F. M., Jalo, N., Thollander, P., Andersson, M. and Johnsson, S. (2021). Energy use categorization with performance indicators for the food industry and a conceptual energy planning framework. *Applied Energy*, 304: 117788. https://doi.org/10.1016/j.apenergy.2021.117788.

Karagöz, Y. and Kösterelioğlu, İ. (2008). Developing the communication skills assessment scale with the factor analysis method. *Dumlupınar University Journal of Social Sciences*, (21).

Karasar, N. (2007). Scientific research method: Concepts, principles, techniques. *Ankara: Nobel Yayınevi* http://hdl.handle.net/123456789/921.

KPMG. (2019). Global Automotive Executive Survey 2019. KPMG Automotive Institute, retrieved on Jun. 2021. [Online]. Available: https://home.kpmg/xx/en/home/insights/2019/01/global-automotive-executive-survey-2019.html.

Mattick, C. S., Williams, E. and Allenby, B. R. (2010). Historical trends in global energy consumption. *IEEE Technology and Society Magazine*, 29(3): 22–30. https://doi.org/10.1109/MTS.2010.938106.

Onifade, S. T., Erdoğan, S., Alagöz, M. and Bekun, F. V. (2021a). Renewables as a pathway to environmental sustainability targets in the era of trade liberalization: Empirical evidence from Turkey and the Caspian countries. *Environmental Science and Pollution Research*, 1–12. https://doi.org/10.1007/s11356–021–13684–1.

Onifade, S. T., Alola, A. A., Erdoğan, S. and Acet, H. (2021b). Environmental aspect of energy transition and urbanization in the OPEC member states. *Environmental Science and Pollution Research*, 28(14): 17158–17169. https://doi.org/10.1007/s11356–020–12181–1.

Önüt, S. and Soner, S. (2006). Energy efficiency assessment for the Antalya Region hotels in Turkey. *Energy and Buildings*, 38(8): 964–971. https://doi.org/10.1016/j.enbuild.2005.11.006.

Onifade, S. T., Gyamfi, B. A., Haouas, I. and Bekun, F. V. (2021). Re-examining the roles of economic globalization and natural resources consequences on environmental degradation in E7 economies: Are human capital and urbanization essential components? *Resources Policy*, 74: 102435. https://doi.org/10.1016/j.resourpol.2021.102435.

Onifade, S.T. (2022). Retrospecting on resource abundance in leading oil-producing African countries: how valid is the environmental Kuznets curve (EKC) hypothesis in a sectoral composition framework? *Environmental Science and Pollution Research*, 29(1): 52761–52774. http://dx.doi.org/10.1007/s11356–022–19575–3.

Onifade, S. T. and Alola, A. A. (2022). Energy transition and environmental quality prospects in leading emerging economies: The role of environmental-related technological innovation. *Sustainable Development*, 30(2): 1–13. https://doi.org/10.1002/sd.2346.

Pituch, K. A. and Stevens, J. P. (2015). Applied multivariate statistics for the social sciences: Analyses with SAS and IBM's SPSS. Routledge.

Shafiee, S. and Topal, E. (2009). When will fossil fuel reserves be diminished? *Energy Policy*, 37(1): 181–189. https://doi.org/10.1016/j.enpol.2008.08.016.

Thollander, P. and Ottosson, M. (2010). Energy management practices in Swedish energy-intensive industries. *Journal of Cleaner Production*, 18(12): 1125–1133. https://doi.org/10.1016/j.jclepro.2010.04.011.

TUIK. (2020). Turkish Statistical Institute. Available at: http://www.tuik.gov.tr Access Date: 20 March 2020.

UETM. (2011). Ulusal Enerji Tasarruf Merkezi. Available at: http://www.eie.gov.tr/turkce/en tasarrufu/uetm/hakkimizda.html.

WorldBank.(2006).Theroadto2050.SustainableMobilityforthe21stCentury.Availableathttps://documents1.worldbank.org/curated/en/192421468341095824/pdf/360210rev0The0Road0to0205001PUBLIC1.pdf Accessed January 2022.

Chapter 6

Role of Middle East and North African Region in Global Hydrogen Energy Transition

Mustapha D. Ibrahim

1. Introduction

The increase in greenhouse gases (GHG) since the industrial revolution in the 18th century in the West has triggered an unsustainable climate change situation. As a result, the government and other stakeholders are forced to redefine their policies toward climate change to limit GHG emissions and implement a series of measures to ensure environmental sustainability. A significant factor that impacts the exponential increase in environmental pollution is the global increase in energy requirements. As a result, there is a growing necessity to rethink our energy consumption and production patterns to combat global warming and its effects. The energy transition is a necessity and a global challenge, and hydrogen has been identified as a catalyst for combating climate change due to its ability to decarbonize in hard-to-abate sectors such as heavy-duty transport, freight, and air transport, and many heavy industries (Griffiths et al. 2021). Most recently, hydrogen has gained significant momentum as one of the pillar solutions with unprecedented business and political support with ground breaking projects and policies to expand utilization. Hydrogen is the smallest, lightest, and most abundant element on earth and the universe (Jain 2009). It can be found everywhere but almost always combined with other chemical elements. Therefore, it needs to be separated in order to obtain it in its pure form. The method in which hydrogen is produced defines the type of hydrogen.

Industrial Engineering Technology, Higher Colleges of Technology, P.O. Box 7947, Sharjah, United Arab Emirates.
Email: mibrahim1@hct.ac.ae

The global uptake of hydrogen as an energy technology to support decarbonization is a result of the convergence of multiple factors, including the cost-competitive adoption of renewable energy, efficiency improvement of electrolyzers, increase in climate change regulation, and a broader application of hydrogen (Noussan et al. 2020). Global demand for hydrogen was about 90 Mt in 2020, and it is expected to reach 140 Mt in 2025 and 210 Mt in 2030 (IEA 2021). Hydrogen contributes in numerous ways to the energy transition. Figure 1 illustrates seven ways hydrogen can support global energy transition. Hydrogen production can be integrated with variable renewable energy sources and carbon capture technology (IRENA 2019). It can be used in large-scale renewable energy integration, as it enables the storage of excess energy production that could otherwise be wasted using power-to-X technologies. Hydrogen could be used to distribute renewable energy across different regions. Countries such as Japan are not positioned to produce renewable energy from wind or solar energy. Therefore, importing renewable energy might be more economical. Hydrogen and its compound are more energy dense and will help redistribute energy effectively and efficiently (IEA 2019). Hydrogen can be used to balance the energy grid. It can also be used on-site or transported to another location. Enhancing the supply chain of hydrogen, from production and storage to distribution, will contribute to the widespread industrial application as a clean energy carrier and for clean mobility (Iida and Sakata 2019). Hydrogen has an important role to play in decarbonizing the transport sector through fuel cell electric vehicles (FCEVs). Electric vehicles are already reducing vehicle emissions. However, to fully decarbonize the transport sector requires an added deployment of zero-emission vehicles like hydrogen-powered FCEVs.

Many industrial processes use natural gas, coal, and oil to provide energy, consequently producing about 20% of global CO_2 emissions (IEA 2019). Thus, the industry must improve energy efficiency and reduce energy consumption. Hydrogen offers a zero-emission alternative for heating. Hydrogen-based chemistry can be used to decarbonize parts of the petrochemical value chain. If Carbon Capture and Utilization (CCU) technology takes off, green hydrogen could be used to convert the captured carbon into valuable chemicals such as methanol, methane, formic acid, or urea.

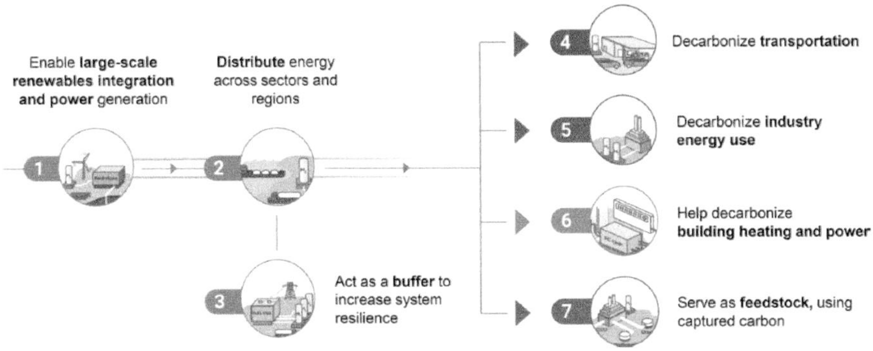

Figure 1: Roles of hydrogen in global energy transition (adapted from Hydrogen Council 2021 (HC 2021)).

Transitioning to a low-carbon economy not only fights climate change but also enhances energy security and boosts local industries' productivity (Yang et al. 2019). Hydrogen has a significant role in the low-carbon economy. This transition comes with certain difficulties. There are technological and geographical perspectives to hydrogen energy adoption driven by technical challenges, economic, and geopolitical implications (Noussan et al. 2020). For low-carbon and renewable energy to fulfill its potential in the future energy mix, there is a need to build new synergies, federate the industrial ecosystem, improve every production link, and build human capacity (Letessier 2021). Given the technological and geographical implication of the global hydrogen energy transition, the Middle East and North Africa (MENA) has a role in an efficient transition given their position in the energy market. This paper explores the challenges of establishing a low-carbon hydrogen transition and MENA's role in mitigating some of the challenges and enhancing the hydrogen transition.

2. Low-Carbon Hydrogen

Hydrogen production processes are often classified according to colors based on the energy sources. Hydrogen-generated fossil fuel such as natural gas, petroleum, and coal is denoted as 'grey,' 'black,' and 'brown,' respectively. About 80% of global hydrogen production is sourced from emission-intensive sources such as natural gas reforming 'grey hydrogen' and coal gasification (CG) 'brown hydrogen' (IEA 2021). Most brown hydrogen comes from China and other parts of Asia (IEA 2019). Utilizing hydrogen from unabated fossil fuels offers minimal environmental benefits and, in some cases, increases global emissions. Therefore, it is critical for low-carbon hydrogen to replace the current hydrogen production in order for hydrogen to contribute significantly to the clean energy transition.

Hydrogen production with natural gas where carbon dioxide emission is captured is referred to as 'blue hydrogen.' Hydrogen purely produced from renewable energy sources such as solar and wind energy is known as 'green hydrogen.' Generally, green hydrogen is produced using renewable energy sources to run an electrolyzer to split water producing H_2 and O_2 (Gandhi et al. 2022). Typically, blue hydrogen and green hydrogen are considered low-carbon hydrogen. Green hydrogen production does not emit CO_2. Nonetheless, the green hydrogen value chain, such as electrolyzer production, presents some CO_2 emissions that could be calculated. Figure 2 illustrates the hydrogen production pathways and description.

There has been a significant advancement in electrolyzer technology. This improvement, couple with the declining cost of renewable energy production (Apostoleris et al. 2019), has positively impacted the hydrogen transition and global hydrogen economy. Currently, green hydrogen cost about two to three times the cost of blue hydrogen (IRENA 2022b), and research shows a continuous decline in green hydrogen production cost, especially in the MENA region (Razi and Dincer 2022). Advancing the green hydrogen initiative is a priority for many countries. The three indicators for tracking clean hydrogen progress are the level at which low-carbon hydrogen meets new applications and replaces conventional hydrogen in industrial applications, the demand growth in new sectors where hydrogen is used to reduce CO_2

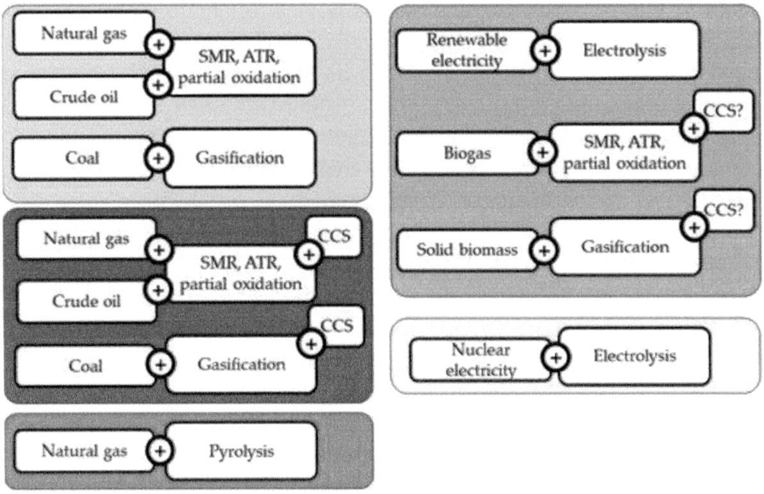

Figure 2: Hydrogen production pathways. SMR: steam methane reforming, ATR: autothermal reforming, CCS: carbon capture and sequestration. Source: Noussan et al. 2020.

emissions, and cost reduction, process integration, scale-up, and improvement in the efficiency of cross-cutting technology such as fuel-cells, electrolyzers, and CCUS-equipped hydrogen production (IEA 2021). These three indicators signify growth in low-carbon hydrogen adoption. Low-carbon hydrogen will account for about 20–40% of hydrogen supply by 2050 (HC 2021). This percentage is equivalent to 140 to 280 MT of hydrogen supply. To reach the net-zero target, relying on a single pathway to low-carbon hydrogen supply will slow down the process. There needs to be an even swifter ramping up of the value chain and clean hydrogen production pathways project development.

3. Challenges in Hydrogen Transition

In the past decade, there has been significant progress in the hydrogen economy. However, there are still challenges in the transition. These obstacles need to be addressed for the full benefits of hydrogen to be materialized. Few studies have looked at the challenges of the Hydrogen economy transition at the national level. Dou et al. (Dou et al. 2017) used strengths, weaknesses, opportunity, and threats (SWOT) analysis to analyze China's hydrogen economy transition. The study summarized the challenges as safety in hydrogen delivery and storage, shortage of hydrogen production, lack of transportation, and insufficient infrastructure. There has been tremendous progress in addressing the safety concerns of hydrogen. The promise of a hydrogen future comes with enormous safety responsibility.

There is a disparity in the attainment of hydrogen economy among the so-called top hydrogen energy countries due to challenges faced by individual countries. The challenges can be categorized into six major factors: technological, economic, infrastructure, environmental, social, and political aspects (Ren et al. 2020). There has been advancement in the technological aspect of the hydrogen energy transition.

This includes improved electrolyzer efficiency to boost hydrogen production and advancement in solar and wind energy technologies. However, hydrogen production at the current industrial levels, for example, SMR and CG, is not environmentally sustainable (Sharma and Ghoshal 2015). Production of hydrogen from water electrolysis using renewable energy is in the medium and long-term (Ping et al. 2018). Consequently, the production of fuel-cell technology for the commercialization of hydrogen is a top priority (Zegers 2006). Technological advances are required to improve fuel cell durability and efficiency. Investment in research and development (R&D) for components is also necessary.

The economic aspect remains a significant challenge to ushering in the hydrogen economy. The cost competitiveness of hydrogen is a primary concern. Regarding cost, the most economically efficient hydrogen production routes are SMR, nuclear pyrolysis, biomass, and coal gasification (Bakenne et al. 2016). Coal and natural gas production costs are estimated at around \$0.36–\$1.83, and 2.48–3.17 per kg, respectively. The production related to other sources exceeds \$4.00/kg (Bartels et al. 2010). Coal gasification is attractive in China, Russia, and the United States. However, this comes with a significant carbon footprint if CCS is not used. Biomass gasification is seen as an economically attractive option for future hydrogen production from renewable energy (Ball and Weeda 2016a). Renewable energy sources are the best choice for hydrogen production (Dincer and Zamfirescu 2012). While in the short and medium term, fossil fuels will still play an essential role in the transition due to their economic and technical reference (Moliner et al. 2016). In both the short and long run, MENA is positioned to contribute significantly to the hydrogen transition due to the region's low energy cost.

The demand for hydrogen has grown more than threefold since 1975. The demand for pure hydrogen is about 70 million tonnes/year, almost entirely supplied from fossil fuel. A significant challenge in the hydrogen energy transition is replacing hydrogen from fossil fuels with renewable and low-carbon hydrogen. Increasing low-carbon hydrogen in the sector application will make the hydrogen energy transition swiftly. IEA estimated global hydrogen demand in a net-zero scenario (Figure 3). A significant amount of hydrogen will still be produced from fossil fuels. Low-carbon hydrogen from electrolysis and CCS/CCU integration must be expanded for an effective and efficient transition.

Infrastructural challenges are a critical factor in expanding hydrogen applications and demand. Although there are many applications of hydrogen today, and many are being explored, the economic and cost implications of the application due to infrastructure are crucial. The doubt about increasing hydrogen demand comes from the criticism of supply chain infrastructure. The technical and cost feasibility, efficiency, and availability of rare earth metals required for the hydrogen supply chain are often highlighted. Shinnar (Shinnar 2003) outlined six misconceptions about hydrogen and why the hydrogen economy is inefficient and electricity is a better alternative. Page and Krumdieck (Page and Krumdieck 2009) analyzed hydrogen energy efficiency and the hydrogen economy, and their results expressed doubts about the environmental benefits of pursuing the hydrogen economy. The practicality of the hydrogen economy was analyzed by Yun (Yun 2016), and his results concluded that

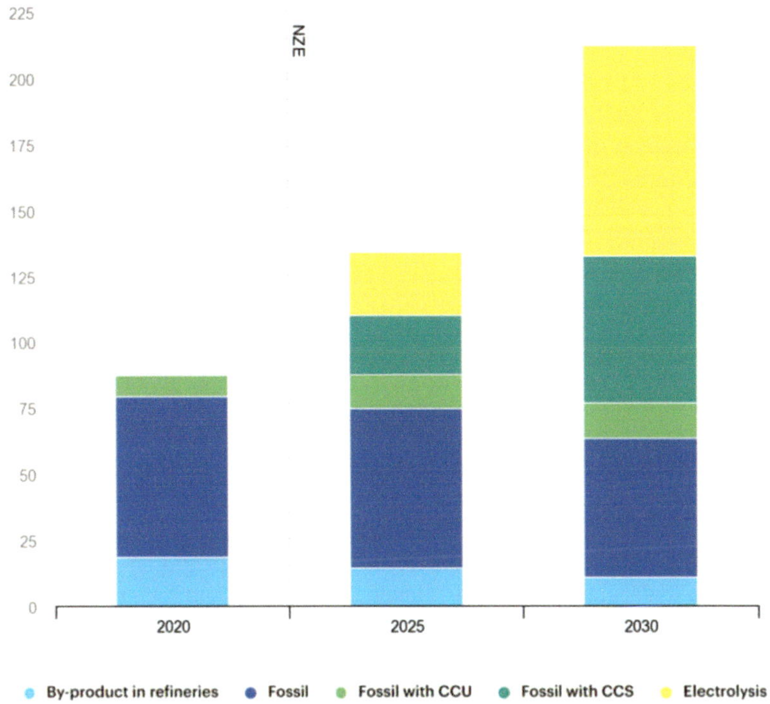

Figure 3: Net-zero scenario of global hydrogen demand by production (IEA 2021).

the hydrogen engine could end as a failure based on the theorem of indirect energy conversion. Based on empirical data and analysis, Flood (Flood 2012) acknowledged the technical feasibility and hydrogen in various transportation and power-generation sectors. However, the cost and additional infrastructural development and investment will limit the application in the future. Few studies, such as Kleijn and Voet (Kleijn and van der Voet 2010), point to the scarcity of resources required for hydrogen infrastructure as a limiting factor to a green hydrogen system rather than the availability of renewable energy. Macdonald (Macdonald 2004) augured that a critical aspect of the hydrogen economy debate is hydrogen production because hydrogen production from fossil fuels does not meet environmental requirements. The pros and cons of the hydrogen economy raise valid points and skepticism. However, the advantages of hydrogen as a future energy source make it an inevitable part of our energy system. Under a distinctive pessimistic and optimistic viewpoint of hydrogen, Simbeck (Simbeck 2004) discussed the feasibility of hydrogen economy in the long run with short-term challenges such as technical feasibility of onboarding of FCVs, inefficient hydrogen production, and cost of production. To appreciate the role of hydrogen in the future energy mix, hydrogen should not be evaluated in isolation (Michael Ball and Wietschel 2009, Ball and Weeda 2016b). The infrastructural requirement for scaling hydrogen has to be assessed from the production, transportation, and demand perspectives.

The environmental challenges of low-carbon hydrogen come mostly from blue hydrogen. For hydrogen produced from fossil fuel to be considered low-carbon (Figure 2), the carbon emitted must be captured, stored, or utilized (CCS/U). The progress so far falls well short of the 2050 target since the technology for utilization of carbon from hydrogen production is still at its early stages and is not scaled. Furthermore, CCS/U comes with environmental challenges such as leakage (Zhang 2021).

People's attitude is a significant huddle towards the adoption of hydrogen. There is a general lack of understanding about renewable energy generation, its advantages, and the fundamental definition of renewable energy. Social barriers arise when implementing new technologies, including hydrogen development (Ingaldi and Klimecka-Tatar 2020). Social resistance may influence investment, delay implementation, and policy adoption of hydrogen energy transition (Astuti et al. 2019). These types of barriers are among the difficult to overcome. Therefore, there needs to be a general ecological awareness and environmental education about the subject of hydrogen energy systems. However, it is crucial to identify the type of social and ecological action needed to improve the population's attitude towards hydrogen implementation. There are interesting studies about hydrogen adoption such as adoption of hydrogen in mobility(Calandra et al. 2023), and sectoral adaptation of hydrogen transition to low-carbon hydrogen energy systems (Ibrahim et al. 2022); however, they are primarily from a technical point of view. A significant increase in the social perspective study is critical to ascertain the people's attitude about this new energy alternative and transition (Astuti et al. 2019).

The adoption of hydrogen strategies has political implications. These strategies are largely catalyzed by the United Nations Climate Change Conference in Glasgow (COP26). Most countries pushing for hydrogen strategies are advanced and emerging economies. There is a clear gap in the hydrogen economy between Europe, Africa, MENA, and Asia. According to a report from the United Nations Environmental Programme (UNEP), the inability of the developing world to participate in research and development to advance the hydrogen economy is primarily due to financial and technical deficiencies [40]. The gap in hydrogen economy attainment could advance the per capita income discrepancies between advanced and developing nations and aggravate the social and malnutrition problem of respective regions. This has political and social implications.

4. Role of MENA in Hydrogen Transition

MENA is a significant player in the global energy market and seeks to maintain its position in the energy transition and hydrogen energy market. The United Arab Emirates (UAE), Saudi Arabia, and Oman are among the countries that have taken bold steps toward the global hydrogen market. UAE aims to capture 25% of the global hydrogen market by 2025 (Arabnews 2022). Similarly, Saudi Arabia plans to invest $36 billion in hydrogen production by 2030 with its hydrogen strategy (ZAWYA 2022). The global hydrogen transition comes with certain challenges and scepticisms, as discussed in Section 3. Given MENA's strategic location and resources, it is positioned to capture a major share of the global hydrogen market and maintain its position as a leading energy provider. More importantly, the long-run clean hydrogen projections and net-zero emission target by many energy bodies such as IEA and the hydrogen council signifies the importance of the MENA region in the energy transition.

The technologies for hydrogen production already exist. What is required is scaling production, in particular green and other forms of low-carbon hydrogen. There are three methods for low-carbon hydrogen production: the first is "steam methane reforming (SMR), utilizing bio-methane, or fusing carbon capture and storage or utilization (CCS/U). The second is through electrolysis using renewable energies such as solar and wind. The third is biomass gasification (Weimann et al. 2021). Among the low-carbon hydrogen production technologies, SMR and electrolysis are mature technologies (Sánchez-Bastardo et al. 2021) that continue to experience significant production efficiency improvement.

On the other hand, SMR with CCS/U and gasification are in the pilot phase with great potential for scaling. However, research shows promising results in both cases. Given the resources and investment required in all three low-carbon hydrogen production technologies, MENA is in a position to scale low-carbon hydrogen production.

The potential of renewable energy production for hydrogen varies geographically (Dehshiri and Dehshiri 2022). A major driver of green hydrogen economics is the location which determines the consistency of power generation for green hydrogen production (de León et al. 2022). Therefore, the need for energy trading will persist. This is compounded by the limited "storability" of electricity. Functional cross-border hydrogen infrastructure is essential to ensure the sustainability of hydrogen transition, such as a secure energy supply. MENA has a geographical advantage of securing viable infrastructure to support the hydrogen transition (Dunn 2002).

Bilateral hydrogen trading relations are imperative in hydrogen transition. Several countries have signed MOUs for trading hydrogen and hydrogen-related technologies. According to IRENAs report, about 30% of the prospective hydrogen exporters to prospective importers such as Belgium, Germany, Japan, Netherlands, and the Republic of Korea are MENA countries (Figure 4) (IRENA 2022a). More importantly, a large amount of global green hydrogen production is expected to come from MENA (Nicita et al. 2020).

The MENA region exhibits great potential for hydrogen infrastructural development based on its robust hydrocarbon production infrastructure. There is substantial research supporting the repurposing of infrastructure to support hydrogen development. The capital cost associated with establishing an entirely new infrastructure limits hydrogen uptake in many countries and regions. The region's robust expertise in hydrocarbon infrastructure is expected to minimize infrastructural investment hurdles. In addition, the strategic location of MENA as refuelling stations for future international logistics using hydrogen is encouraging. However, the broader cost associated with upgrading surrounding infrastructure for hydrogen transportation, handling, and storage is problematic.

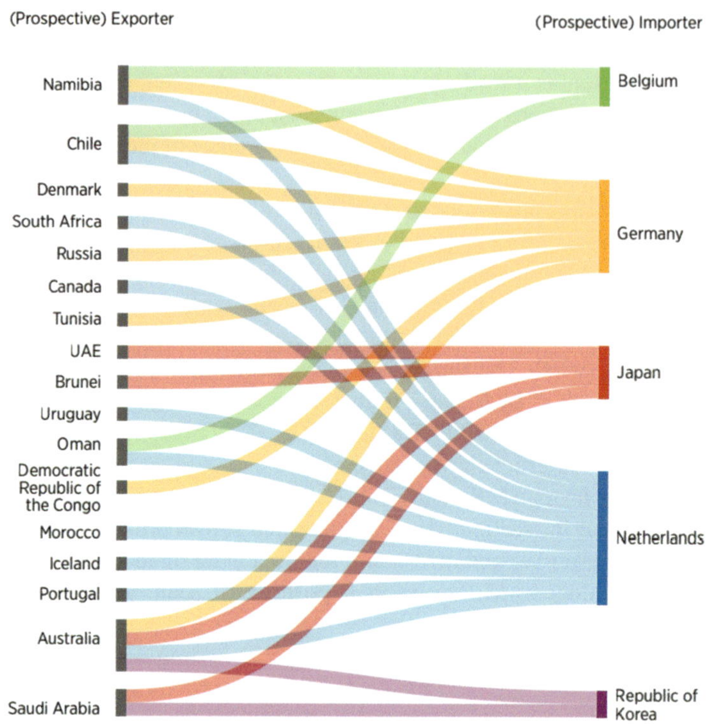

Figure 4: Bilateral hydrogen trade agreements and Memorandum of Understanding (Source: IRENA report (IRENA 2022)).

Nonetheless, the mature ports and gas handling infrastructure could be transformed for hydrogen utilization. This will require significant investment in ports and other associated equipment. The sizable export capacity of the region provides the foundation for the development of export-oriented infrastructure to support the global hydrogen transition. For example, the UEA provides a potential for clusters of hydrogen-related activities to minimize investment risk, diversify hydrogen investment, and mitigate investment risk. The study shows that the MENA region has the potential to mitigate the technological and infrastructural challenges of the global hydrogen transition. In addition to the economic advantage of low hydrogen production cost to boost low-carbon hydrogen production.

5. Conclusion

While hydrogen is considered a game changer in the net-zero emission transition, some challenges limit its uptake at a global level. MENA is a region with enormous potential for enhancing global hydrogen energy transition. This study looks into various challenges affecting the global hydrogen transition, which include technological, economic, infrastructural, environmental, social, and political. At a global level, the MENA regions is expected to have a major impact on the hydrogen transition. The study reveals that the MENA region has a significant advantage in low-carbon hydrogen transition given its mature hydrocarbon infrastructure that can, in theory, be repurposed for blue hydrogen infrastructure. The geographical location also provides sustainable renewable energy for green hydrogen production, which is considered the future of sustainable energy for many industrial activities. The MENA region has an economic advantage in hydrogen production due to its low production cost. On the technological and infrastructural front, MENA country's hydrogen strategies create a friendly ecosystem for investors, bilateral trade agreements, R&D collaboration, and policy backing. While none of these challenges are mutually exclusive, technological and infrastructural challenges are critical, and the MENA region has significant advantages in enabling the hydrogen transition.

References

Apostoleris, H., Sgouridis, S., Stefancich, M. and Chiesa, M. (2019). Utility solar prices will continue to drop all over the world even without subsidies. *Nature Energy*, 4(10): 833–834.

Astuti, S. P., Day, R. and Emery, S. B. (2019). A successful fuel transition? Regulatory instruments, markets, and social acceptance in the adoption of modern LPG cooking devices in Indonesia. *Energy Research & Social Science*, 58: 101248.

Bakenne, A., Nuttall, W. and Kazantzis, N. (2016). Sankey-Diagram-based insights into the hydrogen economy of today. *International Journal of Hydrogen Energy*, 41(19): 7744–7753.

Ball, M. and Weeda, M. (2016a). 11-The hydrogen economy—Vision or reality? *Compendium of Hydrogen Energy*, 237–266.

Ball, M. and Weeda, M. (2016b). 11-The hydrogen economy—Vision or reality? pp. 237–266. *In*: Ball, M., Basile, A. and Veziroğlu, T. N. (eds.). *Compendium of Hydrogen Energy*. Oxford: Woodhead Publishing.

Ball, M. and Wietschel, M. (2009). The future of hydrogen – opportunities and challenges. *International Journal of Hydrogen Energy*, 34(2): 615–627. doi:https://doi.org/10.1016/j.ijhydene.2008.11.014.

Bartels, J. R., Pate, M. B. and Olson, N. K. (2010). An economic survey of hydrogen production from conventional and alternative energy sources. *International Journal of Hydrogen Energy*, 35(16): 8371–8384.

Calandra, D., Wang, T. D., Cane, M. and Alfiero, S. (2023). Management of hydrogen mobility challenges: A systematic literature review. *Journal of Cleaner Production*, 137305.

de León, C. M., Ríos, C. and Brey, J. (2022). Cost of green hydrogen: Limitations of production from a stand-alone photovoltaic system. *International Journal of Hydrogen Energy*.

Dehshiri, S. S. H. and Dehshiri, S. J. H. (2022). Locating wind farm for power and hydrogen production based on Geographic information system and multi-criteria decision making method: An application. *International Journal of Hydrogen Energy*.

Dincer, I. and Zamfirescu, C. (2012). Sustainable hydrogen production options and the role of IAHE. *International Journal of Hydrogen Energy*, 37(21): 16266–16286.

Dou, Y., Sun, L., Ren, J. and Dong, L. (2017). Opportunities and future challenges in hydrogen economy for sustainable development. In *Hydrogen Economy* (pp. 277–305), Elsevier.

Dunn, S. (2002). Hydrogen futures: Toward a sustainable energy system. *International Journal of Hydrogen Energy*, 27(3): 235–264.

Flood, W. (2012). The Hydrogen Economy—Global Solution, or Pie in the Sky? *Energy and Environment*, 23(6-7): 1097–1104.

Gandhi, K., Apostoleris, H. and Sgouridis, S. (2022). Catching the hydrogen train: Economics-driven green hydrogen adoption potential in the United Arab Emirates. *International Journal of Hydrogen Energy*.

HC. (2021). Hydrogen for Net-Zero: A critical cost-competitive energy vector.

Ibrahim, M. D., Binofai, F. A. and Mohamad, M. O. (2022). Transition to low-carbon hydrogen energy system in the UAE: Sector efficiency and hydrogen energy production efficiency analysis. *Energies*, 15(18): 6663.

IEA. (2019). The Future of Hydrogen.

IEA. (2021). Hydrogen: More Efforts Needed.

Iida, S. and Sakata, K. (2019). Hydrogen technologies and developments in Japan. *Clean Energy*, 3(2): 105–113.

Ingaldi, M. and Klimecka-Tatar, D. (2020). People's attitude to energy from hydrogen—from the point of view of modern energy technologies and social responsibility. *Energies*, 13(24): 6495.

IRENA. (2019). gen: A renewable energy perspective.

IRENA. (2022a). Geopolitics of the Energy Transformation: The Hydrogen Factor. Abu Dhabi.

IRENA. (2022b). Green hydrogen cost reduction. https://www.irena.org/publications/2020/Dec/Green-hydrogen-cost-reduction. Accessed Aug. 22 2022.

Jain, I. (2009). Hydrogen the fuel for 21st century. *International Journal of Hydrogen Energy*, 34(17): 7368–7378.

Kleijn, R. and van der Voet, E. (2010). Resource constraints in a hydrogen economy based on renewable energy sources: An exploration. *Renewable and Sustainable Energy Reviews*, 14(9): 2784–2795. doi:https://doi.org/10.1016/j.rser.2010.07.066.

Letessier, O. (2021). The Role of Hydrogen in the Energy Transition: In this paper adapted from his presentation at the IRI Virtual 2021 Annual Conference, Olivier Letessier discusses hydrogen's growing role in the transition toward greener energy and how Air Liquide has been a pioneer in hydrogen technologies. *Research-Technology Management*, 64(6): 56–60.

Macdonald, D. D. (2004). Fueling the hydrogen economy. *Materials Today*, 7(6): 64.

Moliner, R., Lázaro, M. J. and Suelves, I. (2016). Analysis of the strategies for bridging the gap towards the hydrogen economy. *International Journal of Hydrogen Energy*, 41(43): 19500–19508. doi:https://doi.org/10.1016/j.ijhydene.2016.06.202.

Nicita, A., Maggio, G., Andaloro, A. and Squadrito, G. (2020). Green hydrogen as feedstock: Financial analysis of a photovoltaic-powered electrolysis plant. *International Journal of Hydrogen Energy*, 45(20): 11395–11408.

Noussan, M., Raimondi, P. P., Scita, R. and Hafner, M. (2020). The role of green and blue hydrogen in the energy transition—A technological and geopolitical perspective. *Sustainability*, 13(1): 298.

Page, S. and Krumdieck, S. (2009). System-level energy efficiency is the greatest barrier to development of the hydrogen economy. *Energy Policy*, 37(9): 3325–3335. doi:https://doi.org/10.1016/j.enpol.2008.11.009.

Ping, Z., Laijun, W., Songzhe, C. and Jingming, X. (2018). Progress of nuclear hydrogen production through the iodine–sulfur process in China. *Renewable and Sustainable Energy Reviews*, 81: 1802–1812.

Razi, F. and Dincer, I. (2022). Renewable energy development and hydrogen economy in MENA region: A review. *Renewable and Sustainable Energy Reviews*, 168: 112763.

Ren, X., Dong, L., Xu, D. and Hu, B. (2020). Challenges towards hydrogen economy in China. *International Journal of Hydrogen Energy*, 45(59): 34326–34345. doi:https://doi.org/10.1016/j.ijhydene.2020.01.163.

Sánchez-Bastardo, N., Schlögl, R. and Ruland, H. (2021). Methane pyrolysis for zero-emission hydrogen production: A potential bridge technology from fossil fuels to a renewable and sustainable hydrogen economy. *Industrial & Engineering Chemistry Research*, 60(32): 11855–11881.

Saudi Arabia's hydrogen strategy targets $36bln of investments by 2030. (2022).

Sharma, S. and Ghoshal, S. K. (2015). Hydrogen the future transportation fuel: From production to applications. *Renewable and Sustainable Energy Reviews*, 43: 1151–1158.

Shinnar, R. (2003). The hydrogen economy, fuel cells, and electric cars. *Technology in Society*, 25(4): 455–476. doi:https://doi.org/10.1016/j.techsoc.2003.09.024.

Simbeck, D. R. (2004). CO_2 capture and storage—the essential bridge to the hydrogen economy. *Energy*, 29(9): 1633–1641. doi:https://doi.org/10.1016/j.energy.2004.03.065.

UAE seeks to capture 25% of the hydrogen market. (2022), Arabnews.

Weimann, L., Gabrielli, P., Boldrini, A., Kramer, G. J. and Gazzani, M. (2021). Optimal hydrogen production in a wind-dominated zero-emission energy system. *Advances in Applied Energy*, 3: 100032.

Yang, W., Zhao, R., Chuai, X., Xiao, L., Cao, L., Zhang, Z. et al. (2019). China's pathway to a low carbon economy. *Carbon Balance and Management*, 14(1): 1–12.

Yun, T. (2016). Investment in hydrogen engine must be ended with failure. *Journal of Mathematical Finance*, 6(01): 64.

Zegers, P. (2006). Fuel cell commercialization: The key to a hydrogen economy. *Journal of Power Sources*, 154(2): 497–502.

Zhang, H. (2021). Regulations for carbon capture, utilization and storage: Comparative analysis of development in Europe, China and the Middle East. *Resources, Conservation and Recycling*, 173: 105722.

Chapter 7

Examining the Existence of Speculative Bubbles in Oil and Natural Gas Prices

Hakan Yildirim,[1] *Kemal Eyüboğlu,*[2] *Saffet Akdağ*[2] and *Andrew Adewale Alola*[3,4,*]

1. Introduction

Although most energ-dependent countries have consistently aimed at implementing economic diversification policy, the approach is not as easy as it is always suggested. İn addition, the energy market has in recent time been significantly driven by the environmental sustainability and sustainable growth campaign of United Nations Development Programmes (UNDP) and other environmental-oriented inter-govermental agencies such as the United Nations Framework Convention in Climate Change (UNFCCC) (United Nations 2020, UNFCCC 2020). While the global energy outlook by 2050 is attributed to the driving force of the global energy transition, the manner of energy utilization across the main sectors (primary energy consumption by end-use sectors) of the economy will play a significant role (British Petroleum 2021). In specific, the British Petroleum (2021) report noted that the industrial sector accounted for the largest sectoral (45%) of global energy utilization in 2018, while residential and commercial building, transportation, and the non-combusted use of fuels respectively accounted for 29%, 21%, and 5% of global energy use in the same year.

[1] Istanbul Gelisim University, Department of Logistics Management, Istanbul, Turkey.
[2] Tarsus University, Department of Banking and Finance, Tarsus, Turkey.
[3] CREDS-Centre for Research on Digitalization and Sustainability, Inland Norway University of Applied Sciences, 2418, Elverum, Norway.
[4] Adnan Kassar School of Business, Lebanese American University, Beirut, Lebanon.
Emails: hayildirim@gelisim.edu.tr; kemaleyuboglu@tarsus.edu.tr; saffetakdag@tarsus.edu.tr
* Corresponding author: andrew.alola@hotmail.com

Given the forces affecting the dymanics of energy use, the energy market, especially the prices of energy mix have increasingly flunctuated amidst the key uncertainties arising from manmade and natural aspects such as pandemics and climate change-related issues. For instance, the value of the future oil and gas production has been signficantly reduced in recent time by the lower energy prices and downward trend of demand arising from the coronavirus (COVID-19) pandemic (International Energy Agency 2021). The International Energy Agency (2020) further noted that over 50 billion (USD) has been slashed from the assets of leading oil and gas companies, thus leading to a sharp fall in oil and gas investments and prompting an acute fiscal pressure in the oil and gas exporting states especially in the Middle East and Africa. Importantly, the COVID-19 pandemic caused the Brent prices to decline to the lowest (an average of $18 per barrel) monthly average in real terms since the first quarter of 1999 (United States Energy Information Administration 2020). Despite this observation, the United States Energy Information Administration (2020) expressed optimism about the gradual rebound in energy prices, especially oil and gas, starting from the second quarter of 2021.

In line with the aforementioned concern in the global oil and gas market, the current study is timely in the sense that it addresses the contemporary challenges facing the global energy market. In line with this, the current approach carefully examines the evidence of speculative bubbles in crude oil price, brent oil price, and natural gas price from January 2000 to December 2019. Considering the advantage or superiority of the generalized sup augmented Dickey–Fuller (GSADF) test, this investigation then employs the GSAF to illustrate the evidence of ballons in the aforementioned commodities. The significant of this study is it potential to offer relevant policy to stakeholders in global energy market and especially oil and gas market entities.

The next section presents a brief overview of related studies while Section three highlights the data and method employed for the investigation. In Sections four and five, the findings of the investigation and conclusion are presented, respectively.

2. Related Studies

One of the first studies on the existence of bubbles in oil prices is Eckaus (2008). In the implied study, the statistical significant evidence of bubbles were formed periodically between 2000–2008. Another pioneering work with similar results is Khan (2009). The study of Khan (2009) found that after 2008, bubbles were formed in oil prices. The study of Sornette et al. (2009) revealed a significant evidence supporting the existence of bubbles in oil prices between 2006–2008. In the study of Lammerding et al. (2013), in which a longer period was examined, it was stated that their findings revealed significant evident of speculative bubbles in oil prices between 1985 and 2010. Another study in which the price bubble in oil prices was determined by Zhang and Wang (2015). In specific, Zhang and Wang (2015) revealed that there is a significant evidence of price bubbles in the relevant oil prices between 2003 and 2012. In addition, the study of Fantazzini et al. (2016) is one of the first studies about the existence of negative price bubbles in oil prices. Accordingly, the study presented that there were negative bubbles in oil prices between 2014–2015. Furthermore, in the study of Su et al. (2017), the findings support the existence of price bubbles in crude oil prices between 1986 and 2016.

Moreover, Caspi et al. (2018) utilized dataset with a longer period, thus it offers a more reliable outcome in comparison with most of the earlier studies. In specific, Caspi et al. (2018) revealed a statistical significant evidence of the existence of bubbles in oil prices between 1876–2014. In the study of Sharma and Escobari (2018), the study investigated whether there is a price bubble in natural prices as well as oil prices. As a result of the study using weekly data between 1994 and 2015, the findings revealed in a conclusive manner that there were bubbles in oil prices and natural gas prices. In addition, Lawal (2019) employed the crude oil, brent oil, and OPEC oil price indexes, such that the findings indicated a periodic price bubbles in crude oil brent oil and OPEC oil price index between 1986 and 2018. A more recent study of Yıldırım (2020) also found the existence of price bubbles for sixteen different real-time commodities was tested. The findings obtained through the SADF method implied that the price bubble is in a concern for beef cattle and timber, while there is no price bubble for the other fourteen different commodities examined.

3. Data and Methodology

This study uses monthly data of crude oil, brent oil, and natural gas prices in dollar terms between January 2000 and December 2019. Relevant data are taken from an online source known as Investing.com.

In order to test the price balloons, the GSADF test developed by Phillips et al. (2015a) study is used. The following regression Equation (1) is used while calculating the unit root in SADF and GSADF test statistics (El Mentasser et al. 2015):

$$y_t = m + \lambda y_{t-1} + \sum_{i=1}^{p} \alpha_i \Delta y_{t-i} + \epsilon_t, \epsilon_t \sim iid\ N(0, \sigma^2),\ t = 1, ..., T \tag{1}$$

Therefore, the GSADF test calculations can be expressed from the following expressions (Phillips et al. 2015):

$$SADF_{r_2}(r_0) = sup_{r_1 \in [0, r_2 - r_0]} ADF_{r_1}^{r_2}, \tag{2}$$

$$GSADF(r_0) = sup_{r_2 \in [r_0, 1]} SADF_{r_2}(r_0) \tag{3}$$

In the SADF and GSADF unit root tests, the null hypothesis H0: $\lambda = 1$, which states that there is a price bubble, versus the alternative hypothesis H1: $\lambda > 1$, which states that the price bubble does not exist in unit root tests. In iterative regressions, the model expressed in Equation 3 is estimated repeatedly by using subsets of the sample data, which is increased by one observation at each pass (Philips et al. 2015b).

4. Findings

In the study, at first, the descriptive statistics of crude oil, brent oil, and natural gas prices were calculated. Descriptive statistics of the relevant data are given in Table 1.

Table 1: Descriptive statistics.

Variables	Average	Maximum	Minimum	Standard Deviation
Crude oil	62.14396	140.0000	19.44000	26.26820
Brent Oil	65.32417	139.8300	19.14000	29.99795
Natural gas	4.659021	13.92100	1.711000	2.280082

When the descriptive statistics are evaluated, it is seen that the highest volatility in oil prices is in brent oil. On the other hand, it is observed that the volatility in natural gas prices is lower than oil prices. In Table 2, GSADF test results are given in order to detect bubbles in crude oil, brent oil, and natural gas prices.

Table 2: GSADF test results.

Variables	t Statistic	p-Value	Result
Crude oil	3.010147**	0.0200	The H0 hypothesis stating that the price bubble does not exist was rejected.
Brent oil	3.732830*	0.0000	The H0 hypothesis stating that the price bubble does not exist was rejected.
Natural gas	1.709631	0.1400	The H0 hypothesis stating that the price bubble does not exist was accepted.

* 1% is significant at the ** 5% significance level.

According to GSADF test results, a definite result regarding the presence of bubbles in crude oil and brent oil prices were reached. But there is no significant evidence establishing the presence of bubbles in natural gas prices. The results regarding the existence of bubbles in oil prices are consistent with the results of the studies in the literature. However, the results for natural gas prices differ from the results of Sharma and Escobari's (2018) study in the literature. This difference is thought to arise from the period examined and the use of data at different frequencies. Graphics of GSADF test results are given in Figure 1.

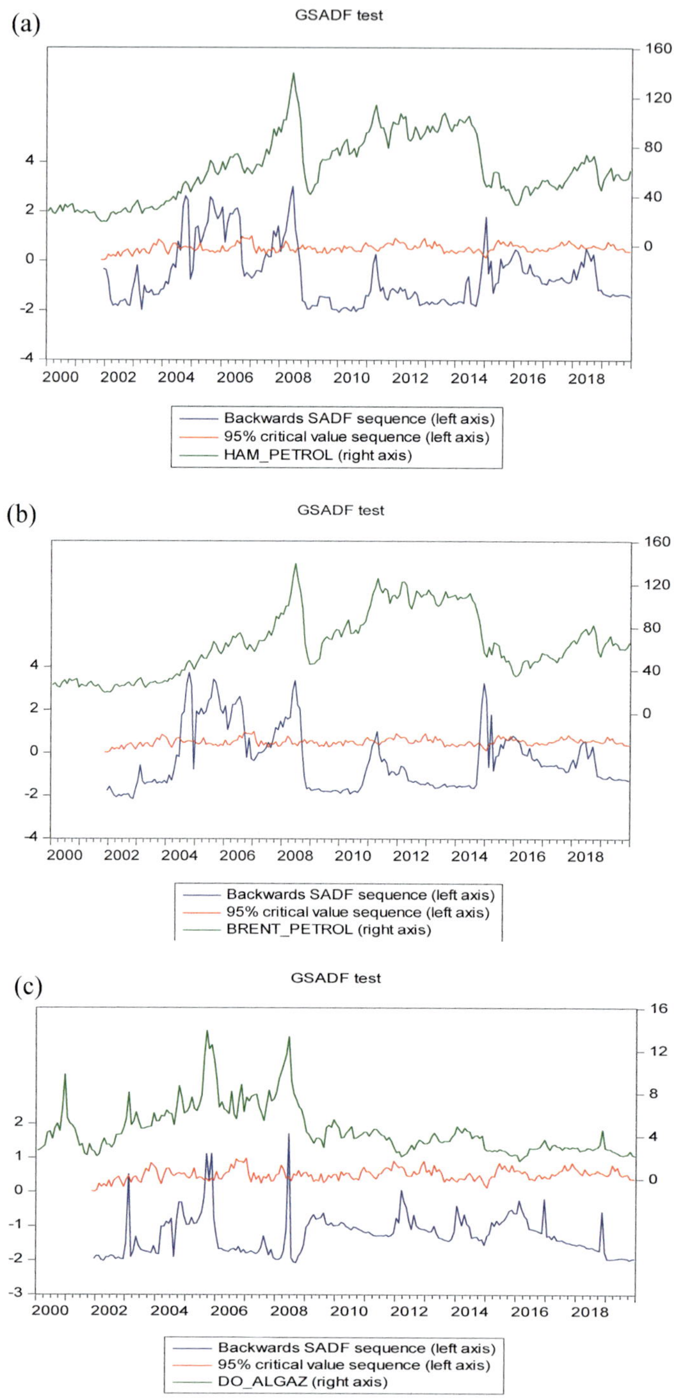

Figure 1: The visual presentation of bubbles in (a) crude oil price, (b) brent oil prices, and (c) natural gas prices.

5. Conclusion

This study test whether there are bubbles in oil and natural gas prices which are the two prominent natural energy resources. In this context, monthly data of crude oil, brent oil, and natural gas prices between January 2000 and December 2019 were used. In order to determine the price bubble of the relevant energy resources, the study employed the GSADF tailed unit root test developed by Phillips et al. (2015a). Given the findings of the investigation, the study found the presence of bubbles in crude oil and brent oil prices, while there is no significant evidence of the presence of bubbles in natural gas prices. It is important to follow up the bubbles in asset prices, which are seen as the cause of many global economic crises, and to take the necessary measures to avoid serious adverse effect related with the indicated bubbles in the implied period. As a matter of fact, oil price bubbles, which are an important cost element for many developing countries, will negatively affect many sectors.

However, in order to prevent the adverse challenge that may occur in the supply of related energy resources on a global scale, the price bubbles should be eliminated. As a recommendation for future study, similar studies may be dedicated to investigate the effects of price bubbles with other sectors in the periods when they occur and to investigate whether they cause similar price bubbles.

References

British Petroleum. (2021). Energy use by sector. https://www.bp.com/en/global/corporate/energy-economics/energy-outlook/demand-by-sector.html. (Accessed 17 January 2021).

Caspi, I., Katzke, N. and Gupta, R. (2018). Date stamping historical periods of oil price explosivity: 1876–2014. *Energy Economics*, 70: 582–587.

Eckaus, R. S. (2008). The oil price really is a speculative bubble. *Center for Energy and Environmental Policy Research*, 8(7): 1–9.

El Montasser, G., Gupta, R., Martins, A. L. and Wanke, P. (2015). Are there multiple bubbles in the ethanol–gasoline price ratio of Brazil? *Renewable and Sustainable Energy Reviews*, 52: 19–23.

Fantazzini, D. (2016). The oil price crash in 2014/15: Was there a (negative) financial bubble? *Energy Policy*, 96: 383–396.

International Energy Agency. (2020). World Energy Outlook 2020. https://www.iea.org/reports/world-energy-outlook-2020. (Accessed 17 January 20201).

Khan, M. S. (2009). The 2008 Oil Price Bubble (No. PB09–19).

Lammerding, M., Stephan, P., Trede, M. and Wilfling, B. (2013). Speculative bubbles in recent oil price dynamics: Evidence from a Bayesian Markov-switching state-space approach. *Energy Economics*, 36: 491–502.

Lawal, A. I., Olayanju, T. M. A., Salisu, A. A., Asaleye, A. J., Dahunsi, S. O., Dada, A. O., Omoju, O. E. and Popoola, O. (2019). Examining rational bubbles in oil prices: evidence from frequency domain estimates. *International Journal of Energy Economics and Polic*, 9(2): 166–173.

Phillips, P. C., Shi, S. and Yu, J. (2015a). Testing for multiple bubbles: Historical episodes of exuberance and collapse in the S&P 500. *International Economic Review*, 56(4): 1043–1078.

Philips, P.C.B., Shi, S. and Yu, J. (2015b). Testing for multiple bubbles: Limit theory of dating algorithms. *International Economic Review*, 52: 1079–1134.

Sharma, S. and Escobari, D. (2018). Identifying price bubble periods in the energy sector. *Energy Economics*, 69: 418–429.

Sornette, D., Woodard, R. and Zhou, W. X. (2009). The 2006–2008 oil bubble: Evidence of speculation, and prediction. *Physica A: Statistical Mechanics and its Applications*, 388(8): 1571–1576.

Su, C. W., Li, Z. Z., Chang, H. L. and Lobonţ, O. R. (2017). When will occur the crude oil bubbles? *Energy Policy*, 102: 1–6.

United Nations Development Programme. (2020). Sustainable Development Goals. https://www.undp.org/content/undp/en/home/sustainable-development-goals.html. (Accessed 17 January 2021).

United Nations Framework Convention iin Climate Change. (2020). https://unfccc.int/. (Accessed 17 January 2021).

United States Energy Information Administration. (2020). SHORT-TERM ENERGY OUTLOOK. https://www.eia.gov/outlooks/steo/report/prices.php. (Accessed 17 January 2021).

Yildirim, H. (2020). Testing bubbles formation at real-time commodity prices. *J. Public Affairs*, 2020: e2243. https://doi.org/10.1002/pa.2243.

Zhang, Y. J. and Wang, J. (2015). Exploring the WTI crude oil price bubble process using the Markov regime switching model. *Physica A: Statistical Mechanics and its Applications*, 421: 377–387.

Environment

Chapter 8

Analysis of Particulate Matter in Street Dust of Middle and Eastern Parts of Northern Cyprus

Emmanuel Akaiso

1. Introduction

Air pollution is the negative shift and change in the composition of the environment comprising of gaseous components and particulate matter (PM). Particulate matter can either be coarse particles, fine particles, and ultrafine particles; the composition of particulate matter depends on the sources from which it came from (Li et al. 2018). Combustion sources such as traffic emission or residential heating produce carbonaceous particles, they are carbon based as they pull along organic chemicals such as reactive metals and polycyclic aromatic hydrocarbons. Desert dust and mineral dust from agriculture produce inorganic particles. Nitrogen dioxide (NO_2), nitric oxide (NO), ozone (O_3), sulfur dioxide (SO_2), volatile organic compounds, and carbon monoxide (CO) are toxic gaseous pollutants that adds to the formation of particle from the complex atmospheric photochemical reactions, they are called secondary particles (Bourdrel et al. 2017).

Seven million premature deaths were recorded in 2010 due to both indoor and outdoor air pollution. Nevertheless, indoor air pollution is more heterogeneous with huge variations in sources and pollutants between countries. Across the globe secondhand smoke is the main indoor air pollution source. Cooking and heating with solid fuels causes great indoor air pollution in Asia, meanwhile in Europe several sources contribute to the indoor pollution which includes particulate matters from wood burning and cooking, household products, and volatile organic compounds from organic solvents. However a recently conducted European study reveals that 60% of the global weight of indoor air pollution related diseases comes from outdoor air pollution infiltrating via air interchange (Sahakyan et al. 2018).

Department of Contemporary Management, Isik University, Maslak, Istanbul Turkey.
Email: akaisoe@gmail.com

The population of the world has been predicted to reach 10 billion by 2050, this changes in the population size would eventually affect the dynamics of the population and structure as well as affecting the consumption patterns. This increase of population and socioeconomic increment in development have been part of the drivers bringing change to the environment thereby magnifying existing problems and producing latest problems for the human and natural environment. Cancer usually starts with the irregular division of cells in the body and this cell spreads across other parts of the body causing great damage (Spiru and Simona 2017). The rise in construction activities in a region is a result of an increase in population and increase in accommodation demand. This demand for more accommodation further leads to an increase in the usage of cement, paints, steel, and household items made from steel. Scientists invented ways by which the carcinogenic risk can be calculated so that people are aware of the areas around them that are polluted and take necessary precautions. Cancer has different causes ranging from poor diet to infections. Generally, the incremental lifetime cancer risk assessment formula is used to calculate this risk, if the result gotten is $ILCR \leq 10^{-6}$ then it is at low risk but if the value gotten is $ILCR > 10^{-4}$ then it is high risk and a lot of precautions should be taken (Rao et al. 2017).

The aim of this study was to assess the origin and composition of particulate matter across Northern Cyprus as to determine which metals and location are more cancerogenic.

2. Literature Review

In Thailand, Srithawirat et al. (2016), assessed ten residential buildings on one side of the Phitsanulok road in Thailand at the dry and wet periods of 2014 was investigated to find out the indoors and outdoors concentrations of PM10. The inductively coupled plasma optical emission spectrometry (ICP-OES) was used to analyze PM10 for seven unknown metallic Elements (Zn, Fe, Pb, Cd, Ni, Cu, and Cr) via inductively coupled plasma optical emission spectrometry (ICP-OES). The PM10 indoor and outdoor concentrations every month were found to be from 41.50 to 105.30 $\mu g\ m^{-3}$ and 95.20 to 145.10 $\mu g\ m^{-3}$, consecutively. There was a comparison of PM10 in the dry season and PM10 in the wet season. The ratio between indoor and outdoor values was lower than 1, which resolves the fact that majority of the particulate matters came from the environment outdoor. Overall, the heavy metals concentrations in PM10 were from 0.2 to 0.2 to 2.7 $\mu g\ m^{-3}$ for indoor environment and 0.5 to 7.1 $\mu g\ m^{-3}$ for the outdoor environment, respectively.

Škrbić et al. (2019) conducted an initial investigation that recognized seasonal occurrence, distribution, and sources of sixteen polycyclic fragrant hydrocarbons (PAHs) in sixty road dust samples amassed inside urban quarter of Novi Sad, the second biggest city in Serbia. The received consequences were similarly used for complete assessment of carcinogenic risk of Serbian population exposed to PAHs present in street dirt by the incremental lifetime most cancers chance technique. The full level of 16 PAHs ranged between 35 $\mu g\ kg^{-1}$ and 2422 $\mu g\ kg^{-1}$ in samples taken in summer season and among 35 $\mu g\ kg^{-1}$ and 587 $\mu g\ kg^{-1}$ in samples taken in wintry weather. In both seasons, four-ring PAHs were the maximum dominant compounds and excessive molecular weight (HMW) PAHs had a similar contribution (55% in summer and65% in winter). The highest content material becomes determined for fluoranthene (Fly) in each season (597 μg kg^{-1} in winter, 301 μg kg^{-1} in summer season). The PAHs source apportionment was analyzed by way of fundamental aspect analysis (PCA) and diagnostic ratios, and combustion of petroleum appeared to be the main assets of the PAHs in road dirt. The most cancers danger stage for youngsters and person have been similar for dermal contact and by ingestion, the` e possibility ranging from 10^{-6} to 10^{-4} μg kg^{-1}. Moreover, the overall incremental lifetime cancer chance (ILCR) become assessed for children and grown-up populations thinking of three viable exposure routs and the median overall cancer danger changed into $> 10^{-5}$, with 7% of the samples having the threat $> 10^{-4}$ that need to be taken into consideration of high concern with potential health hassle. Those effects are the primary of this kind for the complete Serbia and the Western Balkan vicinity and may be taken into consideration because the baseline for destiny studies.

A strong positive correlation in indoor PM10 occurred between Zn and Cu, Zn and Ni, and Cu and Ni. Zn and Ni, Pb and Cu, Cu and Ni, Cd and Ni, and Zn and Cu showcased strong positive correlations in the outdoor environment. The enrichment factors of Zn, Cu, Pb, Cr, and Ni were less than one, suggesting that the metals in indoor PM10 have originated from crustal materials. Cr proved to have the highest excess cancer risk in an evaluation using an integrated risk information system for the health risk assessment. Gustafsson et al. (2018) presented an efficient method for the isolation of dust particles in the respirable size range. The respirable fraction was recovered from vacuum cleaner bags, separated by stepwise sieving, followed by characterization for size, morphology, surface area, organic content, and elemental composition. The dust samples were collected and sieved, then characterization was done using scanning electron microscopy. Surface area was determined using the Braeuer Emmett Teller (BET) method, the dust composition by ICP-MS, and elemental composition by X-ray powder diffraction (XRD). Aluminum and zinc were the dominating metals measured in the dust, whereas the major mineral components were found to be silicon dioxide and calcium carbonate. The fraction of organic matter in the dust was measured to be 69 ± 1%. The organic matrix contained bacterial and fungi and a presence of skin fragments.

Kidak (2017) argues that street surfaces get distinctive number of contaminations by the procedure of natural development, through wet and dry deposition, sedimentation, impaction, and capture attempt. The concentration of all the seven heavy metals were broken down (As, Co, Cr, Cu, Ni, Pb, and Zn) utilizing X-ray fluorescence technique, where nine streets were examined, which includes highways, residential roads and industrial roads. Three residue samples for every street are gathered amid two unique periods of the year (for example, winter and summer). It was exceptionally clear that Cu, Cr, and Zn were the most abundant heavy metals found in all classes of roads in the city (WHO guidelines). In this examination, there are two major supporters of contamination, the lithogenic impacts originate from the starting point of the material while anthropogenic impacts originate from vehicles exhaust and modern waste. Parkways in winter and residential streets in summer had "high contamination."

3. Methodology

3.1 Study area

Northern Cyprus is situated in the eastern constituent of the Mediterranean Sea. It possesses a landmass of around 9,251 km^2 and has a coastline of around 648 km. In this study, the sampling locations include Turkmenkoy, Haspolat, Famagusta, City Center, Kucuk-kaymakli and Quarry site.

3.2 Sample collection

The samples were collected from six distinct functional areas across Northern Cyprus. In each location except for Turkmenkoy and Famagusta, five samples were collected from five points after which they were buked together, and a representative sample was gotten. The samples were collected by sweeping areas, such as pavements and concrete surfaces, and edges of roads with a brush and a dustpan in each sampling site. The samples were then sieved and stored in self-sealed plastic bags.

3.3 XRD analysis

X-ray diffraction (XRD) was used for the phase identification of crystalline material and provides information on unit cell dimensions of the samples. The dust samples were pounded into finely grained texture and placed on a glass slate taken into the XRD machine for analysis. The analysis was Operated at 45 kV and 40 mA, recorded at 5º–90º at 2Ø range with a scanning speed of 0.1 seconds/step with a step width of 0.01945º (Gustafsson et al. 2018, Nowak et al. 2018).

3.4 Heavy metal analysis

3.4.1 Sample digestion

The samples were digested by measuring 1 g of the sample into a volumetric flask containing 2 ml of water, 6 ml of hydrochloric acid, and 2 ml of nitric acid and was placed into the dust solution. The sample solution was heated for 30 minutes until the brown fume leaves. Twenty ml of distilled water was added to the solution and it is allowed to cool before being filtered into a 100 ml Erlenmeyer flask and made up to the 100 ml mark with distilled water (Srithawirat et al. 2016, Das and Ting 2017, Sahakyan et al. 2018).

3.4.2 Analysis

The analysis was done using the inductively coupled plasma—Mass spectrophotometer (ICP-MS) in the Northern Cyprus State Laboratory, Lefkosa—Northern Cyprus. The samples were evaluated for the presence of lead (Pb), cadmium (Cd), and iron (Fe).

3.4.3 Cancer risk assessment analysis

Cancer risk assessment is calculated by adding the incremental lifetime cancer risk assessment (ILCR) of inhalation (Equation 1), digestion (Equation 2) and dermal contact (Equation 3) to determine the cancer risk of a particular location. There are different constants in the equations below and their meanings are fully explained. **Ingestion Rate** (mg/day): Adult (100) Child (200); Exposed skin area, **SA** (cm^2): Adult (5700) Child (2800); Skin adherence Factor, **AF** soil (mg/cm): Adult (0.07) Child (0.2); Exposure frequency (EF) (days/year): Adult (365) Child (365); Exposure duration, **ED** (year): Adult (24) Child (6); Body weight **BW** (kg): Adult (60) Child (18); Averaging time, **AT** (days): Adult (25550) Child (25550); Dermal absorption fraction (**ABS**): Adult (0.13) Child (0.13); **Inhalation ratio** (m^3/day): Adult (20) Child (10); **Particulate emission Factor** (m^3/kg): Adult (1.36×10^9) Child (1.36×10^9); **CSF ingestion** (mg/kg/day): Adult (7.3) Child (7.3); **CSF inhalation** (mg/kg/day): Adult (3.85) Child (3.85); **CSF dermal** (mg/kg/day): Adult (25) Child (25).

$$\boldsymbol{ILCR}_{INGESTION} = \frac{Cs \times \left(CSF_{ing} \times \sqrt[3]{\frac{BW}{70}} \right) IR_{ing} \times EF \times ED}{BW \times AT \times 10^6} \quad (1)$$

$$\boldsymbol{ILCR}_{INHALATION} = \frac{Cs \times \left(CSF_{inhal} \times \sqrt[3]{\frac{BW}{70}} \right) R_{inhal} \times EF \times ED}{BW \times AT \times PEF} \quad (2)$$

$$\boldsymbol{ILCR}_{DERMAL} = \frac{Cs \times \left(CSF_{derm} \times \sqrt[3]{\frac{BW}{70}} \right) SA \times AF \times ABS \times EF \times ED}{BW \times AT \times 10^6} \quad (3)$$

$$\boldsymbol{ILCR}_{TOTAL} = \boldsymbol{ILCR}_{INGESTION} + \boldsymbol{ILCR}_{INHALATION} + \boldsymbol{ILCR}_{DERMAL} \quad (4)$$

Source: (Gope et al. 2018).

The ILCR value that indicates risk or safety according to Petit et al. (2019). Exceptionally minimal risk (ILCR = 10^{-7}); Minimal risk (ILCR $\leq 10^{-6}$); Moderate risk (ILCR = 10^{-5}); High risk (ILCR > 10^{-4}) High Risk (ILCR = 10^{-2}).

4. Results

4.1 Mineral characterization of samples

Turkmenkoy contains calcium carbonate, quartz, dolomite; Haspolat contains calcium carbonate, quartz, dolomite; Nicosia City Centre contains calcium carbonate, dolomite, sodalite, Berlinite; Famagusta contains calcium carbonate, quartz, dolomite; Nicosia surroundings contains calcium carbonate, quartz, dolomite and andesine while Lefkosa Quarry contains calcium carbonate, quartz, dolomite and albite.

4.2 Inductively Coupled Plasma—Mass Spectrometer (ICP-MS)

The samples across the six locations were analyzed by the inductively coupled plasma mass spectrophotometer to detect the presence and the concentration of Cadmium (Cd), Lead (Pb), and Iron (Fe) (see Table 1). The level of presence of these three heavy metals alongside other mineral compounds can be used to provide the origin and composition of particulate matter across Northern Cyprus.

The table below gives the standard of the heavy metals in question at a unit of mg/kg provided by the World Health Organization and the Environmental Protection Agency (see Table 2).

Table 1: Concentrations of each heavy metal detected by the ICP-MS.

LOCATION	CADMIUM (mg/kg)	LEAD (mg/kg)	IRON (mg/kg)
TURKMENKOY	0.102	0	0
HASPOLAT	0.095	10.510	4738.818
NICOSIA CITY CENTER	0.020	8.840	4993.818
FAMAGUSTA	0.295	0	500
NICOSIA'S SURROUNDINGS	0.002	4.362	0
QUARRY	0.009	7.654	8007.092

Table 2: The recommended concentration by WHO & EPA.

Heavy Metals	WHO? Mg/kg	EPA Mg/kg
Cadmium (Cd)	0.39	2
Lead (Pb)	85	8
Iron (Fe)	10,900	4,750

Source: (Opaluma et al. 2012, Ghanavati et al. 2019).

4.3 The cadmium concentration in each location

From the table above, it shows that Famagusta has a cadmium concentration of 0.295 mg/kg (Table 1). This is attributed to the construction activities going on at the Famagusta location where the sample was gathered from. Cadmium is one of the components derived from cement production. Even with the concentration of 0.295 mg/kg (Table 2), it is still not up to the standard set by both World Health Organization (WHO) and the Environmental Protection Agency (EPA); as a result, people living in Famagusta environment are safe as the cadmium concentration is quite below the maximum standard concentration.

4.4 The lead concentration in each location

The table above proves that the Haspolat sample has a lead concentration of 10.51 mg/kg (Table 1), followed by the one from Nicosia City Centre with a value of 8.84 mg/kg (Table 1). The presence of lead is attributed to the exhaust petrol and diesel. Looking at these locations, they are characterized by severe traffic and vehicular activities resulting in the presence of lead in the samples. The lead concentration in the Nicosia City Centre is lower than the standard maximum values set by WHO and EPA at 85 mg/kg (Table 2) and 10 mg/kg (Table 2), respectively. However, the Haspolat samples have a lead concentration of 10.51 mg/kg which is higher than the standard value set by EPA.

4.5 The iron concentration in each location

The Quarry sample has an iron concentration of 8007.092 mg/kg (Table 1), followed by the sample from the Nicosia City Centre at 5672.888 mg/kg (Table 1), and the one from Haspolat at 4738.818 mg/kg (Table 1). Table 1 shows that the three locations with available iron concentrations are higher than the standard values set by WHO and EPA; this is extremely dangerous because everyone in this area is prone to iron contamination if proper precautions are not taken into consideration. There is need for urgent check to the iron concentrations in Quarry, Nicosia City Centre, and Haspolat region because of heavy traffic activities, vehicular activities, drilling, metallic parking of sands, and rusting of iron available in such areas. After all the analysis was conducted, the origin of the particulate matter in street dusts is detected and summarized below (see Table 3).

Table 3: Origin of particulate matter.

LOCATION	ORIGIN OF PARTICULATE MATTER
TURKMENKOY	African origin, sea spray, and cement production
HASPOLAT	African origin, sea spray, exhaust petrol, and industrial activities
NICOSIA CITY CENTRE	African origin, exhaust petrol, and industrial activities
FAMAGUSTA	African origin, sea spray, and cement production
NICOSIA'S SURROUNDINGS	African origin, sea spray, cement production, and exhaust petrol
QUARRY	African origin, sea spray, exhaust petrol, and industrial activities

4.6 Cancer risk assessment

According to Petit et al. (2019), the cancer risk assessment is determined using the incremental life cancer risk assessment formula. If the results gotten that is ILCR $\leq 10^{-6}$, it indicates safety but if it is ILCR $\geq 10^{-4}$, it indicates elevated risk.

4.6.1 Turkmenkoy

In this location, only the ICP-MS. detected the metal Cadmium. The ILCR value was calculated for both the children and adults. The ILCR value for a child is 1.01×10^{-6} while that for an adult is 1.12×10^{-6}; both the values are in the range of 10^{-6} and from the table above, we can see that they are at low risk as the cadmium concentration in Turkmenkoy location does not pose any carcinogenic threat.

4.6.2 Haspolat

The ILCR values for the metal cadmium affecting both children and adults are 9.43×10^{-7} and 1.04×10^{-6}, respectively. The ILCR value for children is at an extremely insignificant risk level while the ILCR value for adults is at the minimal risk level, so both of them are not at a risk of any carcinogenic threat. The lead ILCR values for both children and adult are 1.04×10^{-4} and 1.16×10^{-4}, respectively. Both the ILCR values are at a high-risk level and pose a carcinogenic threat to the residents in such a location; this is attributed to the gasoline release from the heavy traffic activities at the Haspolat region of Northern Cyprus. Iron ILCR values for both children and adults are 4.71×10^{-2} and 5.2×10^{-2}, respectively. Both the ILCR values are very high-risk level and pose a great carcinogenic threat to the residents in such location; this is attributed the heavy traffic activities at the Haspolat region of Northern Cyprus and also the movement of desert dusts from Sahara desert into Northern Cyprus.

4.6.3 Nicosia city centre

The ILCR values for the metal cadmium affecting both children and adults are 1.10×10^{-7} and 2.20×10^{-7}, respectively. The ILCR child value and the ILCR adult value is at an extremely minimal risk level, so they both do not spose any carcinogenic threat. Lead ILCR values for both children and adults are 8.78×10^{-5} and 9.72×10^{-5}, respectively. The ILCR value for both of them is at the moderate risk level and it poses a moderate carcinogenic threat to the residents in such a location, this is due to the renovations and construction going on in various parts of the city centre. So, residents in this location should avoid contact with lead byproducts from these buildings so as to avoid bioaccumulation of lead compound in the body. Iron ILCR values for both children and adults are 6.01×10^{-2} and 5.43×10^{-2}, respectively. The ILCR value for both of them is at a very high-risk level and poses a great carcinogenic threat to the residents in such location; this is attributed the heavy traffic activities at the Nicosia City Center of Northern Cyprus and also the movement of desert dusts from Sahara desert into Northern Cyprus.

4.6.4 Famagusta

The ILCR values for the metal cadmium affecting both children and adults are 2.93×10^{-6} and 3.25×10^{-6}, respectively. The ILCR child value and the ILCR adult value is at the minimal risk level, so they both do not pose any carcinogenic threat. Iron ILCR values for both children and adults are 5.49×10^{-2} and 4.50×10^{-2}, respectively. The ILCR values for both of them is at the very high-risk level and it poses a great carcinogenic threat to the residents in such location, this can be attributed to the domestic waste or household waste from where the sample was taken, resulting in a high iron concentration.

4.6.5 Nicosia's surroundings (kucuk-kaymakli)

The ILCR values for the metal cadmium affecting both children and adults are 1.98×10^{-8} and 2.20×10^{-8}, respectively. The ILCR Child value and the ILCR Adult value are at an exceptionally negligible risk level, so they both do not pose any carcinogenic threat. Lead ILCR values for both children and adults are 4.33×10^{-5} and 4.80×10^{-5}, respectively. The ILCR value for both of them is at the moderate risk level and it poses moderate carcinogenic threat to the residents in such locations; this is due to the release of petroleum products in this location. So, residents in the locations should be watchful and prevent bodily contact with any lead containing product.

4.6.6 Quarry

The ILCR values for the metal cadmium affecting both children and adults are 8.94×10^{-8} and 1.19×10^{-8}, respectively. The ILCR child value and the ILCR adult value are at the exceptionally insignificant risk level, so they both do not pose any carcinogenic threat. Lead ILCR values for both children and adults are 7.60×10^{-5} and 2.80×10^{-5}, respectively. The ILCR value for both of them is at the moderate risk level and it poses moderate carcinogenic threat to the residents in such location, this is due to the gasoline release and heavy industrial activities going on in this location. So, workers and residents in this location should avoid contact with lead byproducts from petrol or diesel to avoid bioaccumulation of lead compounds in the body. Iron ILCR values for both children and adults are 7.95×10^{-2} and 8.80×10^{-2}, respectively. The ILCR value for both of them is at a very high-risk level and poses a great carcinogenic threat to the residents in such locations; this is attributed the industrial activities at Quarry and also the movement of desert dusts from Sahara desert into Northern Cyprus. The iron ILCR value gotten for this quarry location is the highest in comparison with other locations. This calls for serious precautions at quarries in Northern Cyprus.

5. Conclusion and Recommendations

5.1 Conclusion

The results of the analysis revealed that the origin of samples was African because of the high numbers of Africans on the Island, sea spray, cement production, petrol exhaust, and steel production and all samples contained carbon carbonate (calcite), dolomite, and quartz. The cadmium (0.002 mg/kg to 0.295 mg/kg) and lead (2.362 mg/kg to 10.510 mg/kg) concentrations were within the recommended rate (2 mg/kg for cadmium and 85 mg/kg for lead) provided by WHO and EPA but the iron concentration (4738.818 mg/kg to 6007.092 mg/kg) was higher than the recommended rate of 4750 mg/kg provided by the EPA. The incremental lifetime cancer risk (ILCR) assessment values showed that the cadmium concentration was negligible risk because the ILCR value was $\leq 10^{-6}$, the lead concentration was moderate with ILCR value of 10^{-5}, but the iron concentration was high with ILCR value of 10^{-4} to 10^{-2}.

5.2 Recommendations

Car trips should be reduced as residents should be encouraged to use public buses to reduce pollution; wood burning during winter for heating in homes should be reduced to prevent massive indoor pollution. Sustainable materials should be used for constructions of houses in Cyprus. The environmental agencies in Cyprus should post daily air quality tests in newspapers and television so that residents can be aware and take necessary precautions. Once residents notice themselves coughing irregularly, they should visit the doctor for a medical checkup, residents should also be discouraged from smoking. Environmental management plans which include dust management should be put in place across Northern Cyprus so that residents can live safely. There should be regular experiments to determine the components of particulate matter across Northern Cyprus to identify which heavy metal to tackle. The Environmental Protection Agencies in Northern Cyprus should enforce regulations and compliance to environmental policies.

References

Bourdrel, T., Bind, M. A., Béjot, Y., Morel, O. and Argacha, J. F. (2017). Cardiovascular effects of air pollution. *Archives of Cardiovascular Diseases*, 110(11): 634–642.

Chao, Z. (2014). Catalysis over zinc-incorporated berlinite (ZnAlPO4) of the methoxycarbonylation of 1,6-hexanediamine with dimethyl carbonate to form dimethylhexane-1,6-dicarbamate (May).

Das, S. and Ting, Y. -P. (2017). Evaluation of wet digestion methods for quantification of metal content in electronic scrap material. *Resources*, 6(4): 64.

Ghanavati, N., Nazarpour, A. and Watts, M. J. (2019). Status, source, ecological and health risk assessment of toxic metals and polycyclic aromatic hydrocarbons (PAHs) in street dust of Abadan, Iran. *Catena*, 177(February 2018): 246–259.

Gope, M., Masto, R. E., George, J. and Balachandran, S. (2018). Exposure and cancer risk assessment of polycyclic aromatic hydrocarbons (PAHs) in the street dust of Asansol city, India. *Sustainable Cities and Society*, 38(March 2017): 616–626.

Gustafsson, Å., Krais, A. M., Gorzsás, A., Lundh, T. and Gerde, P. (2018). Isolation and characterization of a respirable particle fraction from residential house-dust. *Environmental Research*, 161(November 2017): 284–290.

Joshi, U. M., Vijayaraghavan, K. and Balasubramanian, R. (2009). Elemental composition of urban street dusts and their dissolution characteristics in various aqueous media. *Chemosphere*, 77(4): 526–533

Kariper, A., Üstündağ, İ., Deniz, K., Mülazımoğlu, İ. E., Erdoğan, M. S. and Kadıoğlu, Y. K. (2019). Elemental monitoring of street dusts in Konya in Turkey. *Microchemical Journal*, 148(February): 338–345.

Keshavarzi, B., Tazarvi, Z., Rajabzadeh, M. A. and Najmeddin, A. (2015). Chemical speciation, human health risk assessment and pollution level of selected heavy metals in urban street dust of Shiraz, Iran. *Atmospheric Environment*, 119: 1–10.

Khademi, H., Gabarrón, M., Abbaspour, A., Martínez-Martínez, S., Faz, A. and Acosta, J. A. (2019). Environmental impact assessment of industrial activities on heavy metals distribution in street dust and soil. *Chemosphere*, 217: 695–705.

Kidak, R. (2017). Seasonal assessment of heavy metal pollution in street dust of nicosia city in northern cyprus. *European Journal of Sustainable Development*, 6(4).

Li, T., Hu, R., Chen, Z., Li, Q., Huang, S. and Zhu, Z. (2018). Science Direct Fine particulate matter (PM 2.5): The culprit for chronic lung diseases in China. *Chronic Diseases and Translational Medicine*, 4(3): 176–186.

Ma, J. and Singhirunnusorn, W. (2012). Distribution and health risk assessment of heavy metals in surface dusts of maha sarakham municipality. *Procedia - Social and Behavioral Sciences*, 50(July): 280–293.

Nowak, S., Lafon, S., Caquineau, S., Journet, E. and Laurent, B. (2018). Quantitative study of the mineralogical composition of mineral dust aerosols by X-ray diffraction. *Talanta*, 186(November 2017): 133–139.

Opaluma, O. D., Aremu, M. O., Ogbo, L. O., Abiola, K. A., Odiba, I. E., Abubakar, M. M. and Nweze, N. O. (2012). Heavy metal concentrations in soils, plant leaves and crops grown around dump sites in Lafia Metropolis, Nasarawa State, Nigeria. *Advances in Applied Science Research*, 3(2): 780–784.

Rao, S., Klimont, Z., Smith, S. J., Van Dingenen, R., Dentener, F., Bouwman, L., ... and Tavoni, M. (2017). Future air pollution in the Shared Socio-economic Pathways. *Global Environmental Change*, 42: 346–358.

Sahakyan, L., Tepanosyan, G., Maghakyan, N., Kafyan, M., Melkonyan, G. and Saghatelyan, A. (2018). Contamination levels and human health risk assessment of mercury in dust and soils of the urban environment, Vanadzor, Armenia. *Atmospheric Pollution Research*, (July): 1–9.

Salem, D. M. S. A., Khaled, A., El Nemr, A. and El-Sikaily, A. (2014). Comprehensive risk assessment of heavy metals in surface sediments along the Egyptian Red Sea coast. *Egyptian Journal of Aquatic Research*, 40(4): 349–362.

Spiru, P. and Simona, P. L. (2017). A review on interactions between energy performance of the buildings, outdoor air pollution and the indoor air quality. *Energy Procedia*, 128: 179–186.

Srithawirat, T., Latif, M. T. and Sulaiman, F. R. (2016). Indoor PM10 and its heavy metal composition at a roadside residential environment, Phitsanulok, Thailand. *Atmosfera*, 29(4): 311–322.

Petit, P., Maître, A., Persoons, R. and Bicout, D. J. (2019). Lung cancer risk assessment for workers exposed to polycyclic aromatic hydrocarbons in various industries. *Environment International*, 124(December 2018): 109–120.

Chapter 9

Human Risk Assessment of Heavy Metals Present in Four Motor Park Soils in Lagos State, Nigeria

Emmanuel Akaiso

1. Introduction

The belief that heavies metal pollution is only gotten high intense industries is a common perception of people living in rural areas (Brown et al. 2003). In reality, nowadays, roadways and automobiles are considered to be one of the largest sources of heavy metals. Lead (Pb), copper (Cu), and zinc (Zn) are the most common heavy metals produced from automobiles. Small or little amounts of other metals like nickel and cadmium are found to originate from carparks, bus stops, the roadside, and automobile exhaust (Brown et al. 2003).

Heavy metals are metals between copper to bismuth in the periodic table. They can be defined as elements with a specific gravity greater than four or five, located from atomic numbers 22 to 34 and 40 to 52 on the periodic table, having a specific biological response. Metals have always been present on Earth and a reasonable amount of the metals are good for biochemical processes (Riffat et al. 2011).

However, some heavy metals are dangerous for living organisms and human beings. Heavy metal contamination of soil is an environmental concern due to accumulation of metals which have severe effects on soil ecology, agricultural production, animal and human health as well as groundwater quality. It was reported that there is a correlation between heavy metal concentration and the dust in the area (Amuda et al. 2007). Heavy metals escape into the air in the form of dust from either car exhausts or from generators; this gradually disrupts air quality. Even though the process to atmospheric suspended particles is through inhalation, ingestion is the

Department of Contemporary Business Management, Isik University, Maslak, Istanbul Turkey.
Email: akaisoe@gmail.com

major means of toxic exposure. In developing countries like Nigeria, a large number of less privileged children live by the roadside or the streets with little or no hygiene, eating and picking up varied materials from the soils. This is of great concern as rapid environmental protection is needed and human life and health safety is to be taken up as topmost priority (Amuda et al. 2007).

From the beginning of its existence, the Earth has gone through one stress or the other, but it has not been as bad as it is today. The world has been expressing concerns over certain excess concentration of some heavy metals in the environment; this has been a result of distribution of heavy metals across the earth majorly by anthropogenic activities and the hazard posed to living organisms and its environment. Due to increased developments in various parts of the world, industrialization has also increased, leading to great pollutants production. Fe, Mo, and Mn are classified as micronutrients, harmful metals and trace elements includes Zn, Ni, Cu, Co, Va, and Cr. Chemicals which are toxic to plants and micro-organisms are Ag, As, Hg, Cd, and Pb. Automobiles releases 50% of copper and zinc to the environment. Zinc is released when tires wear out while copper is released from automobile brakes. There is oil leakage when an automobile has a fault, and it needs to be fixed or repaired by the automobile engineer; the oil leaks that occur serve as a pathway by which heavy metals move into the environment. Heavy metal pollution of soils overall becomes dangerous to all living organisms because they can either move into the underground water body. They can also get washed into the surface waterbody leading to eutrophication; they can be released to the air leading to air pollution; they can also affect soils by making it difficult for agricultural activities by rendering the soil infertile because of the impurities the pollution brings to the soils and in return it affects the agricultural processes. Some vegetables we eat might be polluted by bad soils and fertilizers and when such foods are eaten, it can lead to acute poisoning (Garbu et al. 2014).

All heavy metals are toxic to living organisms; even the essential metals like calcium and magnesium if in an excess amount can be toxic. They obstruct biochemical reactions, and they are a significant risk to plant growth, animal existence, and human life. Research proves that these contaminants are dangerous to plants, wildlife, and human existence. Most heavy metals cause gross bioaccumulation and biomagnification of food chains from plants to humans, which is a great environmental and society issue. Comparing heavy metals to organic compounds which are degradable is not viable as extremely minute quantity of heavy metals are very harmful to human health. Lead contamination results in nervous breakdown and unstable behaviors causing retardation and reduction in the activity of the brain.

Copper and zinc are essential for human development, but an un-prescribed dose of these chemicals results in illnesses of various degrees which includes stooling, vomiting, gastrointestinal bleeding, acute tubular necrosis, and hemolytic anemia. A large degree of contaminants is emitted from manufacturing industries, agricultural activities, and domestic activities. In industrialized areas, most of its pollution is from vehicular or mechanical exhaust of various degrees.

The huge population growth in Nigeria attracts a lot of industries due to the high demand for industrialization leading to a great amount of pollution, extreme population, and industrialized explosion (Duruibe et al. 2007). The quest to cater for the overpopulation and feed people, which directly increases the need for food production due to the large population directly increases the heavy metal pollution in the system because of heavy industrialization and high agricultural activities. Little or no research facilities in a country makes it difficult for that country to know its level of pollution and the drastic measures needed to prevent such pollution from further occurring; this has been the case of residents and citizens of Lagos state, Nigeria (Duruibe et al. 2007).

The intentional and unintentional exploitation of soils for decades due to anthropogenic source pollution makes the chosen sites fit for research (Ali et al. 2015). A control sample is gotten from a botanical garden. The Mettler Teledo pH meter was used to determine pH, Walkley Black method was used to analyze the organic matter content, soil grain size was analyzed using hydrometer method, and the aqua regia method was used for digestion of the soil samples so we can extract the liquid sample for Analysis. The Flame Atomic Absorption Spectrophotometer (F.A.A.S.) would help in the quantification and chemical analysis of all soils samples in order to determine the concentration of all the heavy metals present in the soil.

2. Literature Review

The contamination factor and pollution indices were analyzed by comparing the concentration of some heavy metals in soils of selected sites (odo oba, sabo, and general areas) in Ogbomoso to a control site in the Lautech Campus also in Ogbomoso. According to Yekeen (2013), there was a strong indication that most pollutants were from vehicular emissions because of the correlation between the daily average traffic flow and the concentration of the heavy metals. If bioaccumulation of the heavy metals occurs, this would lead to gross impact on the residents of such locations. More experiments need to be conducted to determine the level of heavy metal pollution and the health effects on the people living on the major roads in Ogbomoso.

Suruchi and Pankaj (2011) explained that soils used in farming, or any agricultural processes should contain minute level of heavy metals. In explaining the results of the soil tests, it was noticed that there can be an increase in the natural number of heavy metals in the soils due to industrial wastes or byproducts, thereby causing great hazards to all living organisms. The quantity of contaminants a living organism is exposed to, and the duration of exposure determines the level of health effects the living organism would go through, in essence the amount of polluted soil exposed to, or food consumed per time.

A chemically analyzed review on the level of metal contamination of some cities in China for at least ten years was conducted by Yang and Wei (2010). In this project work, the amount, medium through which pollutants are gotten and the analytical tools of the metals in the road dusts, farm soils, and urban soils underwent serious investigation, and they were analyzed in this study. From the analysis, it was found that the concentrations of Cr, Ni, Cu, Pb, Zn, As, Hg, and Cd was greater than the recommended values in China soils and there was a large variation in the heavy metal pollution levels. The pollution levels of Cu, Pb, Zn, and Cd were greater than that of

Ni and Cr. Cd, Hg, and Pb metals were due to human activities affecting soil used for agriculture (Mmolawa 2011).

Pollutants get into the body through inhalation, dermal contact, and ingestion. After absorption, they get distributed into the body through the blood stream, they metabolize and are eventually partially excreted outside the body and the rest causes great damage to the body. When present in such polluted environment, the air inhaled is polluted with heavy metals carried from the soil, the heavy metals also seep down into the soil thereby affecting the groundwater leading to dangerous levels of water pollution or even contaminating the oceans and causing a lot ofdamage (Oyeku and Eludoyin 2010). The concentrations provided by the World Health Organization are illustrated in Table 1.

The gastrointestinal tract, central and peripheral nervous systems, kidneys, immune and reproductive system, cardiovascular system, and every organ and system in the body gets affected by lead contamination because lead is a multi-target toxicant (see Table 2). Lead concentration above 30–60 mg/kg in the human body causes great damage to the liver system of the body, it also leads to the irreversible damage of the nervous system in children because they are very prone to the effects of pollution (Eguwrugwu et al. 2007).

Fernando et al. (2012) explains that the recommended concentration of zinc is from 4 mg/kg/day (in children) to 20 mg/kg/day (in adult); too little or excess concentration of zinc leads to health problems such as kidney and brain damage. Nevertheless, zinc is an essential element for human growth and survival and should be taken in the right proportion as this can help with the proper growth of the bones.

Cu(II) is the chemical state in which copper compounds are mostly present in topsoil even though some copper salts exist in two chemical valence states Cu(I) or Cu(II) ions. When the uptake of copper concentration by human beings exceeds the

Table 1: The discussion of the WHO recommended concentration of heavy metals, pH values, organic carbon, and matter and particle size value (Caspah et al. 2016).

Metals	pH	Organic Carbon and Organic Matter	Particle Size Value
Cadmium 0.9 mg/kg	pH 6.5	0.5%	Clay 40%
Copper 30 mg/kg	to	to	Sand 45%
Lead 32 mg/kg	pH 7.0	3.0%	Silt 40%
Zinc 20 mg/kg			

Table 2: Fernando et al. (2012) studied the WHO Values of hazard quotients (HQ) and hazard index (HI) for the daily oral intake value of both adult and child. The values are represented in the table below.

	Concentration (mg/kg)	Concentration (mg/kg)
METALS	ADULT	CHILD
Cadmium	0.107	0.130
Copper	0.103	0.016
Lead	0.417	0.499
Zinc	0.57	0.68

recommended concentration of 0.103 mg/kg/day, such a human body is exposed to copper toxicity.

When a human body is exposed to cadmium, there is a significant risk of complication which is from cadmium toxic concentration. The recommended daily oral intake is 0.107 mg/kg/day and a concentration of 0.9 mg/kg/day (Eguwrugwu et al. 2007).

In determining the level of heavy metals present in the sample soils in this research project, the following research objectives will be achieved:

1. What is the present maximum heavy metal concentration for motor park soils?
2. What happens to the heavy metals present in motor park soils when not controlled or treated?
3. Does the heavy metal present in soils have effects on plant and animal life?
4. Is the human population affected by heavy metal pollution since they get their nutrition from plants and animals?
5. Can the health challenges we discussed earlier be traced back to the uncontrolled heavy metal pollution?

3. Methodology

3.1 Study area

The soil samples used for analysis were collected from motor parks in Obalende, Oshodi, Ketu and Iyana ipaja, all in Lagos state, Nigeria (see location in Figure 1). The study area is very suitable for research purposes as various vehicular and automobile maintenance activities take place in the motor parks leading to gross soil pollution. The soil samples used for the chemical analysis were collected from this location and the controlled sample was gotten from a botanical garden.

Figure 1: An overview map of Lagos state (Source: Lagosmap.com).

3.2 Equipment/materials used

Analytical grade chemicals (such as HCL, HNO3, H2SO4, NaF), properly calibrated glasswares and the analytical instrument flame atomic absorption spectrophotometer (F.A.A.S) is used for qualification and quantification of heavy metals.

3.3 Sample collection

The soil samples were collected using a small hand shovel into clear polythene bags for further analysis. The collected topsoil samples were air dried for five days, crushed and a portion of the soil was properly sieved through a 2-mm stainless steel sieve mesh to remove non-soil impurities and soil materials greater than 2 mm which can be categorized as gravel and the sieved soils were stored in clean polyethene bags (Da Silva et al. 2016).

Representative soil samples were collected into analytical bags for chemical analysis. The soils were all subjected to air drying for five days, they are homogenized, and a sieve is used to remove the non-soil impurities by sieving the soils. It was then stored into a chemical analysis container.

3.4 Soil analysis/digestion

3.4.1 Particle size analysis

From the soil samples, 150 g was dried in the oven, then into a beaker and filled with water was placed 100 g of the oven-dried soil. Sodium hexametaphosphate (50 ml) was dropped into the soil and water solution, the solution was allowed to rest for 2–40 minutes (sodium hexametaphosphate serves as a dispersing agent). Sieving of the remaining soil samples were done using sieve of sizes sixty-three micron and 425 microns, a bowl was placed at the bottom to receive the resulting soil sample from the sieving procedure. The soil sample was then washed with 100 ml distilled water, it was removed, cooled, and weighed again to know the quantity of particles retained (Eleni et al. 2015).

$$\text{percentage retained} = \frac{\text{weight retained}}{\text{initial dry weight of sample}} \times 100$$

The above equation is used to determine the percentage composition of the soil.

3.4.2 Determination of pseudo total metal concentration in soil

Into a 20 ml aqua regia solution in a Teflon beaker, 2 g of soil sample was dissolved in it and allowed to rest for 20–30 minutes; it was then placed on a hot plate for two hours after which the solution was cooled for 30 minutes. The solution underwent filtering to get the digested soil sample (Ming and Lena 2001).

3.4.3 Organic matter content (walkey blackey method)

The Walkley Black method was used to determine the organic matter content of the soil; 2 g of the dried soil sample was weighed and placed into a 50 ml flask, potassium dichromate (10 ml) was placed using pipette into the flask and sulfuric acid (20 ml) was added using dispenser; the solution was stirred gently and allowed to rest for 30 minutes. Ferroin indicator (10 drops) was placed into the solution before titration to avoid deactivation by adsorption into soil surfaces. With about 100 ml of distilled water the solution was diluted to give a visible precipitation so the endpoint can be seen. Orthophosphate (10 ml) was placed into the solution using a dispenser, after which sodium fluoride (0.2 g) was added to complex the ferrous ion that stands as a hindrance to the visibility of the titration end point (Juliana et al. 2014).

The solution was titrated against 0.5 M Fe^{2+} to a brownish red end point. During initial titration process a yellowish orange to dark green color is observed, relying on the concentration of unreacted $Cr_2O_7^{2-}$ the color moves to a turbid gray before the end point of the titration is reached and then the color changes sharply to a wine red or reddish brown at the end point. With all the gotten end point titer values for each soil samples, the percentage organic carbon of the soil is determined using the formula below:

$$\text{Total organic carbon} = \frac{(\text{ml blank} - \text{ml sample}) \times (M_{fe}^{\;2+}) \times 0.003 \times F \times 100}{\text{Mass of soil}}$$

ml = titre value for blank

ml sample = titre value fors sample

$M_{fe}^{\;2+}$ = Molarity of the Fe^{2+} ion

F = Correction factor 1.33

Total organic matter = Total organic carbon ×1.73.

3.5 Quality control

- All samples were collected into analytical chemical containers used for analysis.
- The glass wares were soaked in 5% HNO_3 and rinsed with distilled water.
- For precision sake, all titrations were done in triplicate.

3.6 Data analysis

The data analysis is a progressive step-by-step procedure, which includes collecting the soil sample heavy metal analyzed concentration data from the flame atomic absorption spectrophotometer laboratory (Nnaemeka et al. 2017). Estimating the amount of living organism exposures, assessment of toxicity level, and heavy metal risk characterization is all done using the formula below.

$$\text{Daily oral intake} \left(\frac{\frac{DI}{mg}}{kg}\right) = \frac{C \times IR \times EF \times ED}{BW \times AT}$$

C = concentration of the contaminants in the soil $\left(\frac{mg}{kg}\right)$

IR = Ingestion rate $\left(\frac{mg}{kg}\right)$; Child – 0.0002 kg/day, Adult – 0.0001 kg/day.

ED = Exposure Duration (Yr); Child – 6 years, Adult – 30 years.

EF = Exposure frequency(day/yr); 350 days/year.

AT = Average time for non-carcinogens; 365 days/year.

BW = Body weight (kg); Child – 15 kg, Adult – 70 kg.

4. Results/Discussion

Starting with the pH values of the soils, the recommended pH value of soils giving by WHO is between pH 6.5 to pH 7.0. The soil samples from Ketu and Oshodi Bus Park was found to have a neutral pH value of 7.5 compared with that of Obalende and Iyana Ipaja sample from the bus park, both of which had a pH value of 8.4, showing that they are very alkaline in nature (see Table 3). pH is an important soil property, having great effects on solute concentration and absorption in soil. Both the pH values of the soils in Obalende and Iyana Ipaja axis are higher than the WHO recommended value; this means that the soils in such areas are unsuitable for agricultural processes. Agricultural activities on such soils are unproductive because soil nutrients strive in a neutral environment. Botanical garden soil gives a pH value of 6.54 showing that it is neutral and contains plant nutrients which aids good agricultural environment, and this pH value is in line with the WHO recommended value. Also, the neutral pH value is due to the fact that the soils in botanical gardens are the least polluted soils in the region of research interest.

The recommended organic carbon and organic matter percentage by WHO is between 0.5% and 3.0%. The percentage organic carbon and organic matter helps to

Table 3: pH value of various soil samples in separate locations.

Parameter/ Samples	pH Value	Organic Carbon %	Organic Matter %	% gravel	% sand	% silt	% clay
Iyana Ipaja A	8.00	0.10	0.20	1.00	59.00	18.00	22.00
Iyana Ipaja B	7.97	0.30	0.50	1.00	46.00	22.00	34.00
Oshodi A	6.77	0.60	0.96	1.00	61.00	18.00	20.00
Oshodi B	6.78	0.50	0.90	1.00	63.00	14.00	22.00
Obalende A	8.32	0.50	0.90	1.00	57.00	17.00	25.00
Obalende B	8.47	0.60	1.04	1.00	59.00	18.00	22.00
Ketu A	7.45	0.30	0.51	0.00	75.00	11.00	14.00
Ketu B	7.58	0.20	0.34	2.00	70.00	11.00	17.00
Botanical Garden	6.54	0.70	1.21	0.00	28.00	18.00	54.00

determine the quantity of carbon present in any organic compound. Botanical garden soil sample has the highest organic carbon value of about 0.70% and organic matter value 1.21%, indicating that it would contain more clay particles than others. Oshodi and Obalende have an organic carbon content of about 0.60% and organic matter content of 0.90%. Iyana Ipaja contains the lowest organic carbon value of 0.10% and organic matter content of 0.20%. All the soils samples are in the recommended range given by WHO and this shows that they have good proportion of organic carbon and organic matter but looking at the Iyana Ipaja sample its organic carbon value is lower than the WHO recommended value. A direct effect of poor soil organic carbon is reduced biomass, activity, and nutrients mineralization due to shortage of energy sources in non-calcareous soils, aggregate, stability infiltration, drainage, and airflow are reduced. The organic matter and organic carbon values are greater than the blank values signifying a good property for all the soil samples.

Grain size analysis uses wet and dry methods to analyze and check various components of the soil. It helps for the identification of soil samples from different motor parks in Lagos state. This helps to tell the percentage quantity of gravel, sand, silt and clay in the different soil samples. The WHO recommended the value for Clay 40%, sand 45% and silt 40%.

Ketu sample B has the highest percentage of gravel of 2%, other soil samples contain minimal quantity of gravel of 1%, except Ketu sample A and the botanical garden sample, which does not contain gravel. The soil samples are all sandy, especially Ketu and Oshodi samples, which contain the highest percentage of sand of about 70%, and this is higher than the WHO recommended value, this means that the soil would not be good for agricultural processes. They all contain on average the same amount of silt of about 18% only for Ketu which is about 11%. In terms of clay composition, they contain averagely about 20% except botanical garden that contains about 54% which is higher than the WHO recommended value of 40%. Botanical garden soil sample is more clayey than all the other soil samples; this could be attributed to the high organic matter content.

The concentration of Zn present in Obalende and Ketu samples were found to be 142.5 mg/kg and 88 mg/kg, respectively while Iyana Ipaja and Oshodi has a concentration of 39 mg/kg and 45 mg/kg, respectively which is higher than the WHO recommended concentration (20 mg/kg) of Zn in soils. The high Zn concentration could be linked to the fact that there exist a lot of oil and grease and corrosion of galvanized parts in the motor park.

Samples in Ketu and Obalende was also observed to contain high concentration of Pb of 129 mg/kg and 78.6 mg/kg, respectively while Oshodi and Iyana Ipaja has a concentration of 48.6 mg/kg and 45.8 mg/kg, respectively, which is both higher than the WHO recommended concentration at 32 mg/kg. This irregularity could be due to gasoline, tire wear etc. in the motor park.

A higher concentration 125.18 mg/kg of Cu was found in Obalende which is far higher than the WHO recommended concentration at 30 mg/kg while Oshodi, Ketu, and Iyana Ipaja have a concentration of 30.6 mg/kg, 26.5 mg/kg, and 18.6 mg/kg, respectively. So Oshodi, Ketu, and Iyana Ipaja motor park does not contain high concentration of Cu. The high concentration of Cu in Obalende is due to engine parts repairs, brake emulsions e.t.c in the motor park.

The concentrations for cadmium in all locations are very minimal and some locations even have no trace of cadmium. Ketu has the highest cadmium concentration of 0.7 mg/kg, which is lower than the WHO recommended concentration of 0.9 mg/kg. This indicates that all the locations possess little or no trace of cadmium. The results of the concentrations of various metals in various soil samples in separate locations are shown below (see Tables 4 and 5).

Pb has been shown to affect every organ and system in the human body. Evidence has shown that Pb is a multi-target toxicant which causes effects in the gastro-intestinal tract, cardiovascular system, central and peripheral nervous systems, kidneys, immune system and reproductive system. The results for adults and children with WHO recommendation are provided in Table 6, 7, and 8. Children are most prone to the toxic effects of lead, causing breakdown of central nervous system irreversible, which damage is reported to occur when blood lead level exceeds the WHO recommended values of 30–60 mg/kg. In this study, Ketu soil sample contains an extremely elevated level of Pb with a concentration of 129 mg/kg which is very toxic and dangerous to the human health, followed by Obalende 78.6 mg/kg, Oshodi 37.1 mg/kg, and Iyana Ipaja 45.8 mg/kg. Also, the daily oral intake in Ketu 0.0053 mg/kg/day in adult and 0.0098 mg/kg/day in children is highly dangerous because it is higher than the WHO recommended daily oral intake estimate of 0.0020 mg/kg/day. This clearly shows that the users of Ketu motor park are prone to lead poisoning and people in such environment would experience constant illnesses such as fatigue, vomiting, and skin irritation.

Table 4: Concentrations of various metals in various soil samples in separate locations.

Locations	Zinc (mg/kg)	Lead (mg/kg)	Copper (mg/kg)	Cadmium (mg/kg)
Iyana Ipaja A	43.2	45.8	18.6	0.2
Iyana Ipaja B	30.2	46.3	7.65	ND
Oshodi A	37.1	48.6	12.2	1.0
Oshodi B	53.5	35.1	30.6	ND
Obalende A	91.4	52.9	72.6	1.2
Obalende B	143	78.6	125	0.23
Ketu A	83.5	129	27.4	0.70
Ketu B	88.7	108.7	23.6	0.60
Botanical	4.53	13.4	5.88	0.60

Note: ND is not detected.

Table 5: WHO Recommended concentration of heavy metals.

Metals	Concentration (mg/kg)
Cadmium	0.9
Copper	30
Lead	32
Zinc	20

Table 6: Adult daily oral intake.

Locations	Zinc (mg/kg)	Lead (mg/kg)	Copper (mg/kg)	Cadmium (mg/kg)
Iyana Ipaja A	0.002	0.0018	0.0076	0.000008
Iyana Ipaja B	0.0012	0.0019	0.00031	ND
Oshodi A	0.0015	0.0020	0.00049	0.00004
Oshodi B	0.0022	0.0015	0.0013	ND
Obalende A	0.0038	0.0022	0.0029	0.00005
Obalende B	0.0059	0.0032	0.0051	0.00009
Ketu A	0.0034	0.0053	0.0011	0.00003
Ketu B	0.0036	0.0045	0.0009	0.00002
Botanical	0.00019	0.00055	0.00024	ND

Note: ND is not detected.

Table 7: Children daily oral intake.

LOCATIONS	Zinc (mg/kg)	Lead (mg/kg)	Copper (mg/kg)	Cadmium (mg/kg)
Iyana Ipaja A	0.0035	0.0014	0.003	0.000002
Iyana Ipaja B	0.0023	0.0036	0.0005	ND
Oshodi A	0.0028	0.0040	0.0093	0.00008
Oshodi B	0.041	0.0027	0.00234	ND
Obalende A	0.0070	0.0040	0.0056	0.00009
Obalende B	0.010	0.0060	0.0096	0.00002
Ketu A	0.0064	0.0098	0.0021	0.00005
Ketu B	0.0068	0.0083	0.018	0.000046
Botanical	0.00035	0.0010	0.0005	ND

Note: ND is not detected.

Table 8: WHO Recommended daily oral intake.

	Concentration (mg/kg)	Concentration (mg/kg)
METALS	ADULT	CHILD
Cadmium	0.000037	0.000069
Copper	0.00123	0.0023
Lead	0.0013	0.0025
Zinc	0.00082	0.0015

Zn is an essential element that helps the human growth. The WHO Recommended daily intake ranges from 4 mg in children to 20 mg in adults. Too little and too much Zn concentration can cause health problems, but the WHO recommended daily oral intake is 20 mg/kg/day. In this study, Zn contents in soil samples have alarming values; hence, there is toxic risk of Zn exposure to the inhabitants. Obalende contains the highest level of Zn content ration of about 142.5 mg/kg followed by Ketu 88.7 mg/kg, Oshodi 53.5 mg/kg, and Iyana Ipaja 43.2 mg/kg, which is bad, dangerous and toxic to the users of such motor park. The daily oral intake is also the highest at Obalende with a concentration of 0.005 mg/kg/day in adults and 0.010 mg/kg/day in children. This also shows that users of Obalende motor park are more exposed to zinc poisoning.

Copper is usually present in the topsoil as Cu(II) though most Cu salts occur in two valence states Cu(I) or Cu(II) ions. Users of the motor parks in this study happen to be prone to health risk from Cu toxicity due to the daily oral intake that is greater than the WHO recommended daily oral intake of 0.0023 mg/kg/day. It was discovered that Obalende has the highest level of Cu concentration of about 125 mg/kg/day as against the WHO recommended concentration of 32 mg/kg/day and also the daily oral intake is 0.0096 mg/kg/day for children in Obalende and 0.0051 mg/kg/day for adults.

Cadmium is an elevated risk of complications resulting from Cd toxicity in the exposed population of the inhabitants. The WHO recommended daily oral intake is 0.000037 mg/kg/day and a concentration of 0.9 mg/kg/day (see Table 8). Iyana Ipaja and Oshodi soil samples contain a great amount of Cd. Obalende shows an appreciable concentration of about 1.2 mg/kg/day and a daily oral intake of 0.00009 mg/kg/day in children and 0.00005 mg/kg/day in adults. This proves that the users of Iyana Ipaja, Oshodi, and Obalende motor park are at risk of cadmium poisoning.

5. Conclusion

This journal has proven that motor parks in Lagos state, Nigeria are polluted with heavy metals according to various degrees. In most areas, the level of pollution has actually reached a level where its concentration is alarming and calls for concern. The mean concentration of Zn in the four-motor park is 57.1 mg/kg, Pb is 62.0 mg/kg, Cu is 40 mg/kg, and Cd is 0.43 mg/kg. Obalende possesses the highest concentration of copper 125 mg/kg, 0.12 mg/kg of cadmium, while Ketu possesses the highest concentration of lead which is 129 mg/kg. The pH values also explain the alkalinity and acidity of the different soil samples: Iyana Ipaja and Obalende have a soil pH of about 8.0 showing that it is very basic and useful for agricultural activities, the total organic matter helps to determine the amount of organic carbon present in the soils; Obalende and Oshodi have the highest total organic carbon of 0.60%, and a total organic matter of 1.00%. The analysis also showed that children are more prone to heavy metal pollution compared to adults.

5.1 Recommendation

Oil-producing companies like Shell NG and Chevron NG should derive means of proper disposal of their waste. The oil-producing companies should also reduce the content of heavy metals in their products like engine oil, brake oil, etc. Burning should be prevented, especially burning petroleum product derivatives. Heavy metal pollution from road runoff is dangerous for aquatic wildlife, but other types of non-point source pollution like excess nutrients and sediment erosion are just as harmful to habitat, So there should be proper waste treatment before disposal. Carpooling, public transportation, and bicycling should be introduced to the transportation sector. The usage of lead petrol should be banned and discouraged in all parts of Lagos state.

References

Ali, S., Mustafa, S., Erkan, Y., Selehattin, Y. and Hasan, O. (2015). Characterization of heavy metal fractions in agricultural soils by sequential extraction procedure: The relationship between soil properties and heavy metals fractions. *An International Journal of Soil and Sediment Contamination*, 24(1): 1–15.

Amuda, O., Giwa, A. and Bello, I. (2007). Removal of heavy metals from industrial wastewater using modified activated coconut shell carbon. *Biochemical Engineering Journal*, 36(1): 174–181.

Brown, S., Chaney, R., Hallfrisch, J. and Xue, Q. (2003). Effects of bio solids processing on lead bioavailability in an urban soil heavy metal in the environment. *Journal of Environmental Quality*, 32(1): 8–100.

Caspah, K., Manny, M. and Morgan, M. (2016). Health risk assessment of heavy metals in soils from witwatersrand gold mining basin, South Africa. *International Journal of Environmental Research and Public Health*, 13(7): 663.

Da Silva, M., Montanarella, L., Hermann, T. and Tóth, G. (2016). Heavy metals in agricultural soils of the European Union with implications for food safety. *Environment International Journal*, 88(1): 299–309.

Duruibe, J., Ogwuebu, M. and Egwurugwu, J. (2007). Heavy metal pollution and human bio toxic effects. *International Journal of Physical Sciences*, 2(5): 112–118.

Egwurugwu, J., Ufearo, C., Abanobi, O., Nwokocha, C., Duruibe, J., Adeleye, G., Ebunlomo, A., Odetola, A. and Onwufuji, O. (2007). Effects of ginger on cadmium heavy metals. *African Journal of Biotechnology*, 6(18): 2078–2082.

Eleni, A., Maria, J., Susan, B., Josee, D., Luc, C., Sophie, L., John, P., Agnieszka, K. and Nele, B. (2015). Determination of particle size, surface area, and shape of supplementary cementitious materials by different techniques. *Materials and Sturctures Journal*, 48(1): 3687–3701.

Fernando, G., Anderson, T., Takashi, M., Nericlenes, M. and Solange, C. (2012). Heavy metals in vegetables and potential risk for human health. *Science Agricultural Journal*, 69(1): 54–60.

Garbu, S., Ahmed, I., Akan, C. and Dauda, B. (2014). Distribution pattern and level of metals in soil samples within maiduguri metropolis. *International Research Journal of Pure and Applied Chemistry*, 4(5): 486–493.

Iqbal, A., Hayat, S., Ahmad, A., Inam, A. and Samilulah. (2005). Effects of heavy metal on survival of certain groups of Indigenous soil microbial population. *International Journal of Applied Sciences and Environmental Management*, 9(1): 115–121.

Juliana, H., Figueiredo Cícero, C., Marchão Robélio, L., Madari Beáta, E., Benedito Luiz, E., Busato Jader, G. and Souza Diego, M. (2014). Methods of soil organic carbon determination in Brazilian savannah soils. *Scientia Agricola*, 71(4): 302–308.

Ministry of Works and Housing. (2013). The construction of settlements for easy movement. Lagosmap.com.

Ming, C. and Lena, Q. (2001). Comparison of three aqua regia digestion methods for twenty florida soils. *Soil Science Society of America Journal*, 65(2): 491–499.

Mmolawa, K., Likuku, A. and Gaboutloeloe, G. (2011). Assessment of heavy metal pollutants was conducted along major roadsides soils of Botswana. *African Journal of Environmental Science and Technology*, 5(3): 186–196.

Nnaemeka, U., Zelinjo, I., Rose, A. and Orish Ebere, O. (2017). Health risk assessment and dietary exposure to polycyclic aromatic hydrocarbons (PAHs), lead and cadmium from bread consumed in Nigeria. *Publicae Sanitati Augendae*, 68(3): 269–280.

Oyeku, O. and Eludoyin, A. (2010). Heavy metal contamination of ground water resources in a Nigerian urban settlement. *African Journal of Environmental Science and Technology*, 4(4): 201–214.

Riffat, N., Zahoor, H. and Ishfaq, N. (2010). Heavy metal contamination and accumulation in soil and wild plant species from industrial area of Islamabad, Pakistan. *Pakistan Journal of Botany*, 42(1): 291–301.

Suruchi, H. and Pankaj, K. (2011). Assessment of leaves metal contamination in different vegetables grown in and around urban areas. *Research Journal of Environmental Toxicology*, 5(3): 162–179.

Upredra, K. (2006). Agricultural products and by-products as a low-cost absorbent for heavy metal removal from water and wastewater: A review. *Journal of Scientific Research and Essay*, 1(2): 033–037.

Yang, L. and Wei, B. (2010). A review of heavy metal contaminations in urban soils, urban road dusts and agricultural soils from china. *Micro chemical Journal*, 94(2): 99–107.

Yekeen, Z. (2013). The evaluation of the concentrations of soils metals from selected sites along major road in Ogbomosho was made in comparison with control site, lautech campus (LC) based on contamination factor and pollution indices. *Journal of Environment and Earth Science*, 2.

Chapter 10

Another Look at EKC in the Context of Sub-Saharan African Countries

Bright Akwasi Gyamfi

1. Introduction

Scholarly conversation has been preoccupied for a considerable amount of time with the impact that natural resources and energy has on economic growth. An excess of natural resources in a nation has the potential to drive economic growth and development through the facilitation of exports for the generation of foreign exchange earnings, the promotion of local industrialization, creation of jobs, and the attraction of foreign direct investment. In addition to this, they are recognized for contributing to the growth of both infrastructure and human capital (Gyamfi et al. 2022a, Onifade et al. 2021). The counterargument contends that resource-rich nations do not function well as a result of high levels of corruption, political instability, rent-seeking behavior, and lower levels of investment in human capital (Adebayo et al. 2021). Energy consumption is another factor that contributes to economic growth since energy is an essential factor in the production processes of all economies. Energy is essential to the operation of all aspects of a modern economy, including but not limited to the manufacturing sector, the transportation sector, the educational sector, and the healthcare sector, among others. Because of this, the governments and policymakers of every country are working hard to ensure that there is a sufficient supply of energy for the residential and industrial activities. On the other hand, there are worries that rising energy use, particularly of fossil fuels, is related to higher levels of carbon emissions (Adedoyin et al. 2021). The amount of carbon dioxide that is present in the air is one of the primary contributors to both global warming and climate change (He et al. 2021). Since the usage of fossil fuels causes increased carbon dioxide emissions, it is considered unfriendly to the environment.

School of Management, Sir Padampat Singhania University, Bhatewar, Udaipur, Rajasthan, India.
Email: brightgaymfi1987@gmail.com

The exploitation of natural resources has been expanding throughout the years, regardless of the fact that there are contradictory viewpoints regarding the impact natural resources have on economic growth. The World Resources Forum (2020)[1] says that between 1980 (40 billion tons) and 2005 (58 billion tons), the amount of natural resources that were taken from the earth rose by about 25%. In light of this, scholars in recent years have been increasing the focus of research, policymakers, and environmentalists on the impact that the exploitation of natural resources has on ecological sustainability. The procedure of obtaining natural resources (for the purpose of making economic advancement) results in the pollution of water bodies and the deterioration of the environment, as well as the degradation of land and forest resources. Furthermore, the extraction of natural resources and the transformation of these resources into high-value products both need the use of energy. Therefore, the more energy that is utilized for the exploitation of natural resources and the processing of their products, the greater the amount of carbon emissions (Ahmed et al. 2020). It is therefore reasonable to anticipate that the extraction and processing of natural resources will create an adverse effect on the surrounding environment if energy efficiency and environmental consciousness are prioritized during these processes and vice versa. If the process of extracting natural resources employs renewable energy for its activities, then there is a good possibility that their negative effect in terms of carbon emissions will be modest. This is due to the fact that renewable energy sources are better for the environment than traditional energy sources because they produce less carbon dioxide (Alola et al. 2019).

Africa is a continent that is richly blessed with natural resources such as mineral reserves, oil reserves, and forest resources. Although the influence of natural resources on economic growth has been scientifically investigated for nations on the planet (Hayat and Tahir 2021, Yasmeen et al. 2021, Rahim et al. 2021), very few analyses have been undertaken to evaluate the influence of natural resources on the environment (Khan et al. 2022, Erdoğan et al. 2021, Nathaniel et al. 2021 among others). According to the findings of the few studies, the process of extracting natural resources leads to a rise in carbon emissions, which in turn lowers the health of the environment (Onifade et al. 2021, Gyamfi et al. 2022b, Gyamfi 2022). This is something that appears to be true in the majority of the studies. Because revenues from natural resources over a certain level can lead to high ecological degradation. Moreover, there is a chance that the connection between the two is not linear, which is due to the fact that natural resource profits can trigger lower levels of environmental degradation. As a result, the initial negative effect of extracting natural resources might be reversed after a specific threshold has been reached. If enough money is made off natural resources, industry participants, policymakers, and the government agencies might be able to repair the damage that has been done to the environment.

[1] https://www.bing.com/ck/a?!&&p=91cfcb349a0f43e5f9b62bfeae426c46JmltdHM9MTY1Nzc0NDM0MiZpZ3VpZD00OTYyNmIzZC1kZWYzLTQ3MTItOGU3My0yZjNiYWYzNTBiZDUmaW5zaWQ9NTIxMA&ptn=3&hsh=2&fclid=e5a38890-02ea-11ed-a870-740fb12d7848&u=a1aHR0cHM6Ly93cmYyMDIwLndyZm9ydW0ub3JnL3dwLWNvbnRlbnQvdXBsb2Fkcy8yMDIwLzAxL0NhbGw0YWJzdHJhY3RzMjAyMC5wZGY&ntb=1.

After significant economic advantages, actors in the sector and legislators may become environmentally sensitive. As a result, energy-efficient technology or more stringent rules may be implemented to ensure that the extraction of natural resources does not negatively impact the environment. In addition to some gains, extractive companies can buy equipment that uses less energy and helps them run their businesses better.

Moreover, analysis that looks at environmental deterioration within the context of pollutants has been split into three categories. In the first category is the research investigates whether or not the Environmental Kuznets Curve (EKC) assumption can be applied in real-world situations (Ozturk and Al-Mulali 2015, Agboola et al. 2022, Bamidele et al. 2022, Ohajionu et al 2022). The next category composed of analysis that uses the EKC equations and takes into account the effect that other coefficients have on the environment. These other factors include, but are not limited to, energy usage, globalization, and industrialization (Alola et al. 2022, Akadırı et al. 2021, Alola and Donve 2021). The third category of researchers includes those who stated the link between economic growth and carbon emissions but did not precisely account for the EKC concept's projections (Steve et al. 2022, Appiah et al. 2022).

Furthermore, the management of the environment has proven a challenging task for many nations that are still developing, especially in SSA. The environmental challenges plaguing the sub-region include but are not limited to water contamination, loss of forest cover, and improper waste management. Recently, there has been a rise in the level of worry among experts on air pollution, namely CO_2 emissions in the sub-region. Even though the region's carbon emissions are lower in comparison to other places, the upward trend over the past few years has led many people to question whether or not the region is prepared to assist in the fight against the threat posed by climate change and global warming, the effects of which are more severe for the countries in the region. Despite the fact that China, India, and the United States are among the renowned nations in terms of carbon emissions, nations in sub-Saharan Africa are in a vulnerable position (Gyamfi et al. 2022b, c). This is due to the fact that a significant number of people in these countries rely on agriculture-related operations for their livelihoods, which are also significantly affected by climatic factors. As a result, in order for the globe to achieve sustainable development, countries that produce high amounts of carbon dioxide emissions need to be dedicated to bringing the level of carbon emissions down. In light of the fact that carbon dioxide emissions in sub-Saharan Africa have increased from 399 thousand (kt) in 1990 to 823 thousand (kt) in 2018, it is essential to determine the factors that are driving this increase in order to assist in the formulation of appropriate policies to minimize emissions. While there are increasing examinations on the determinants of CO_2 emission in the sub-region and globally (Bekun et al. 2021b, Ntarmah et al. 2021, Iheonu et al. 2021, Yameogo et al. 2021), the impact of resource exploitation, clean energy as well as other variables on carbon emission in SSA is very few. However, similar to the study of Gyamfi (2022), this study utilized carbon emission instead of consumption-based carbon emission. Again, there is an interaction term between renewable energy and natural resource which look at how the usage of clean energy with resource rent impact the eco-system.

In light of the preceding, the purpose of this analysis is to investigate how the utilization of natural resources and alternative forms of energy source affects carbon dioxide emissions in sub-Saharan Africa (SSA). This research attempts to accomplish the following key goals: (a) empirically determine the linear and non-linear impacts of natural resources on CO_2 emissions in sub-Saharan Africa; (b) estimate the influence of alternative energy usage on CO_2 emissions in sub-Saharan Africa; and (c) investigate the interaction impacts of alternative energy usage in the natural resources-CO_2 emission nexus in sub-Saharan Africa.

In order to make an addition to the existing body of research, the purpose of this study is to investigate the effect that is caused by the interacting dynamic connecting the usage of green energy and the depletion of natural resources. Second, it investigates the quadratic influence that the economic growth has on the emission of CO_2. Prior analyses (Alola and Donve 2021, Gyamfi et al. 2022c, Bekun 2022) have made the assumption that the link between carbon and economic growth is linear which may not be the case. Thirdly, it makes use of the ARDL technique in order to investigate the influence of the extraction of natural resources and the use of renewable energy on CO_2 emissions. Fourth, it presents empirical findings from a location that has not gained much consideration in the energy and climate economics literature; thus, SSA.

2. Methodology

2.1 Data source and estimation framework

The study makes use of panel data from World Bank (2022) for Oil producing sub-Saharan Africa countries (see Table 5 of Appendix) from 1990 to 2020 based on availability of data. Carbon dioxide emissions is measured by carbon dioxide (kt), population is measured by the total population and income is measured by the GDP per capita (2015 USD). Moreover, urban population is measured by urban population (% total population), natural resources extraction is also measured by total natural resources rent (% GDP) and renewable energy consumption is measured by combustible renewables and waste. These measurements follow what has been commonly used by previous studies. This study draw strength from the study of Gyamfi (2022) by incorporating population growth and renewable energy in to the manuscript. As a result, the main effect function is provided as;

Main effect function; $CO_2t = f(GDPCt, PUPt, URBt, NRRt, RUEt)$ (1)

where CO_2 is the dependent variable carbon emission, GDPC is GDP per capita, PUP is population, URB been urban population, NRR for natural resources rent, and RUE been renewable energy usage while t is the time frame. However, taking the natural logarithm of Equation 1 for a panel data results in:

$$LCO2t = \beta 0 + \beta LGDPCt + \beta 2LPUPt + \beta 3LURBt + \beta 4LNRRt + \beta 5LRUEt + \varepsilon t \quad (2)$$

Moreover, the study factor EKC into the equation to access the square of GDPC which is stated in Equation 3 as;

$$LCO2t = \beta 0 + \beta 1LGDPCt + \beta 2GDPC2t + + \beta 3LPUPt + \beta 4LURBt + \beta 5LNRRt + \beta 6LRUEt + \varepsilon t \quad (3)$$

where CO_2, GDPC, PUP, URB, NRR and URB have been defined above but $GDPC^2$ is the square of income. Nevertheless, since on the argument that green energy intake can moderate the environmental impact of natural resource mrent, an interactive term between the two (LNRRxLREU) is introduced to Equation 4:

$$LCO2t = \beta 0 + \beta 1LGDPCt + \beta 2GDPC2t ++ \beta 3LPUPt + \beta 4LURBt + \beta 5LNRRt + \beta 6LRUEt + \beta 7LNRR^{*}LREUt + \varepsilon t \quad (4)$$

Time series and panel data are found to contain unit root which usually leads to getting spurious regression results. It, therefore, has become almost a ritual for studies that use such data for analysis to conduct unit root. If unit root is present a variable is usually differenced to remove the unit root. Once unit root is absent the variable becomes stationary. However, it is argued that interdependency among panel data renders these tests ineffective. In such instances, the Pesaran (2007) Panel Unit Root test thus, cross-sectionally augmented IPS (CIPS) is used to test for the stationarity. This makes it crucial to therefore test for the cross-sectional dependence among the variables before the unit root test is conducted.

Following this is the Kao (1999) cointegration test to ascertain the long-run relationship between the dependent variable and the regressors. Once cointegration is established among the variables the autoregressive distributed lag (ARDL) regression technique is used to determine the effects of population, urbanization, income, natural resources rent, and renewable energy consumption on CO_2 emissions. The ARDL technique which generates short-run and long-run coefficients simultaneously is chosen for this estimation since it deals with endogeneity problems-associated panel data. It is also useful when variables for analysis are stationary at levels and/or first difference; and also when the sample size is small. Following Kollie (2020), the basic ARDL model is often specified as:

$$Y_{it} = \sum\nolimits_{j-1}^{p} \lambda_{tj} Y_{i,t-j} + \sum\nolimits_{j=0}^{q} \delta_{ij}^{i} X_{i,t-j} + \mu_i + \in_{it} \quad (5)$$

where Y_{it} is the dependent variable; $Xi, t-j$ is the vector of independent coefficients; λ_{ij} is the coefficient on the lags of the dependent variable; $\delta_{ij'}$ is the coefficients on the current and lags of the explanatory variables; μ_i is the fixed effect; and ϵ_{it} is the error term. A precondition to establish the long-term relationship among the variables is that they must be cointegrated. An error correction model (ECM) is also introduced to capture short-run deviations that are corrected in the long run. Consequently, Equation 7 is re-parameterized to account for the ECM to arrive at the panel ARDL model given below:

$$\Delta lnY_{it} = \phi_1(lnY_{i,t-1} - \theta_1' lnX_{it}) + \sum\nolimits_{j=1}^{p-1} \lambda_{ij} \Delta lnY_{i,t-1} + \sum\nolimits_{i=0}^{q-1} \delta_{ij} \Delta lnX_{i,t-1} + u_1 + \in_{it} \quad (6)$$

The variables Y and X in Equation 10 remained as already explained; $\theta i'$ is a vector of long-run coefficients; λ_{ij} and δ are short run coefficients; and ϕi captures the speed of adjustment which should be negative and statistically significant.

3. Empirical Results and Discussion

This section provides the discussion of all empirical results and discussion accordingly. From Table 1 below it was proven that, urbanization has the highest mean, median, maximum as well as minimum which GDP has the lowest. Moreover, from the correlation, GDPC, population growth, and urbanization have positive connection with the dependent variable carbon emission whiles clean energy and natural resources have a negative correlation with carbon emission for the-oil producing SSA countries. However, Table 2 proves that all the three cross-sectional techniques used shows there is cross-section among the variables, which allowed us to use the CIPS

Table 1: Descriptive statistic and correlation matrix.

	LCO_2	LGDPC	LREU	LPUP	LNRR	LURB
Statistic						
Mean	1.7674	0.7983	1.0965	1.6631	1.5163	2.4827
Median	1.5845	0.7541	1.7065	1.0130	1.3618	2.5519
Maximum	2.4630	1.6962	3.3923	2.2420	1.9703	3.4931
Minimum	0.0209	–2.7132	–0.0634	–0.7146	0.0568	1.3456
Correlation						
LCO_2	1					
LGDPC	0.4113	1				
LRUE	–0.9112	–0.8342	1			
LPUP	0.4474	0.7001	0.4224	1		
LNRR	–0.3064	–0.0954	0.6117	0.2667	1	
LURB	0.5141	0.4535	–0.4324	–0.7425	0.7667	1

Table 2: Cross-sectional dependence and unit root test results.

Cross-sectional Dependence Test		
Test	**Statistic**	**Prob.**
Breusch-Pagan LM	1370.0123***	(0.000)
Pesaran scaled LM	146.0057***	(0.000)
Pesaran CD	13.1053***	(0.000)
CIPS Unit Root Test		
Variable	0(I)	1(I)
LCO_2	–3.0011**	N/A
LGDPC	–0.1060	–4.6734***
LPUP	–3.2191***	N/A
LURB	–2.9192***	N/A
LNRR	–3.3708**	N/A
LRUE	–2.6495**	N/A

Note: *** and ** denote 1% and 5% levels of significance, respectively.

unit root test. The outcome of the unit root shows there are mix stationarity among the variables, that is, carbon emission, population growth, urban population, natural resources, and clean energy all have stationarity at level while GDP has stationarity at first difference. From there, the Kao cointegration technique was used to access the long-run relationship among the variables. From Table 3 below, the result shows that there is cointegration among the variables.

The ARDL technique were used to determine the short-term and long-term effects of the coefficients on environmental degradation within the oil producing countries of SSA which was presented in Table 4. This was done so that the analysis could produce accurate and reliable results. From the analysis in Table 4 below it was observed that, from all the four models, there is a positive significant linkage involving GDPC and CO_2 in the long run whiles models 3 and 4 shows a negative relationship between the square of GDPC and CO_2 emission that shows the presence of U-Shape EKC hypothesis for the oil-producing SSA countries, which affirms the findings of Jian et al. (2022), Mehmood (2022), and Zeraibi et al. (2022). This means that an increase in income will initially increase CO_2 emissions but beyond a certain threshold, income will dampen carbon dioxide emissions. The reason may be as a result of the technique effect associated with higher income levels (Gyamfi et al. 2021, Bekun et al. 2021a).

Moreover, from all the models, the population has a positive significant connect with emissions, which affirms the outcomes of Gyamfi et al. (2020), Khan et al. (2021), and Rehman et al. (2021). This implies that more land is need for activities like building, farming, and industrial development which produce a lot of emissions. On the other hand, urban population has an adverse influence on the environment for all the four models. According to the compact theory, urbanization brings about a reduction in the amount of energy consumed and, as a result, a reduction in the amount of carbon emissions. Because of the increased concentration of people and businesses in metropolitan areas, there has been an increase in the effectiveness with which resources, most notably energy, are utilized, which has resulted in a decrease in the amount of carbon emissions. The discovery leads one to believe that the sub-region may not be suffering from the negative effects of urbanization with regard to carbon-dioxide emissions. It should not come as a surprise that this result occurred, given the efforts that have been made in recent decades to minimize the amount of traffic congestion and slum conditions in the sub-region. Efforts have been made to make urban households more aware of climate change; this may have led to environmental consciousness on the part of urban households and business managers, which in turn may have resulted in a reduction of carbon emissions in urban regions. This finding is in line with that of Hashmi et al. (2021) and Liu et al. (2021).

Table 3: Kao cointegration test results.

Kao Test		**Statistic**	**Prob.**
	ADF	–3.6510***	0.007

Note: *** and ** denote 1% and 5% levels of significance, respectively.

Table 4: ARDL regression results.

	Model 1	Model 2	Model 3	Model 4
Variable	Coefficient (P-value)	Coefficient (P-value)	Coefficient (P-value)	Coefficient (P-value)
LONG-RUN ESTIMATES				
LGDPC	0.1462*** (0.001)	0.6429*** (0.000)	0.9961*** (0.001)	0.8903*** (0.000)
LGDPC2			–0.1404*** (0.000)	–0.1076** (0.020)
LPUP	0.4671** (0.035)	0.5058* (0.075)	1.0079*** (0.000)	0.8449** (0.000)
LURB	–1.1122*** (0.003)	–1.0050*** (0.000)	–1.0025*** (0.000)	–1.0416*** (0.000)
LNRR	0.1007*** (0.004)	0.1101** (0.000)	0.1508*** (0.000)	0.2807*** (0.000)
LRUE	–0.4385*** (0.000)	–1.0287*** (0.000)	–0.8190** (0.012)	–0.8554*** (0.000)
LRUE x LNRR		–0.1427 (0.1456)		-0.3009 (0.232)
SHORT-RUN ESTIMATES				
ECM	–0.7505*** (0.001)	–0.3054** (0.035)	–0.5539*** (0.00)	–0.4002** (0.015)
D(LGDPC)	0.5238 (0.120)	0.4461** (0.017)	1.0915 (0.804)	2.4382 (0.526)
D(LGDPC2)			0.4007 (0.168)	0.5133*** (0.004)
D(LPUP)	1.2532 (0.937)	0.8957 (0.645)	0.9353 (0.660)	1.0096 (0.396)
D(LURB)	2.2006 (0.1110)	2.8617 (0.161)	2.0235 (0.386)	2.4090 (0.291)
D(LNRR)	–0.1025 (0.243)	–0.7065 (0.128)	–0.6087 (0.292)	–1.0325** (0.023)
D(LRUE)	–0.3667 (0.568)	–0.4348 (0.292)	–0.5001 (0.295)	–1.1377 (0.264)
D(LRUE x LNRR)		0.0930 (0.463)		0.1004** (0.045)

Note: *** and ** denote 1% and 5% levels of significance, respectively.

Nevertheless, natural resources are seen to have positive connection with emissions for all the models. Although it is arguable that governments in sub-Saharan Africa have benefited from natural resources in terms of income mobilization, currency trading, and employment prospects, amongst other things, the same cannot be said about the health of the environment. This is what this means. As a result, the natural resources in the sub-region have been exploited, which has had a negative impact on the ecosystem. This leads one to believe that the harvesting of natural resources, such as oil, minerals, and forests, has not been handled in a more sustainable manner. This finding is in line with Shittu et al. (2021), Jahanger et al. (2021), and Nathaniel (2021). Furthermore, renewable energy is said to have negative significant connection with carbon emission. The implication is that more of such energy can be used for production purposes in an economy and still protect the environment. Many researchers have reported renewable energy enhances environmental sustainability (Gyamfi et al. 2021a, b, Gyamfi et al. 2022c). An interactive term connecting the intake of renewable energy and the depletion of natural resources has been developed in order to investigate the potential role that renewable energy may play in the relationship between natural resources and carbon dioxide emissions. The findings of the regression on all variables demonstrate that the interaction term has a significantly negative effect that is supported by statistics. Inference leads us to believe that the use of carbon-neutral, renewable energy sources reduces the beneficial influence of natural resources on carbon dioxide emissions. Therefore, the amount of carbon dioxide emissions will decrease as the companies that extract natural resources increase their use of renewable energy in the extraction and refining stages of their operations. The utilization of renewable energy results in a slower rate of carbon release than there otherwise would have been. For the short-run analysis, it was observed that, GDPC in model two, $GDPC^2$, and the interaction between clean energy and natural resources all are positively significant with emissions whiles natural resource had a negative connection with emissions.

4. Conclusion and Policy Recommendation

4.1 Conclusion

This study investigates the connection regarding economic growth, natural resources, urban population, population growth, and clean energy on CO_2 emissions for the oil-producing SSA countries from 1990 to 2020 by accounting for the presence of EKC hypothesis. The study utilized the three cross-sectional dependency techniques (Breusch-Pagan LM, Pesaran scaled LM and Pesaran CD), the CIPS unit root test, Kao cointegration test and the ARDL regression techniques. From the analysis, there is a long run cointegration among the under-study variables. Moreover, GDPC and its square shown a positive and negative significant connection with emission confirming the presence of U-Shape EKC hypothesis. Again, population growth and natural resources have a positive connection with carbon emissions while urban populations, renewable energy, and the interaction between clean energy and natural resources all have a negative connection with carbon emissions for the oil-producing SSA countries. Lastly, for the short run analysis it was observed that, GDPC in model two, GDPC2, and the interaction between clean energy and natural resources all are positively significant with emissions while natural resources have a negative connection with emissions.

4.2 Policy recommendation

The decision-makers in this sub-region ought to make a concerted effort to eliminate or reduce the impact of variables that affect the emergence of low-carbon economies. It is important to instill environmentally responsible practices in the sub-increasing region's numerous inhabitants. Because leading a healthier approach will contribute to a reduction in CO_2 emissions, governments need to get serious about educating the public about topics such as the efficient use of electricity and water, the preservation of food, and the reduction of waste. The findings provide credence to the contention that renewable energy sources are less harmful to the environment. As a result of this, increasing the use of renewable energy sources in the sub-region will contribute significantly to the reduction of carbon-dioxide emissions. Therefore, it is absolutely necessary for policymakers to work toward attracting investors to the renewable energy sector. In actuality, the provision of tax refunds and subsidies on renewables as forms of incentives will enhance both the production and consumption of renewable energy. In the same way, if the tax rate on nonrenewable energy sources goes up, people will use less of that kind of energy.

It will be easier for companies in the natural resources sector to generate financial prosperity without negatively impacting the health of the environment if they switch to using renewable sources of energy. Therefore, it is essential for policymakers to ensure that the extraction of natural resources is carried out in a manner that is more environmentally friendly. Therefore, actions such as cutting taxes on technology is beneficial to the environment and can improve the ecological exploitation and processing of forest resources, mineral resources, and oil resources.

References

Adebayo, T. S., Awosusi, A. A., Kirikkaleli, D., Akinsola, G. D. and Mwamba, M. N. (2021). Can CO_2 emissions and energy consumption determine the economic performance of South Korea? A time series analysis. *Environmental Science and Pollution Research*, 28(29): 38969–38984.

Adedoyin, F. F., Bein, M. A., Gyamfi, B. A. and Bekun, F. V. (2021). Does agricultural development induce environmental pollution in E7? A myth or reality. *Environmental Science and Pollution Research*, 28(31): 41869–41880.

Agboola, P. O., Hossain, M., Gyamfi, B. A. and Bekun, F. V. (2022). Environmental consequences of foreign direct investment influx and conventional energy consumption: Evidence from dynamic ARDL simulation for Turkey. *Environmental Science and Pollution Research*, 1–14.

Ahmed, Z., Asghar, M. M., Malik, M. N. and Nawaz, K. (2020). Moving towards a sustainable environment: The dynamic linkage between natural resources, human capital, urbanization, economic growth, and ecological footprint in China. *Resources Policy*, 67: 101677.

Akadırı, S. S., Alola, A. A. and Usman, O. (2021). Energy mix outlook and the EKC hypothesis in BRICS countries: A perspective of economic freedom vs. economic growth. *Environmental Science and Pollution Research*, 28(7): 8922–8926.

Alola, A. A. and Donve, U. T. (2021). Environmental implication of coal and oil energy utilization in Turkey: is the EKC hypothesis related to energy? *Management of Environmental Quality: An International Journal*.

Alola, A. A., Adebayo, T. S. and Onifade, S. T. (2022). Examining the dynamics of ecological footprint in China with spectral Granger causality and quantile-on-quantile approaches. *International Journal of Sustainable Development and World Ecology*, 29(3): 263–276.

Alola, A. A., Yalçiner, K., Alola, U. V. and Saint Akadiri, S. (2019). The role of renewable energy, immigration and real income in environmental sustainability target. Evidence from Europe largest states. *Science of the Total Environment*, 674: 307–315.

Appiah, M., Gyamfi, B. A., Adebayo, T. S. and Bekun, F. V. (2022). Do financial development, foreign direct investment, and economic growth enhance industrial development? Fresh evidence from Sub-Sahara African countries. *Portuguese Economic Journal*, 1–25.

Bamidele, R., Ozturk, I., Gyamfi, B. A. and Bekun, F. V. (2022). Tourism-induced pollution emission amidst energy mix: evidence from Nigeria. *Environmental Science and Pollution Research*, 29(13): 19752–19761.

Bekun, F. V. (2022). Mitigating emissions in India: Accounting for the role of real income. *Renewable Energy Consumption and Investment in Energy*, 670216917.

Bekun, F. V., Alola, A. A. and Sarkodie, S. A. (2019). Toward a sustainable environment: Nexus between CO_2 emissions, resource rent, renewable and nonrenewable energy in 16-EU countries. *Science of the Total Environment*, 657: 1023–1029.

Bekun, F. V., Alola, A. A., Gyamfi, B. A. and Ampomah, A. B. (2021b). The environmental aspects of conventional and clean energy policy in sub-Saharan Africa: Is N-shaped hypothesis valid? *Environmental Science and Pollution Research*, 28(47): 66695–66708.

Bekun, F. V., Gyamfi, B. A., Onifade, S. T. and Agboola, M. O. (2021a). Beyond the environmental Kuznets Curve in E7 economies: Accounting for the combined impacts of institutional quality and renewables. *Journal of Cleaner Production*, 314: 127924.

Danish Baloch, M. A., Mahmood, N. and Zhang, J. W. (2019). Effect of natural resources, renewable energy and economic development on CO_2 emissions in BRICS countries. *Sci. Total Environ.*, 678: 632–638.

Demissew Beyene, S. and Kotosz, B. (2020). Testing the environmental Kuznets curve hypothesis: An empirical study for East African countries. *International Journal of Environmental Studies*, 77(4): 636–654.

Destek, M. A. and Sinha, A. (2020). Renewable, non-renewable energy consumption, economic growth, trade openness and ecological footprint: Evidence from organisation for economic Co-operation and development countries. *Journal of Cleaner Production*, 242: 118537.

Erdoğan, S., Çakar, N. D., Ulucak, R. and Kassouri, Y. (2021). The role of natural resources abundance and dependence in achieving environmental sustainability: Evidence from resource-based economies. *Sustainable Development*, 29(1): 143–154.

Gyamfi, B. A., Bein, M. A., Ozturk, I. and Bekun, F. V. (2020). The moderating role of employment in an environmental Kuznets curve framework revisited in G7 countries. *Indonesian Journal of Sustainability Accounting and Management*, 4(2): 241–248.

Gyamfi, B. A. (2022). Consumption-based carbon emission and foreign direct investment in oil-producing Sub-Sahara African countries: The role of natural resources and urbanization. *Environmental Science and Pollution Research*, 29(9): 13154–13166.

Gyamfi, B. A., Adebayo, T. S., Bekun, F. V. and Agboola, M. O. (2022c). Sterling insights into natural resources intensification, ageing population and globalization on environmental status in Mediterranean countries. *Energy and Environment*, 0958305X221083240.

Gyamfi, B. A., Adedoyin, F. F., Bein, M. A., Bekun, F. V. and Agozie, D. Q. (2021a). The anthropogenic consequences of energy consumption in E7 economies: Juxtaposing roles of renewable, coal, nuclear, oil and gas energy: Evidence from panel quantile method. *Journal of Cleaner Production*, 295: 126373.

Gyamfi, B. A., Agozie, D. Q., Bein, M. A., Bekun, F. V. and Adedoyin, F. F. (2021b). Unlocking the investment impact of biomass energy utilization on environmental degradation for an isolated island. *International Journal of Energy Sector Management.*

Gyamfi, B. A., Agozie, D. Q. and Bekun, F. V. (2022b). Can technological innovation, foreign direct investment and natural resources ease some burden for the BRICS economies within current industrial era? *Technology in Society*, 102037.

Gyamfi, B. A., Onifade, S. T., Nwani, C. and Bekun, F. V. (2022a). Accounting for the combined impacts of natural resources rent, income level, and energy consumption on environmental quality of G7 economies: A panel quantile regression approach. *Environmental Science and Pollution Research*, 29(2): 2806–2818.

Hashmi, S. H., Fan, H., Habib, Y. and Riaz, A. (2021). Non-linear relationship between urbanization paths and CO_2 emissions: A case of South, South-East and East Asian economies. *Urban Climate*, 37: 100814.

Hayat, A. and Tahir, M. (2021). Natural resources volatility and economic growth: Evidence from the resource-rich region. *Journal of Risk and Financial Management*, 14(2): 84.

He, X., Adebayo, T. S., Kirikkaleli, D. and Umar, M. (2021). Consumption-based carbon emissions in Mexico: An analysis using the dual adjustment approach. *Sustainable Production and Consumption*, 27: 947–957.

Iheonu, C. O., Anyanwu, O. C., Odo, O. K. and Nathaniel, S. P. (2021). Does economic growth, international trade, and urbanization uphold environmental sustainability in sub-Saharan Africa? Insights from quantile and causality procedures. *Environmental Science and Pollution Research*, 28(22): 28222–28233.

Jahanger, A., Usman, M., Murshed, M., Mahmood, H. and Balsalobre-Lorente, D. (2022). The linkages between natural resources, human capital, globalization, economic growth, financial development, and ecological footprint: The moderating role of technological innovations. *Resources Policy*, 76: 102569.

Jian, L., Chuimin, K., Jijian, Z., Yusheng, K. and Ntarmah, A. H. (2022). The relationship between economic growth and environmental degradation: Could West African countries benefit from EKC hypothesis? *Environmental Science and Pollution Research*, 1–19.

Kao, C. (1999). Spurious regression and residual-based tests for cointegration in panel data. *Journal of Econometrics*, 90: 1–44.

Khan, I., Han, L., Bibi, R. and Khan, H. (2022). Linking natural resources, innovations, and environment in the Belt and Road Initiative countries using dynamic panel techniques: The role of innovations and renewable energy consumption. *Environmental Science and Pollution Research*, 1–10.

Khan, I., Hou, F., Irfan, M., Zakari, A. and Le, H. P. (2021). Does energy trilemma a driver of economic growth? The roles of energy use, population growth, and financial development. *Renewable and Sustainable Energy Reviews*, 146: 111157.

Kollie, G. B. (2020). Export-led growth hypothesis in ECOWAS: A panel data analysis. *African Journal of Economic Review*, 8(2): 258–275.

Liu, W., Zhan, J., Zhao, F., Wei, X. and Zhang, F. (2021). Exploring the coupling relationship between urbanization and energy eco-efficiency: A case study of 281 prefecture-level cities in China. *Sustainable Cities and Society*, 64: 102563.

Mahalik, M. K., Mallick, H. and Padhan, H. (2021). Do educational levels influence the environmental quality? The role of renewable and non-renewable energy demand in selected BRICS countries with a new policy perspective. *Renewable Energy*, 164: 419–432.

Mehmood, U. (2022). Examining the role of financial inclusion towards CO_2 emissions: Presenting the role of renewable energy and globalization in the context of EKC. *Environmental Science and Pollution Research*, 29(11): 15946–15954.

Nathaniel, S. P. (2021). Environmental degradation in ASEAN: Assessing the criticality of natural resources abundance, economic growth and human capital. *Environmental Science and Pollution Research*, 28(17): 21766–21778.

Nathaniel, S. P., Yalçiner, K. and Bekun, F. V. (2021). Assessing the environmental sustainability corridor: Linking natural resources, renewable energy, human capital, and ecological footprint in BRICS. *Resources Policy*, 70: 101924.

Ntarmah, A. H., Kong, Y. and Manu, E. K. (2021). Investigating the dynamic relationships between credit supply, economic growth, and the environment: Empirical evidence of sub-regional economies in Sub-Saharan Africa. *Environmental Science and Pollution Research*, 28(5): 5786–5808.

Ohajionu, U. C., Gyamfi, B. A., Haseki, M. I. and Bekun, F. V. (2022). Assessing the linkage between energy consumption, financial development, tourism and environment: Evidence from method of moments quantile regression. *Environmental Science and Pollution Research*, 29(20): 30004–30018.

Onifade, S. T., Gyamfi, B. A., Haouas, I. and Bekun, F. V. (2021). Re-examining the roles of economic globalization and natural resources consequences on environmental degradation in E7 economies: Are human capital and urbanization essential components? *Resources Policy*, 74: 102435.

Ozturk, I. and Al-Mulali, U. (2015). Investigating the validity of the environmental Kuznets curve hypothesis in Cambodia. *Ecological Indicators*, 57: 324–330.

Pesaran, M. H. (2007). A simple panel unit root test in the presence of cross-section dependence. *Journal of Applied Econometrics*, 22(2): 265–312.

Rahim, S., Murshed, M., Umarbeyli, S., Kirikkaleli, D., Ahmad, M., Tufail, M. and Wahab, S. (2021). Do natural resources abundance and human capital development promote economic growth? A study on the resource curse hypothesis in Next Eleven countries. *Resources, Environment and Sustainability*, 4: 100018.

Rehman, A., Ma, H., Ahmad, M., Irfan, M., Traore, O. and Chandio, A. A. (2021). Towards environmental Sustainability: Devolving the influence of carbon dioxide emission to population growth, climate change, Forestry, livestock and crops production in Pakistan. *Ecological Indicators*, 125: 107460.

Shittu, W., Adedoyin, F. F., Shah, M. I. and Musibau, H. O. (2021). An investigation of the nexus between natural resources, environmental performance, energy security and environmental degradation: Evidence from Asia. *Resources Policy*, 73: 102227.

Steve, Y. S., Murad, A. B., Gyamfi, B. A., Bekun, F. V. and Uzuner, G. (2022). Renewable energy consumption a panacea for sustainable economic growth: Panel causality analysis for African blocs. *International Journal of Green Energy*, 19(8): 847–856.

Ulucak, R. and Ozcan, B. (2020). Relationship between energy consumption and environmental sustainability in OECD countries: The role of natural resources rents. *Resources Policy*, 69: 101803.

World Bank. (2022) World Development Indicators. https://databank.worldbank.org/source/world-development-indicators#.

Yameogo, C. E., Omojolaibi, J. A. and Dauda, R. O. (2021). Economic globalisation, institutions and environmental quality in Sub-Saharan Africa. *Research in Globalization*, 3: 100035.

Yasmeen, H., Tan, Q., Zameer, H., Vo, X. V. and Shahbaz, M. (2021). Discovering the relationship between natural resources, energy consumption, gross capital formation with economic growth: Can lower financial openness change the curse into blessing. *Resources Policy*, 71: 102013.

Zeraibi, A., Ahmed, Z., Shehzad, K., Murshed, M., Nathaniel, S. P. and Mahmood, H. (2022). Revisiting the EKC hypothesis by assessing the complementarities between fiscal, monetary, and environmental development policies in China. *Environmental Science and Pollution Research*, 29(16): 23545–23560.

Appendix

Table 5: Countries under study.

Algeria	Angola
Cameroon	Chad
Cong. Rep.	Cote d'lvoir
Egypt	Ghana
Gabon	Libya
Morocco	Sudan
Tunisia	

Chapter 11

Dynamic Relationships between Environmental Degradation, Urbanization, and Agricultural Productivity in Nigeria

Obekpa Hephzibah Onyeje,[1,*] *Adejo Moses Adejo,*[2] *Ali Ayuba,*[2] *Abu Orefi*[2] and *Dzever Donald Denen*[2]

1. Introduction

Environmental degradation is the most visible threat to humanity's survival. It is probably the greatest threat to human life, and it poses a significant threat to the sustainability of economic growth (Nathaniel et al. 2019, Nwodo et al. 2017). Climate change and agricultural productivity have a relationship that produces a dilemma, and both topics have remained global problems since the 1970s (Nwokoro and Chima 2017). It is imperative to manage, conserve, and protect the world's biodiversity, especially in light of the increasing worldwide awareness of the need for cleaner, more ecologically friendly energy sources (Emir and Bekun 2018). Excessive use of specific arable land over a long time contributes significantly to environmental emissions by releasing pollutant gases into the atmosphere (Adedoyin et al. 2020). With increased agricultural activity through mechanization, the greenhouse gases have been shown to increasingly contribute to global warming (Olasehinde-Williams et al. 2020, Alola et al. 2019a). Demands for water, infrastructure, electricity, and food, among other things, put pressure on the environment, causing pollution, resource depletion, and ecological disruptions (WWF 2016, Ahmed and Wang 2019).

[1] Federal University of Agriculture Makurdi, Benue State, Nigeria. Social Standards Node, Center for Innovation in Procurement, Environmental and Social Standards.
[2] Federal University of Agriculture Makurdi, Benue State, Nigeria. Email:
Emails: moses.adejo@uam.edu.ng; aliayuba90@gmail.com; orefi@yahoo.com; dzeverdenen@gmail.com
* Corresponding author: obekpa.hephzibah@uam.edu.ng

Undoubtedly, the world is becoming increasingly urbanized, and the twentieth century saw rapid and unparalleled urbanization of the global population (Aliyu and Amadu 2017). There is no universal agreement among the numerous studies on the effects of urbanization on the environment (Farhani and Ozturk 2015, Al-Mulali et al. 2015, Dogan and Turkekul 2016, Adams and Klobodu 2017, Chia et al. 2018). Nigeria, the most populous nation in Africa and the seventh-most populated nation in the world, has been projected to be the third-most populous country by 2050, surpassing the United States of America, which is currently number three (United Nations 2017). Although there is no standardized definition of an urban settlement, one or more of the following characteristics may be used to classify an area as urban: human population limit, population density, nonagricultural workers, modern infrastructure such as paved roads, electricity, piped water, or utilities, and availability of education and healthcare (Cohen 2006). In 1950, only 10.1 percent of Nigeria's population lived in urban areas; this grew to 20.0 percent by 1970, 43.3 percent by 2000, and 51.16 percent by 2019 (Onibokun and Faniran 1995, Pelcher 2020). In Nigeria, urbanization has become a big problem with the increase in rural-urban migration caused by poverty, infrastructural decay, and low economic activities in rural areas with no corresponding socioeconomic dividends or benefits to the urban environment (Aliyu and Amadu 2017, Nathaniel 2020). Deforestation is another consequence of urbanization. Tropical deforestation accounts for about 20 percent of global greenhouse gas emissions (Derouin 2019). With over 60 percent of Nigeria's over 200 million people (Statista 2022) being in the working class, the amount of labor productivity generated must affect the environment (Nathaniel 2020). Furthermore, urbanization may be linked to a wide range of food consumption patterns in the city. Processed, convenient, fast, and similar foods are becoming more common (FAO-ONU 2017).

With the spate of population growth in Nigeria, there is also a growing demand for feeding the population with increased agricultural productivity (Adedoyin et al. 2020). Nigeria is considered an agrarian society with a huge quantity of fertile farmland and agricultural export potential. Nigeria's agriculture sector accounts for more than 60% of the country's GDP, implying that agriculture is the pillar of the nation's economy (Gokmenoglu et al. 2016). However, the trade-off between increased productivity and environmental pollution is the bane of agricultural mechanization. Agriculture is one of the most significant contributors to global warming, accounting for around one-third of all anthropogenic greenhouse gas emissions (FAO 2016).

In the light of the above, this study aims to contribute to this growing area of research in the following ways; firstly, unlike previous studies (Agboola and Bekun 2019, Adedoyin et al. 2020), which used CO_2 to measure environmental degradation in Nigeria and sub-Saharan Africa, this study measures environmental degradation using Ecological footprint (EFP) in the agricultural-environment nexus for Nigeria. Ecological Footprint (EFP) measures how producing goods impacts the environment (Rashid et al. 2018, Hassan et al. 2019). In addition, EFP looks at global warming in a broader context, looking at how carbon emissions, land utilization, and deforestation affect global warming (Bilgili et al. 2019). Unlike CO_2, EFP is a more robust and comprehensive indicator of environmental status (Ulucak and Bilgili 2018, Danish et al. 2019).

In addition, EFP has been shown by previous studies to capture better environmental degradation than CO_2 (Alola et al. 2019b). Finally, unlike Akomolafe et al. (2021), who used the Vector Error Correction Model (VECM) to model urbanization, CO_2, and agricultural productivity in Nigeria, our study employs the Fully Modified Ordinary Least Squares (FMOLS) method to capture the adverse effects of agricultural activities on the environment using recent time-series data. Our study goes further to test the environmental Kuznet Curve hypothesis for Nigeria. The rest of the paper is organized as follows. A review of the literature is presented first, followed by a discussion of the data and methodology employed, followed by an outline and explanation of our research findings in the results and discussion section.

2. Literature Review

The negative relationship between per capita income and environmental deterioration is conceptualized as the Environmental Kuznets Curve (EKC) hypothesis. These nexus explain the inverse effects on economic growth and environmental quality trajectory. Three avenues—scale, composite, and technique—have been used to observe the EKC phenomenon. The scale effect represents the early period of an economy's development and resembles a pre-industrialization stage; the composite effect, which symbolizes the industrialization stage; and the technique effect, which represents the post-industrialization stage (Grossman and Krueger 1991, Shahbaz and Sinha 2018). Kuznets (1955), who claimed that there is a connection between per capita income and inequality, first raised this topic. Due to the growing environmental consciousness towards healthier and cleaner ecosystems and renewable energy sources, environmental sustainability has been a top priority for most economies worldwide (Agboola and Bekun 2019). Grossman and Krueger's (1991) study, which looked at how the North American Free Trade Agreement (NAFTA) may affect the environment and national income, was the first of its kind in this area. Grossman and Krueger carried out a groundbreaking examination into the inverted U-shaped relationship in 1991. According to Grossman and Krueger (1991), sulphur dioxide and smoke levels rise in nations with low national income levels, whereas the converse was shown in countries with higher income levels.

2.1 Environment and agriculture

Yasin and Sattar (2022) evaluated the impact of forests, agricultural production, domestic investment, and energy consumption on carbon emissions and ecological footprint using data from the BRICS countries. The time covered in this study is 1995 to 2017. The Panel Spatial Correlation Consistent Least-Squares Dummy Variables (PCSS-LSDV) and Panel Quantile Regression (PQR) methodologies were used to achieve this goal. The quantitative research shows that using renewable energy, agricultural production, and financial growth lower carbon emissions, whereas the rest of the variables increase emissions and worsen environmental quality. Furthermore, the transformation of 50 percent of the earth's land surface to crop production has exacerbated humanity's ecological problems for three key reasons: (1) In the form of plants and soil, agriculture is responsible for 25% of all human greenhouse gas emissions. (2) The world's most significant threat to biodiversity is habitat loss, primarily caused by agricultural activities. (3) More than 70% of the agriculture sector's freshwater has been used (Beyer et al. 2018).

2.2 *Environmental policies and agriculture*

Previous studies have found that greenhouse gas emissions (GHG) created by the improper use of nonrenewable resources are to blame for environmental degradation (Daigle and Vasseur 2019). In their work, Apergis and Payne (2009) noted that modifying agricultural regulatory laws, production costs, and economic growth minimizes environmental deterioration. Agboola and Bekun (2019) used annual time series data from 1981 to 2014 to empirically test the EKC hypothesis in an agrarian context in Nigeria. The recently developed Bayer and Hanck combined cointegration and autoregressive distributive lag techniques were used (Agboola and Bekun 2019). In addition, the Granger causality test was employed as an estimation method for causality direction. Their research supported Nigeria's inverted U-shaped EKC pattern, confirming that the country is still in the scale-effect phase of its growth trajectory. Based on their findings, some policy recommendations on the Nigerian energy mix and agricultural operations in response to environmental quality were made for decision-makers, stakeholders, and environmental economists who create environmental rules and strategies to achieve sustainable development target seven (SDG-7).

2.3 *Urbanization and agriculture*

Salari et al. (2021) investigated the impact of explanatory variables like globalization, renewable energy usage, and agricultural production on the ecological footprint in emerging nations from 2002 to 2016. They used fixed-effect panel quantile regression of the 10th, 25th, 50th, 75th, and 90th percentiles. The results show that globalization has the most significant impact among the explanatory variables, negatively and substantially impacting the ecological footprint in all quantiles. However, at the 25th and 50th quantile levels, the agricultural production variable had the most substantial and beneficial impact on the ecological footprint. As a result, greater agricultural production and population contribute to increased environmental deterioration. In contrast, Dogan and Turkekul (2016) examined the relationship between CO2 emissions, actual production, energy consumption, international commerce, urbanization, and financial development in the United States using the Autoregressive Distributed Lag (ARDL) model. The results show that urbanization accelerates environmental damage. Their findings concurred with those of the investigations by Marques et al. (2018), Liu and Bae (2018), and Zhang et al. (2017). Globalization is expected to increase external and internal migration, raise poverty levels, and alter the climate (Leal et al. 2020).

3. Methodology

3.1 Data

The study used time series data from 1981 to 2020. The variables, their measurements and sources of data used is presented in Table 1.

Table 1: Variables, units of measurement, and data sources.

Indicators	Abbreviation	Unit of Measurement	Source
Ecological footprint	EFP	Global hectares	Global footprint network
Agricultural gross domestic product	AGDP	US dollars	Indexmundi
Urbanization	URB	% of total population	Indexmundi
Fossil Fuel Consumption	FFC	% of total energy consumption	Indexmundi

3.2 Data analysis techniques

To achieve the study objectives, we employed both descriptive and inferential statistics. Descriptive statistics were found, including mean, kurtosis, and the Jarque-Bera test, alongside Q-Q plots for the variables' normal distribution (normality). At the same time, lag length selection criteria of the model, the Augmented Dickey-Fuller test (Dickey and Fuller 1981), Philips-Perron (Phillips and Perron 1988), and Fully Modified OLS Model (FMOLS) were used. The FMOLS was employed in the study because variables were stationary at different levels other than I(0) and I(1). The measurement and sources of data used is presented in Table 1.

3.3 Model specification

The impact of FFC, AGDP and URB on EFP using Fully Modified OLS (FMOLS) is specified below:

$$lnEFP_{t-i} = a_0 + a_1\, lnAGDP_{t-i} + a_2\, lnURB_{t-i} + a_3\, lnFFC_{t-i} + u_{1t}$$

where:

$lnEFP_{t-i}$ = Ecological foot print

$lnAGDP_{t-i}$ = Agricultural GDP

$lnURB_{t-i}$ = Urbanization

$lnFFC_{t-i}$ = Fossil Fuel Consumption

u_t = error term.

3.4 Descriptive statistics of data

The result of the descriptive statistics of variables used in the model is presented in Table 2. The result shows the mean value of agricultural gross domestic product

Table 2: Descriptive statistics of variables.

	AGDP	EFP	FFC	URB
Mean	4.31E+10	1.45E+08	19.47375	36.52175
Median	1.97E+10	1.43E+08	19.01500	35.25500
Maximum	1.14E+11	1.98E+08	22.84000	50.38000
Minimum	9711930.	81297473	15.85000	22.67000
Std. Dev.	3.66E+10	44519428	1.576806	8.527811
Skewness	0.543200	-0.114953	0.317028	0.148825
Kurtosis	1.652560	1.393212	2.583468	1.831129
Jarque-Bera	4.993100	4.391041	0.959208	2.424758
Probability	0.082369	0.111301	0.619028	0.297489
Sum	1.72E+12	5.78E+09	778.9500	1460.870
Sum Sq. Dev.	5.22E+22	7.73E+16	96.96634	2836.219
Observations	40	40	40	40

Table 3: Correlation between variables.

	AGDP	EFP	FFC	URB
AGDP	1.00			
EFP	0.88	1.00		
FFC	–0.46	–0.44	1.00	
URB	0.89	0.96	–0.50	1.00

is highest while that of urbanization is the lowest. All the included variables were positively skewed except for EFP. Likewise, the kutorsis shows all the variables have slim tails, while all the variables follow normal distribution with their Jarque-Bera value greater than 5% probability levels. We used a Q-Q plot in addition to the normality test to show the graphical distribution of the data. In the Q-Q figure, the normal distribution is shown by the linear, diagonal blue line. In contrast, as shown in Figures 1 to 4, the dotted line denotes the divergence from the normal distribution. The pairwise correlation between each of the variables used for the study is displayed in Table 3. The correlation between urbanization and ecological footprint is the highest (0.96). On the other hand, the correlation between ecological footprint and fossil fuel consumption is the lowest (0.44).

4. Results and Discussion

To avoid a misleading regression trap and determine the highest possible order of integration among the interest variables, stationarity tests are carried out among them (Agboola and Bekun 2019). Table 4 presents the result of the stationarity test of the properties of variables before regression using Augmented Dickey–Fuller (ADF) test and Phillips-Perron (PP) test. The result is given for AGDP, EFP, FFC, and URB. The ADF test result indicates that all the variables were not stationary at levels I(0),

Table 4: Unit root test for variables used in the analysis.

Variables	ADF			PP		
	Level	1st Difference	2nd Difference	Level	1st Difference	2nd Difference
	T-statistic (Prob.)	T-statistic (Prob.)	T-statistic (Prob.)	T-statistic (Prob.)	T-statistic (Prob.)	T-statistic (Prob.)
LNAGDP	–3.3658** (0.0185)			–3.3566 ** (0.0189)		
LNEFP	–0.9430 (0.7636)	–6.1681*** (0.0000)		–0.9461 (0.7626)	–6.1865*** (0.0000)	
LNFFC	–2.7992 (0.0676)	–6.6644 *** (0.0000)		–2.8195 (0.0648)	–8.4210*** (0.0000)	
LNURB	–1.5323 (0.5067)	–5.5552 (0.8687)	–6.0209 *** (0.0000)	–3.2531** (0.0243)	–0.5552 (0.8687)	–6.0209*** (0.0000)

***, ** rejection of the null Hypothesis at 1% and 5% level of probability respectively.
NOTE: LNEFP = Logarithmic value of Ecological Foot Print, LNFFC = Logarithmic value of Fossil Fuel Consumption, LNURB = Logarithmic value of Urbanization, LNAGDP = Logarithmic value of Agricultural Gross Domestic Product.

Table 5: Lag length selection.

Lag	LogL	LR	FPE	AIC	SC	HQ
0	33.88120	NA	2.34e-06	–1.615200	–1.441047	–1.553803
1	225.4951	331.4403	1.77e-10	–11.10784	–10.23708*	–10.80086
2	250.6211	38.02846*	1.12e-10*	–11.60114*	–10.03376	–11.04856*
3	262.4452	15.33942	1.53e-10	–11.37542	–9.111424	–10.57725

* indicates lag order selected by the criterion.
LR: sequential modified LR test statistic (each test at 5% level).
FPE: Final prediction error.
AIC: Akaike information criterion.
SIC: Schwarz information criterion.
HQ: Hannan-Quinn information criterion.

but AGDP, EPF, and FFC were stationary at the first difference I(1) while URB was stationary at the second difference I(2). The result implies that the level form of these variables has multiple means and covariance and does not revolve around zero. This means these variables in their level form would produce spurious regression results unsuitable for inference. The different levels of unit root stationarity of the variables necessitated the use of cointegration regression, otherwise known as Fully Modified OLS (FMOLS). Table 5 presents the result of lag length from six different selection criteria, AIC was chosen because of its lowest value mi –11.6011 at lag 2. The appropriate lag for the model is 2.

A co-integration test inquiry was conducted on the variables using the Johansen co-integration test to ascertain if a long-run linear combination of non-stationary variables is stationary. This presupposes that a linear combination of non-stationary variables can be static (Engle and Granger 1987). Table 6 displays the Johansen

Table 6: Cointegration test.

Hypothesized No. of CE(s)	Eigenvalue	Trace Statistic	0.05 Critical Value	Prob.**
None *	0.488979	49.32532	47.85613	0.0361
At most 1	0.327271	24.48557	29.79707	0.1807
At most 2	0.212324	9.818275	15.49471	0.2950
At most 3	0.026337	0.987529	3.841466	0.3203

Trace test indicates 1 cointegrating eqn (s) at the 0.05 level.
* denotes rejection of the hypothesis at the 0.05 level.
** MacKinnon-Haug-Michelis (1999) p-values.

Co-integration test results. The result revealed that the combination of these variables has one co-integrating equation using trace statistics, which means that the linear combination of these variables has a single long-run linear combination or relationship.

The result in Table 7 shows the long-run impact of FFC, AGDP, and URB on EFP, respectively, using Fully Modified OLS (FMOLS). The coefficient of adjusted (R^2) of the model was 0.863, indicating that 86.3% of the variation in EFP was explained by EFP, FFC, AGDP, and URB in the previous year. The result further showed that the AGDP and URB significantly affected the EFP in the long run. Specifically, the coefficient of AGDP (0.034) is positive and significant at the 10% level of probability which is in line with the *a priori* expectations. This implies that a unit percentage increase in AGDP will increase the EFP by 0.034. This increase in AGDP could be due to the attractive macroeconomic policy of the government that encourages AGDP through increased production; this means that policies implemented by the government of Nigeria to increase agricultural production also helped increased EFP. This could be attributed to the fact that most agricultural practices in Nigeria are not fully mechanized, thereby leading to less emission due to heavy mechanized equipment. This disagrees with Appiah et al. (2018) and Agboola and Bekum (2019), who found that agricultural production over the decades is mainly responsible for significant adverse changes in EFP and carbon emission in the atmosphere. This result is in harmony with Karl et al. (2020), who found similar results. The result further invalidates the EKC hypothesis for the case of Nigeria because of the positive relationship between AGDP and EFP. This study thus adds to the body of research that rejects the tradeoff between a nation's income level and environmental quality. Given the increased understanding of the effects of climate change, this is not surprising for Nigeria.

The coefficient of URB is positive (1.35) and significant at a 1% probability level. This means that a percentage increase in URB will increase EFP by 1.35. This outcome agrees with the findings of Wang et al. (2019). They found that urbanization and natural resources rent have a positive and significant impact on carbon dioxide emissions in the sub-Saharan African countries. Wang et al. (2013) and Liu and Bae (2018) further confirmed the positive relationship between urbanization and carbon emission. This result is contrary to the findings of Bekun et al. (2019). They found that the real gross domestic product, agricultural value-added, urbanization, and total natural rent are insignificant at a 10% significance level.

Table 7: Fully modified ordinary least square (FMOLS).

Variable	Coefficient	Std. Error	t-Statistic	Prob.
LNAGDP	0.034232	0.018660	1.834468	0.0751
LNFFC	–0.434789	0.336660	-1.291478	0.2050
LNURBN	1.353026	0.130888	10.33727	0.0000
C	14.35194	1.312630	10.93373	0.0000
R-squared	0.874490	Mean dependent var		18.75045
Adjusted R-squared	0.863732	S.D. dependent var		0.326125
S.E. of regression	0.120387	Sum squared resid		0.507258
Durbin-Watson stat	0.575237	Long-run variance		0.019783

5. Conclusion

The study looked into the dynamic relationships between Nigeria's ecological footprint, agricultural GDP, urban population, and fossil fuel consumption from 1981 to 2020. The time series data were stationary at various levels and lacked a unit root. FMOLS and the Johansen co-integration test are used. The environmental Kuznets curve (EKC) theory is reviewed in this work from a fresh angle. Including agricultural gross domestic product as a stand-in for agriculture in conventional EKC. This is done to determine how agriculture affects the environment in Nigeria. The findings of this study are consistent with those of Gökmenolu and Taspinar (2016), which support the Pollution Haven Hypothesis (PHH). The current investigation supports the validity of the PHH in Nigeria. This suggests that AGDP contributes to rising environmental degradation.

According to the current study, agriculture hurts the environment, which calls for government managers and legislators to be aware of agriculture's role in Nigeria's economic growth and its environmental consequences (s). The combustion of fossil fuels during the production stage is a feature of the agricultural sector, which accounts for the pollution caused by AGDP. The constant bush burning practiced by Nigerian farmers, as well as other pollutants like generator fume, and CO_2 emissions from cars are to blame for this rise in EFP (Agboola and Bekun 2019). Farmers' awareness of effective energy sources must be raised to address this problem, especially now that the need for cleaner, more sustainable energy is being recognized globally (Emir and Bekun 2018). Research and development are needed to find new, cleaner energy sources, such as renewable energy (such as solar energy, biofuel, and wind energy). This investment is essential and would be beneficial for the Nigerian economy.

Furthermore, EFP was also observed to rise with urbanization. Therefore, the Nigerian government must slow urbanization. Nigerians' movement from rural to urban areas can be reduced by increasing economic possibilities in rural areas and building quality infrastructure there (Iheonu et al. 2021). This study's inability to consider some factors that influence EFP emissions in Nigeria is one of its flaws due to data accessibility. Future research might want to look at how technological advancement has affected EFP.

References

Adams, S. and Klobodu, E. K. M. (2017). Urbanization, democracy, bureaucratic quality, and environmental degradation. *Journal of Policy Modeling*, 39(6): 1035–1051.

Adedoyin, F. F., Alola, A. A. and Bekun, F. V. (2020). The nexus of environmental sustainability and agro-economic performance of sub-Saharan African countries. *Heliyon*, 6(9): e04878. https://doi.org/10.1016/j.heliyon.2020.e04878.

Agboola, M. O. and Bekun, F. V. (2019). Does agricultural value added induce environmental degradation? Empirical evidence from an agrarian country. *Environmental Science and Pollution Research*, 26(27): 27660–27676. https://doi.org/10.1007/s11356–019–05943-z.

Ahmed, Z. and Wang, Z. (2019). Investigating the impact of human capital on the ecological footprint in India: An empirical analysis. *Environmental Science and Pollution Research*, 26(26): 26782–26796.

Akomolafe, K. J., Madu, A. C. and Ebenezer, A. K. (2021). Urbanization, carbon dioxide emission, and agricultural productivity in Nigeria. *Asian Journal of Agricultural Extension, Economics and Sociology*, 57–72. https://doi.org/10.9734/ajaees/2021/v39i130503.

Aliyu, A. A. and Amadu, L. (2017). Urbanization, cities, and health: The challenges to Nigeria—A review. *Annals of African Medicine*, 16(4): 149–158. https://doi.org/10.4103/aam.aam_1_17.

Alola, A. A., Akadiri, S. S., Akadiri, A. C., Alola, U. V. and Fatigun, A. S. (2019a). Cooling and heating degree days in the US: The role of macroeconomic variables and its impact on environmental sustainability. *Science of the Total Environment*, 695: 133832. https://doi.org/10.1016/j.scitotenv.2019.133832.

Alola, A. A., Yalciner, K., Alola, U. V. and Saint Akadiri, S. (2019b). The role of renewable energy, immigration and real income in environmental sustainability target. Evidence from Europe largest states. *Science of the Total Environment*, 674: 307–315.

Al-Mulali, U., Ozturk, I. and Lean, H. H. (2015). The influence of economic growth, urbanization, trade openness, financial development, and renewable energy on pollution in Europe. *Natural Hazards*, 79(1): 621–644.

Apergis, N. and Payne, J. E. (2009). CO_2 emissions, energy usage, and output in Central America. *Energy Policy*, 37(8): 3282–3286. https://doi.org/10.1016/j.enpol.2009.03.048.

Appaih, K., Du, J. and Poku, J. (2018). Causal relationship between agricultural production and carbon dioxide emission selected energy economies. *Environ. Sci. Pollut. Res.*, 25: 24764–24777.

Bayer, C. and Hanck, C. (2013). Combining non-cointegration tests. *Journal of Time Series Analysis*, 34(1): 83–95.

Bekun, F. V., Alola, A. A. and Sarkodie, S. A. (2019). Toward a sustainable environment: Nexus between CO_2 emissions, resource rent, renewable and non-renewable energy in 16-EU countries. *Sci. Total Environ.*, 657: 1023–1029.

Beyer, R. M., Manica, A. and Rademacher, T. T. (2018). Relocating agriculture could drastically reduce humanity's ecological footprint [Preprint]. *Ecology*. https://doi.org/10.1101/488841.

Bilgili, F., Ulucak, R. and Kocak, E. (2019). Implications of environmental convergence: Continental evidence based on ecological footprint. pp. 133–165. *In*: Shahbaz, M. and Balsalobre, D. (eds.). *Energy and Environmental Strategies in the Era of Globalization*. Cham: Springer.

Chai, J., Liang, T., Lai, K. K., Zhang, Z. G. and Wang, S. (2018). The future natural gas consumption in China: Based on the LMDI-STIRPAT-PLSR framework and scenario analysis. *Energy Policy*, 119: 215–225.

Cohen, B. (2006). Urbanization in developing countries: Current trends, future projections, and key challenges for sustainability. *Technology in Society*, 28(1-2): 63–80. https://doi.org/10.1016/j.techsoc.2005.10.005.

Daigle, C. and Vasseur, L. (2019). Is it time to shift our environmental thinking? A perspective on barriers and opportunities to change. *Sustainability*, 11(18): 5010. https://doi.org/10.3390/su11185010.

Danish, Z., Baloch, M. A., Mahmood, N. and Zhang, J. W. (2019). Effect of natural resources, renewable energy and economic development on CO_2 emissions in BRICS countries. *Science of the Total Environment*, 678: 632–638.

Derouin, S. (2019). Deforestation: Facts, Causes and Effects. Retrieved March 10, 2021, from. https://www.livescience.com/27692-deforestation.html.

Dickey, D. A. and Fuller, W. A. (1981). Likelihood ratio statistics for autoregressive time series with a unit root. *Econometrica: Journal of the Econometric Society*, 49(4): 1057–1072.

Dogan, E. and Turkekul, B. (2016). CO_2 emissions, real output, energy consumption, trade, urbanization, and financial development: Testing the EKC hypothesis for the USA. *Environmental Science and Pollution Research*, 23(2): 1203–1213.

Emir, F. and Bekun, F. V. (2018). Energy intensity, carbon emissions, renewable energy, and economic growth nexus: New insights from Romania. *Energy Environ.*, 1–17. 0958305X18793108.

Engle, R. and Granger, C. (1987). Co-integration and error correction: Representation, estimation and testing. *Journal of Econometrica*, 55(2): 251–276.

Farhani, S. and Ozturk, I. (2015). Causal relationship between CO_2 emissions, real GDP, energy consumption, financial development, trade openness, and urbanization in Tunisia. *Environmental Science and Pollution Research*, 22(20): 15663–15676. https://doi.org/10.1007/s11356–015–4767–1.

FAO-ONU. (2017). The Future of Food and Agriculture. Trends and Challenges. Food and Agriculture Organization of the United Nations.

FAO. (2016). http://www.fao.org/3/a-i6030e.pdf.

Gökmenolu, K. K. and Taspinar, N. (2016). The relationship between CO_2 emissions, energy consumption, economic growth and FDI: The case of Turkey. *The Journal of International Trade and Economic Development*, 25(5): 706–723.

Gökmenolu, K. K., Bekun, F. V. and Taspinar, N. (2016) Impact of oil dependency on agricultural development in Nigeria. *Int. J. Econ. Perspect.*, 10(2): 151–163.

Grossman, G. M. and Krueger, A. B. (1991). Environmental impacts of a North American Free Trade Agreement (No. w3914). *National Bureau of Economic Research.*

Hassan, S. T., Xia, E., Khan, N. H. and Shah, S. M. A. (2019). Economic growth, natural resources, and ecological footprints: Evidence from Pakistan. *Environmental Science and Pollution Research*, 26(3): 2929–2938.

Iheonu, C. O., Anyanwu, O. C., Odo, O. K. and Nathaniel, S. P. (2021). Does economic growth, international trade, and urbanization uphold environmental sustainability in sub-Saharan Africa? Insights from quantile and causality procedures. *Environmental Science and Pollution Research*, 28(22): 28222–28233. https://doi.org/10.1007/s11356–021-12539-z.

Kuznets, S. (1955). Economic growth and income inequality. *The American Economic Review*, 45(1): 1–28.

Leal, P. H., Marques, A. C. and Shahbaz, M. (2021). The role of globalisation, de jure and de facto, on environmental performance: Evidence from developing and developed countries. *Environment, Development and Sustainability*, 23(5): 7412–7431. https://doi.org/10.1007/s10668–020–00923–7.

Lin, D., Hanscom, L., Murthy, A., Galli, A., Evans, M., Neill, E., ... and Wackernagel, M. (2018). Ecological footprint accounting for countries: updates and results of the National Footprint Accounts, 2012–2018. *Resources*, 7(3): 58.

Liu, X. and Bae, J. (2018). Urbanization and industrialization impact of CO_2 emissions in China. *Journal of Cleaner Production*, 172: 178–186. https://doi.org/10.1016/j.jclepro.2017.10.156.

Marques, A. C., Fuinhas, J. A. and Leal, P. A. (2018). The impact of economic growth on CO_2 emissions in Australia: The environmental Kuznets curve and the decoupling index. *Environmental Science and Pollution Research*, 25(27): 27283–27296.

Nathaniel, S. P. (2019). Modelling urbanization, trade flow, economic growth and energy consumption with regards to the environment in Nigeria. *GeoJournal*, 85(6): 1499–1513. https://doi.org/10.1007/s10708–019–10034–0.

Nwodo, O. S., Ozor, J. O., Okekpa, U. E. and Agu, V. C. (2017). Environmental degradation and Nigeria's macroeconomic space. *Environmental Management and Sustainable Development*, 7(1): 37. https://doi.org/10.5296/emsd.v7i1.12155.

Nwokoro, C. V. and Chima, F. O. (2017). Impact of environmental degradation of on agricultural production and poverty in rural Nigeria. *American International Journal of Contemporary Research*, 7(2): 6–14.

Olasehinde-Williams, G., Adedoyin, F. and Bekun, F. (2020). Pathway to achieving sustainable food security in sub-saharan Africa: The role of agricultural mechanization. *Journal of Labor and Society*, 23(3): 349–366. https://doi.org/10.1111/lands.12483.

Onibokun, A. and Faniran, A. (1995). Urbanization and urban problems in Nigeria, 5–17. Available from: http://www.books.openedition.org/ifra.

Pelcher, H. (2020). Urbanization in Nigeria 2019. Retrieved from https://www.statista.com/statistics/455904/urbanization-in-nigeria/.

Phillips, P. C. and Perron, P. (1988). Testing for a unit root in time series regression. *Biometrika*, 75(2): 335–346.

Rashid, A., Irum, A., Malik, I. A., Ashraf, A., Rongqiong, L., Liu, G., ... and Yousaf, B. (2018). Ecological footprint of Rawalpindi; Pakistan's first footprint analysis from urbanization perspective. *Journal of Cleaner Production*, 170: 362–368.

Salari, T. E., Roumiani, A. and Kazemzadeh, E. (2021). Globalization, renewable energy consumption, and agricultural production impacts on ecological footprint in emerging countries: Using quantile regression approach. *Environmental Science and Pollution Research*, 28(36): 49627–49641. https://doi.org/10.1007/s11356-021-14204-x.

Shahbaz, M. and Sinha, A. (2018). Environmental Kuznets curve for CO_2 emission: A literature survey. Working Paper Series.

Statista. (2022). Retrieved from https://www.statista.com/statistics/1288348/number-of-people-employed-in-nigeria/.

Ulucak, R. and Bilgili, F. (2018). A reinvestigation of EKC model by ecological footprint measurement for high, middle- and low-income countries. *Journal of Cleaner Production*, 188: 144–157.

United Nations, Department of Economic and Social Affairs, Population Division (2017). World Population Prospects: The 2017 Revision, Key Findings and Advance Tables. Working Paper No. ESA/P/WP/248.

Wang, Q. W., Zhou, P., Shen, N. and Wang, S. S. (2013). Measuring carbon dioxide emission performance in Chinese provinces: A parametric approach. *Renewable and Sustainable Energy Reviews*, 21: 324–330. https://doi.org/10.1016/j.rser.2012.12.061.

Wang, H. and Chen, Q. (2014). Impact of climate change heating and cooling energy use in buildings in the United States. *Energy and Buildings*, 82: 428–436. https://doi.org/10.1016/j.enbuild.2014.07.034.

Wang, Q., Su, M. R. and Pounce, P. (2019). The effect of energy prices, urbanization and economic growth on energy consumption per capita in 186 Countries. *J. Clean Prod.*, 225: 10 17–1032A.

WWF. (2016). Living planet: Report 2016: Risk and resilience in a new era. Worldwide fund for nature. Retrieved March 10, 2021.

Yasin, I. and Sattar, N. (2022). Environmental Impact of Agriculture and Forestation on the Carbon Emission and Ecological Footprint of BRICS Economies. 1st International Conference on Interdisciplinary Business and Management Research at the Islamia University of Bahawalpur, Bahawalpur, Pakistan.

Zhang, N., Yu, K. and Chen, Z. (2017). How does urbanization affect carbon dioxide emissions? A cross-country panel data analysis. *Energy Policy*, 107: 678–687.

Zivot, E. and Andrews, D. W. K. (1992). Further evidence on the Great Crash, the oil price shock, and the unit root hypothesis. *J. Bus. Econ. Stat.*, 10(3): 251–270.

Appendix

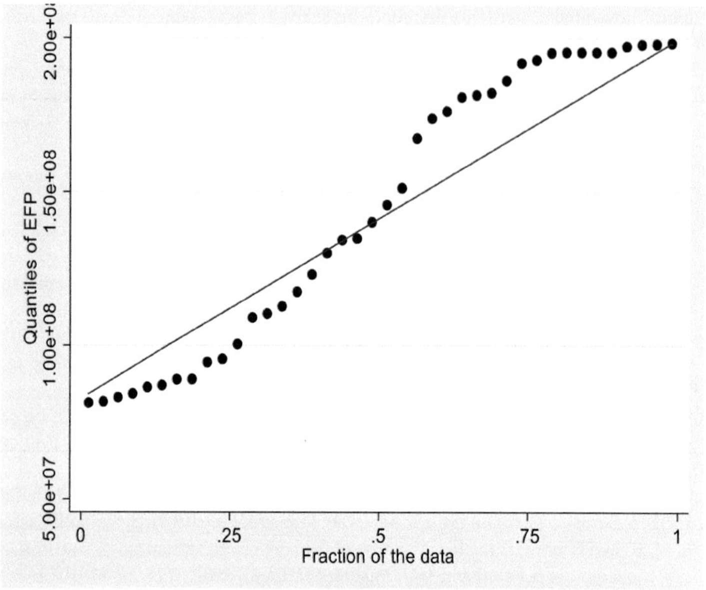

Figure 1: Q-Q plot of EFP.

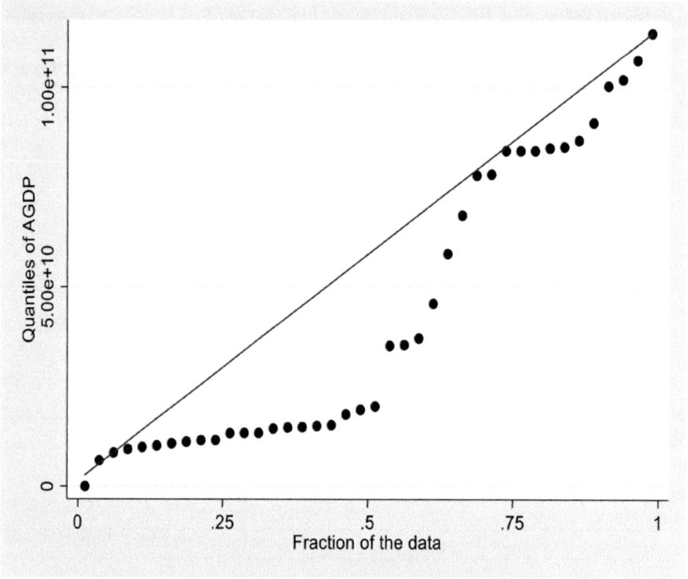

Figure 2: Q-Q plot of AGDP.

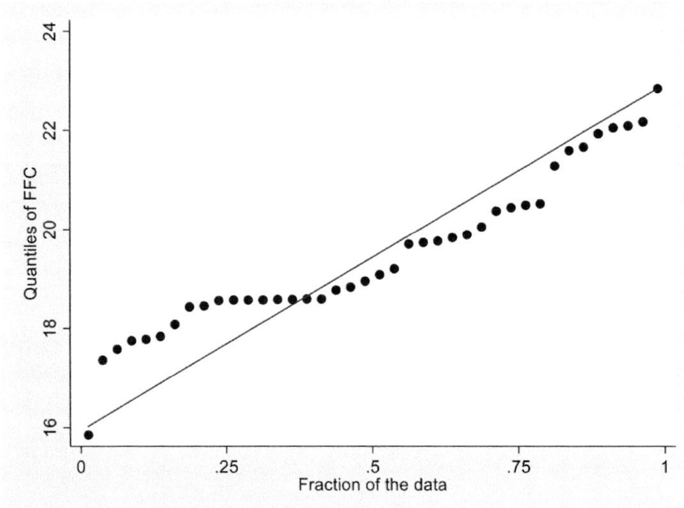

Figure 3: Q-Q plot of FFC.

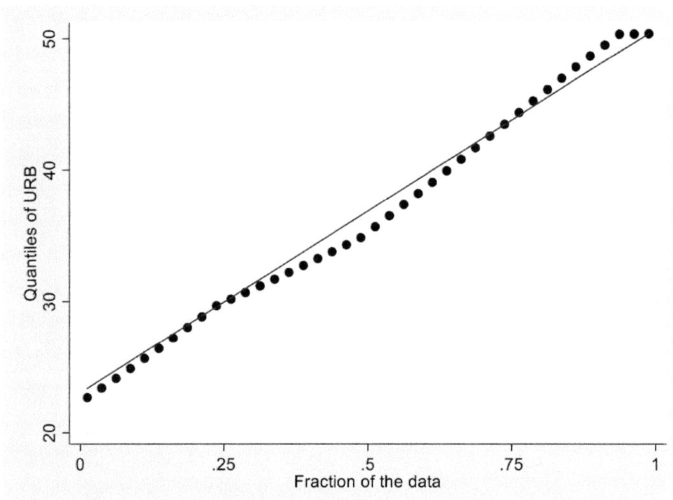

Figure 4: Q-Q plot of URB.

Economy

Chapter 12

Inequality of Income Distribution and Hopes for Democratic Consolidation in Nigeria

Oluwaseyi Mike Bamigboye

1. Introduction

Given the experiences of many countries seemingly stuck between authoritarian and democratic systems, momentary success proved transient and failed to achieve consolidation (Tonta 2016). While factors such as legitimacy crisis, the rise of militant sub-nationalist agitations, ethnoreligious and identity conflicts, corruption, institutional failure, electoral crime and violence, insecurity, injustice, and political apathy pose challenges to the formation of political culture and have been heavily researched, the role of income inequality in the processes of democratic consolidation remains under-researched. This begs the question: Does income inequality expressed in weak economic growth, rising inflation, and high unemployment, especially among young people, have far-reaching implications for the survival of Nigeria's democratic system? Using this West African country as a case study, the author argues that income inequality threatens democratic consolidation and tends to a regression into an authoritarian regime. The choice of Nigeria is informed by doubts which have repeatedly crystallized into debates on the likelihood of consolidation since its democratic transition in 1999.

This paper which bases its assessment on a minimalist concept of democracy uses an analytical strategy and case studies to establish a clear link between democracy and rising income inequality in the country and concludes with an examination of the likelihood of democratic consolidation from a macroeconomic perspective (Przeworski et al. 1996).

Graduate Student, Department of International Relations, Isik University, Istanbul, Turkey.

2. Conceptual Clarification

Democratic transition and consolidation are concepts commonly used in political science literature to explain a dual-phased change from an authoritarian regime to the formation and entrenchment of a democratic political culture. Thus, on the one hand, it involves elite choices and negotiations playing a crucial role in this transition. On the other hand, an adaptation phase in which citizens and leaders fully accept and entrench a democratic system. This presupposes the debate on democratic consolidation on its process and why it is valuable. Invariably, once a transition from authoritarian rule or dictatorship climaxes into a stage where free, fair, and competitive elections are held, democratic actors are often unwilling to regress into authoritarianism. Instead, they shift into institutionalizing what they have achieved. This institutionalization removes the likelihood of a breakdown to a point where there is democratic continuity, entrenchment, survival, and irreversibility (O'Donnell 1992).

3. Relationship between Democratization and Income Inequality

The survival of a young democracy depends on three vital issues, one of which is relat ed to economics, that is, the effectiveness of the state in addressing socio-economic problems (Linz and Stepan 1996). Thus, centrally constituent to democratic consolidation are the challenges of economic underdevelopment. While economic underdevelopment is often analyzed in terms of the impact of inequality on income distribution, the role income inequality plays in democratic consolidation has received less attention.

Modernization theorists pioneered the development-democratization nexus. Emphasizing the emergence of a West-European style democracy irrespective of time and place, they argued that economic development leads to democratization. In other words, as countries become more economically developed, they are likely to become more democratic. Through analytic strategy, correlational findings, and posited mechanisms, a relationship was forged between democracy and development using increased wealth (per capita income), industrialization, urbanization, education, and media participation as indicators (Lipset 1959). They hypothesize that the stability of developed democracies results from their developed economies expressed in a high standard of living and fair income distribution. Thus, the generally accepted argument (which remains unproven) is that the higher the standard of living in a given country and the more socially insured its citizens are, the more stable the democratic system. Conversely, in countries with a low standard of living, where citizens are on the brink of poverty and survival, democracy is unlikely to thrive Przeworski et al. (1996).

Research suggests that countries with a high standard of living are characterized by a large 'middle class,' a disparate group that constitutes about 6080 percent of the population (Mahler et al. 2013). This class is considered the social base of democratic systems and proponents of liberal democracy. Because this social group enjoys a stable income and is guaranteed a certain degree of welfare, it maintains a consciousness that rejects extreme and radical political views. According to the international study "Democratic Values" led by Professor Dr. Detlef Pollack, which examined the relationship between economic status and democratic consciousness and values in eleven post-communist countries in 2000, professionals tend to recognize democratic principles as normative and to identify with them. In other words, the study suggests that people in professional roles often recognize and align themselves with democratic norms.[1] The survey contributes significantly to the analysis of the growing retention of democratic values and the consolidation of Bulgarian society, tacitly pointing to the existence of status hierarchies, which often correspond to the attachment to democratic values. Thus, the higher the group's status, the more significant the proportion of people who "fully agree" that democratic principles are normative. Conversely, low-status groups have significantly lower levels of identification with the premise that democratic principles are always a good thing (European Commission 2005). In this instance, a democracy with a weak middle class will be characterized by more populism, internal conflicts, and an inability to resolve distribution problems civilly.

4. How Does Income Inequality Threaten Democratic Consolidation?

Because of the recent rise in general income inequality, understanding its impact on democracy has become beneficial (United Nations Department of Economic and Social Affairs 2020, Piketty 2014). Empirical evidence shows that rapid economic growth has never translated into economic growth in many countries, including Nigeria, due to the highly unequal income distribution, which has continued to worsen (Aigbokhan 1999, Oyakhilomen and Zibah 2014).

One could talk about income inequality vis-à-vis the top and bottom gap. For explanatory purposes, clarification on general income inequality is expedient. General income includes all forms of income such as wages, salary, gifts, commission, interests, and dividends, especially for those at the top since the poor are almost always wage earners. To calculate, one can find the decile ratio by dividing the income of the top 10 percent of households by the income of the poorest 10 percent (De Maio 2007). The result shows that wage income is not equal to income inequality, and the difference between them tends to be less (Hacker and Pierson 2010). Democracy, when defined in its primary sense, is the equality of individuals before the law. It is often expressed in terms of individuals having an equal vote, the right to express public opinion, and the equal right to influence policy. Within this construct of democracy, commonly referred to as procedural equality, which may also characterize Nigerian democracy, one's financial capability should not be a determinant of one's legal and political standing. Procedural equality is, therefore, of particular importance to society, as it guarantees the individual freedoms necessary for individuals to act independently.

[1] The study was as part of a major international initiative of the same name funded by the European Commission's 5th Framework Programme (HPSE-CT-2001-00062). The project was led by Professor Dr. Detlef Pollack of the European University Viadrina in Frankfurt (O), Germany.

How does inequality threaten democratic consolidation? At the unit level, democracy without a sizeable measure of economic equality is only a mirage because a society with a high concentration of the poor threatens democracy. Rising inequality exacerbates volatility, especially when redistributive transfers are low. According to Acemoglu and Robinson (2006), uprisings and revolutions are natural in unequal societies because the more unequally wealth is distributed in society, the more likely it is that the masses will demand redistributive transfers, which often crystallizes into violence. Levin-Waldman (2013) explains that democratization assumes that society is so unequal that a revolutionary threat is credible. Therefore, elites wishing to avoid the threat of a revolution that could perpetuate their power base might be willing to initiate the transition by expanding voting rights because it is in their interests to do so. However, elite democratization is only necessary to the extent required to reduce the threat of revolution because the elites effectively limit majority power by diluting popular pressure and weakening majority capacity. In this case, income inequality threatens democratic consolidation as it increases redistributive pressures, often leading to protests, mass revolutions, and the subsequent establishment of dictatorships (Boix 2003).

Similarly, income inequality increases the demands for redistribution and the levels of final taxation in a democracy. In this instance, the quantum of tax burdens on those at the top and the bottom gap becomes a vital consideration, particularly when the political system does not respond to supposedly democratic tax rates. Levin-Waldman (2013) argues that an ideal democratic regime would not only eliminate the wealthy elite for the benefit of the masses but could also impose higher tax rates for redistributive purposes. He noted that "in a democracy, everybody votes on the tax rate in accordance with what is known as the median voter theorem. This holds that the more inequality there is, the greater will be the distance between the median income and society's average income. The greater the distance, the more calls there are for redistribution, and the distance itself effectively determines the tax rate." Consequently, the greater the gap, the greater the demand for redistribution by any means, with the gap determining the tax rate.

At the individual level, however, income inequality negatively impacts the ability of individuals to participate equally in democratic processes. Firstly, wealth provides elites leverage in achieving their political goals because it empowers them to pursue their political goals and popularize their ideas even if they are not ideal (Bachrach and Botwinick 1992). Secondly, as studies have found, it creates an unjust system in which institutions and political actors prioritize and serve the interest of those at the top-end of the income gap, thus distorting democracy within the context of classical political thought (Bartels 2008, Gilens 2012).

According to Uslaner and Brown (2005) and Uslaner (2008), rising income inequality creates distrust and disillusionment with the political process and system because those at the lower threshold are alienated; thus, they conclude that the system is unjust and unfavorable. A 2008 study, Current Population Survey, examined the relationship between income inequality and civic engagement. Variables used for engagement were daily discussion of politics, daily newspaper reading that expressed personal knowledge of and interest in politics, participation in protests, attendance at political rallies, visits to public officials, and participation in civil society organizations. Using this study, Levin-Waldman (2013) found that civic engagement varied by household income. The study found that individuals in households that earned more than $100,000 per year participated more in all engagement activities than individuals in households that earned less than $30,000 per year. However, there was no evidence to show that those at the top end of distribution were more involved than those earning between $30,000 and $99,999. These findings indicate that the middle class has a higher level of civic engagement and that people outside the class experience greater social alienation, reducing their participation mainly due to the deep-seated perception of inequality with those at the top. In another study that tested political engagement and human rights activism in emerging democracies such as Nigeria, Obiagu and Ajaps (2022) analyzed the impact of five socio-demographic factors on 372 participants using a modified theory developed by the SDT lens application to generate a community engagement report. Male participants with higher education and incomes were likelier to be involved in politics and activism. With this, one can conclude that income inequality not only negatively affects social engagement but can also lead to political anomie.

5. Nigeria as a Democratizing State

Although democracy in Nigeria was nothing new in the pre-colonial, colonial, and post-colonial periods, its current democratic status has been explained from a Western perspective, as most democratization studies have focused on Latin America and Southern Europe, and more recently, Eastern Europe, paying less attention to Africa (Bratton and Walle 1997). Notwithstanding, scholars posit that the world has experienced increasing democratic change in the past four decades (Huntington 1993). However, an essential international dynamic for the transition to democracy is the "snowballing effect" that brings Africa into the wave of democratization (Huntington 1993, Buhari 2015).[2] Whereas in the pre-colonial era, democracy existed not so much in terms of the selection or election of leaders as in terms of community responsibility, the dynamics of the colonial era were characterized by the nationalist struggle for a place in the management of the state affairs that led to its founding political parties and the struggle for independence. For example, between 1993 and 1999, Nigeria experienced an unprecedented military dictatorship that resulted in a stifled democratic transition and the cancellation of the presidential election of 12 June 1993, widely considered free and fair. Nevertheless, the official 11-month transitional program launched in 1998 leading to the Fourth Republic (which began in 1999 till date) was the product of overwhelming internal consensus and Nigerian determination to democratize (Obiora 2005, Esebonu 2012, Isma'ilaa and Madu 2016).

[2] The proportion of African countries classified by Freedom House as not free fell from 59% in 1983 to 35% in 2003.

6. Dimensions of Inequality in Nigeria

According to statistics, the middle class in Nigeria who have obtained post-secondary education or have studied at an institution of higher learning amounts to about 23% of the Nigerian population (World Economic Forum 2015). This figure constitutes a barely visible middle class and is too inadequate to dictate the course of Nigerian society for an extended period. In the end, this class may eventually be a "loser" from the transition because the disenchanted and disillusioned low class, which constitutes about 77% of the population, and whose income levels are insufficient to satisfy their economic demands, breed negativity about Nigerian democracy due to its practical interpretations that discourage social equality. Thus, the democratic transition is considered the bases of income stratification, resulting in extreme polarization between the country's poor and wealthy citizens.

Attempts at measuring income distribution in Nigeria between 1986 and 2010 estimate its Gini coefficient range between 0.39 and 0.6. The Gini coefficient serves as a quantitative tool for assessing economic inequality within a population. This metric quantifies the deviation of a country's income or wealth distribution from an ideal equal distribution. The values of the Gini coefficient range from 0 to 1, with 0 representing complete equality and 1 representing the highest degree of inequality. The calculation involves constructing the Lorenz curve, a graphical representation of the distribution of income or wealth, and estimating the area separating this curve from the line signifying absolute equality.Bakare (2012), using national data, found Nigeria's Gini coefficients in 1992, 1996, and 2000 to be 0.50, 0.53, and 0.60, respectively. In a related study documented by the World Bank in its 2014 World Development Indicators, Nigeria's Gini coefficient was 0.39, 0.45, 0.47, 0.40, and 0.43 in 1986, 1992, 1998, 2004, and 2010, respectively. While these coefficients show income polarization in Nigeria, the index calculation also notes that 20% of Nigerian income earners earned half of Nigeria's income and received 52.1% of total income in 1998. The lowest 20% of income earners earned a meager 5.0%, and the next 20% of Nigerian workers earned about 10% (World Bank 2014).

According to Oxfam (2017), Nigeria's average economic growth in the 2000s was about 7%, while income inequality, as measured by the Gini Index, grew from 40% in 2003 to 43% in 2009; 39% in 2018, down by 0.00% from the previous year (see Figure 1). Notably, the proportion of people living below the national poverty line increased significantly between 2004 and 2010. The raw numbers rose from 69 million people living in poverty in 2004 to a staggering 112 million in 2010. This increase represents about 69% of the country's total population living in these conditions.

Also, the 2018 National Bureau of Statistics report using either the Gini coefficient or Theil shows that income inequality in Nigeria worsened between 2004 and 2013 but improved in 2016. The Theil index is a statistical tool used to assess economic inequality and other socioeconomic phenomena such as racial segregation. This index is essentially a variant of the generalized entropy index and serves as a measure of various factors such as lack of diversity, redundancy, segregation, isolation, and inequality. In Theil T's version, it is intended to measure the entropic

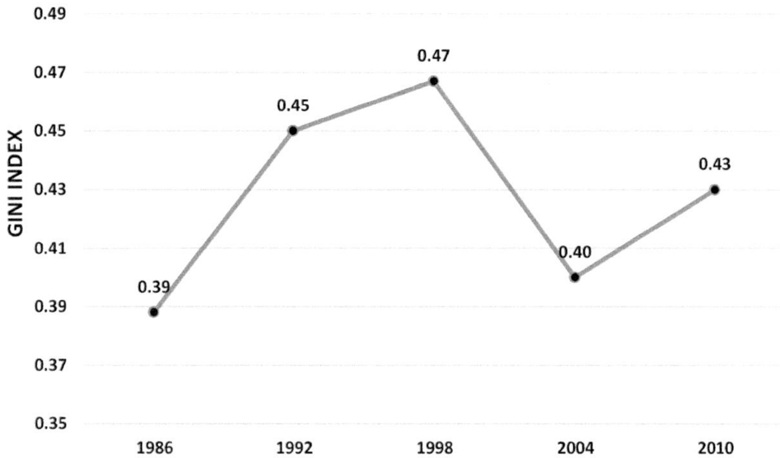

Figure 1: Nigeria's income inequality index between 1986 and 2010.

"distance" separating a population from an ideal state of complete equality in which the income distribution is equal. The result of the calculation is a negative entropy number, meaning that this number increases with an increasing degree of order and further deviation from complete equality. Furthermore, the range of values that the Theil index can take ranges from zero to infinity. Zero represents perfect equality in the distribution, while higher values correspond to an increasing degree of inequality. In the case of Nigeria, where income inequality was measured during the period under study, inequality as measured by the Gini coefficient declined from 0.356 in 2004 to 0.41 in 2013 but improved to 0.391 in 2016. Using Thiel, inequality worsened from 0.217 in 2003 to 0.395 in 2013 but improved to 0.31 in 2016. As a result, the upper class's income share constituted 58.39% of total national income and expenditures, up from 59.42% in 2013. It increased between 2004 and 2013 before declining in 2016.

In contrast, the middle-class share of national income and expenditures increased to 30.26% in 2016, compared to 29.14% in 2013. The middle-class share fell in favor of the upper class between 2004 and 2013 but reversed in 2016. Finally, the percentage of the lower class's income share in total national income/expenditures was 11.35% in 2016, up from 11.43% in 2013. Thus, the middle class gained the highest percentage share of the income/expenditure from 2013 to 2016, while the percentage of the lower class remained the same and that of the upper class declined. In measuring consumption poverty, as used in many countries around the world, the National Bureau of Statistics estimated the Gini coefficient at 0.347 and the index at 35.1 in its 2019 report, underscoring high-income inequality. The calculation of the sample from Borno State, the main theatre of the terrorist group Boko Haram, is assumed to be non-random and non-representative, as houses in very few places in the state were safe during the exercise.

In addition, there is high disparity in regional inequality in Nigeria (see Figure 2). This is teeming polarization is responsible for higher poverty rates in certain regions

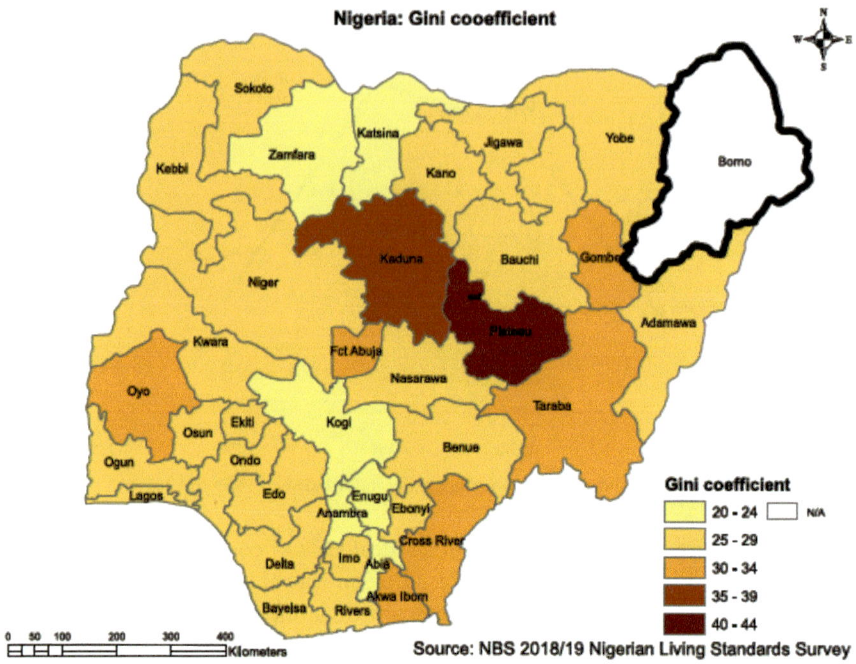

Figure 2: Nigeria's Gini coefficient according to the different states of the federation. Source: National Bureau of Statistics (2020).

such as the Niger Delta and particularly the North (North East and North West) when viewed from the distribution lens of household consumption expenditure. Gender inequality is another driver of this income stratification. Many women are victims of discriminatory traditional and socio-cultural practices that put them at a disadvantage in several areas compared to men. For example, most women are employed in casual, low-paid jobs and are less likely than men to own land. The index shows that 75.8% of the poorest women have never been to school, compared to 28% of the wealthiest men. As a result, women are more likely to be poor than men. They also remain excluded from full participation in the country's economic, social, and political life. Furthermore, a largely regressive tax system that puts the tax burden mostly on poorer companies and individuals equally widens inequality (Oxfam 2017).

7. A Case of the Niger Delta Region

The Niger Delta region of Nigeria provides a good instance of rising income inequality and a threat to democratic consolidation. Incidentally, oil wealth from the region is mainly responsible for sustaining the Nigerian Federation. According to the UNDP (2006), oil extracted from the region between 2000 and 2004 constituted 79.5% of total government revenues. This is about 97% of foreign exchange revenues. Essentially, the region's capacity to generate wealth for the Nigerian Federation and the multinational oil companies that extract oil from it does not translate to human development in the region. The Niger Delta Human Development Report shows that state and regional HDI scores mask inequalities in human development among oil-producing communities and that the decline in the HDI has been steeper for the Niger Delta states than for the rest of Nigeria (UNDP 2006, Higgins 2009).

The crisis in the region began in the 1990s due to tensions between foreign oil companies and minority groups in the region, such as Ogoni and Ijaw, who feel exploited. Militants soon after began to target and destroy pipelines, while local gangs also masked themselves to launch raids (Humphrey 2014). To threaten their agitations for fair wealth redistribution, militant groups threatened to disrupt the 2019 elections if the federal government did not meet their demands on developing the Niger Delta region. The threats were not ignored as the 2019 elections were held amidst hope and fear, and many boycotted elections due to fear, consequently affecting the democratic process. One can conclude that the crisis in Nigeria's Niger Delta region is a product of underdevelopment and inequity in the distribution of resources giving rise to alienation and exclusion from their affairs.

8. A Case in Northern Nigeria

As mentioned earlier and shown in Figure 3 below, the North East and North West states of Nigeria make good examples of income polarization when viewed from the spatial distribution lens of household consumption expenditure. The findings and analysis of a World Bank Assisted research conducted by Clementi et al. (2017) on income polarization in Nigeria provide the author with ample evidence for the thesis. The authors argue thus:

> "For what concerns the North East and the North West, the conflict-stricken areas, had the location effect been the only operating force, we would have seen a disproportionate increase of people in these regions occupying the lower national deciles compared to the rest of the country—they basically lagged behind. The increase of polarization in the rest of the country helped to offset this effect, filling the lower deciles of households from other regions, too, whereas, for the rest of the distribution, we observe in practice a generalized decline of the relative position of these regions in the national distribution. Finally, while the North Central improves relative to the rest of the country in lower deciles, the South East shows a more articulated pattern of distributional change."

How does this impact democratization negatively? At the heart of the link between democratization and income inequality is that it essentially leads to the disappearance of the middle class, the social group that forms the basis of elite power. In this case, as in other societies characterized by income inequality, the "disappearing" middle class is far from the mean or median of the overall distribution, mainly because one earner has many dependents to feed. The middle class is, therefore, unable to consolidate its position. However, in addition to the dire economic impact on the country, rising income inequality in the region has also led to mounting redistributive pressures and, thus, increased political instability. For example, the Boko Haram insurgent group

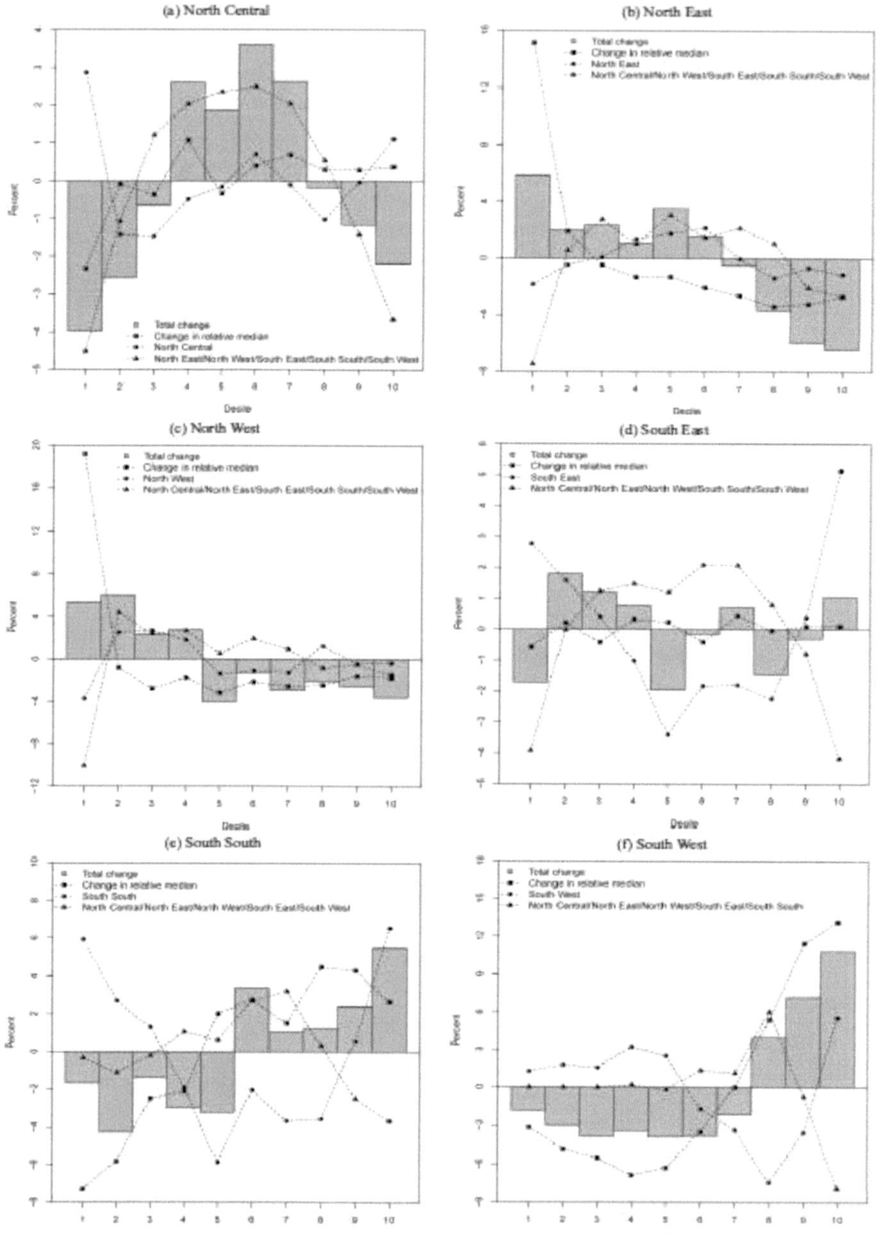

Figure 3: Sources of distributional change in the 2012/2013 to 2003/2004 relative distribution of consumption expenditures by zone. Source: Clementi et al. (2017).

sees northern Nigeria as its main theater of operations. Given the shifting recruitment tactics that use monetary inducements (exploiting the income-inequality induced poverty prevalent in the region), membership of the violent Islamic sect has expanded to include both educated and non-educated actors (Kuznar 2019).

The insurgency has heightened security tensions after it spread to other parts of the country, including cities such as Abuja, Lagos, and, most recently, Ondo. This fuels disillusionment with the nascent democracy beyond the northern part of the country and calls on national leaders to step up their efforts to address security concerns, among many other issues bedeviling the country. For example, Biafran separatists and those who clamor for the Yoruba nation also claim marginalization on the grounds of regional inequality. At the root of these agitations are regional demands for equal wealth distribution and representation in government to drive policies. Overall, the ability to tackle the security scourge, revive the weak economy, resolve the regional polarization, and entrench true federalism have become key competencies whoever becomes Nigeria's next president in 2023 must possess. The author argues that if these issues are adequately resolved, Nigeria will be on the right track to forming a democratic political culture.

9. Findings

All in all, the agitations in these regions underscore the doubts about democratic consolidation in Nigeria on macroeconomic grounds. We conclude that underdevelopment and inequity in the distribution of resources giving rise to alienation and exclusion from their affairs birthed the crisis in Nigeria's Niger Delta and North region. While most of the northern populace groans under the rising income inequality, an unvoiced system of ethnic patronage exacerbates inequality and fuels the instability in the region. As Kuznar (2019) puts it, "the elites retain their constituency, however, [despite the citizens' dissatisfaction with their poor economic state] because of pervasive distrust between ethnic and religious groups that prevent defection from the elites who represent them" Clearly, the people have no way to influence and direct ruling elites and structures in a way that reflects their plight, basic needs, and aspirations. This negates the principles of democracy and threatens the formation of a political culture in which there is democratic continuity, entrenchment, survival, and irreversibility.

10. Regression to Military Dictatorship

One might ask why the "loser class" would find a military dictatorship suitable. Nigeria's underdevelopment is traceable to the current political elites' mismanagement and corruption. Nigeria has experienced military and democratic regimes. The past military regimes, which called themselves "corrective regimes," used civilians' mismanagement of the country's resources, resulting in poor civil society and economic development, as the basis for their intervention in governance (Eshikena 2012). Statistics show that the military is better at managing the economy, despite accusations of mismanagement of public resources and disregard for human dignity. Using data from the Nigerian junta era, Odetola supports military-led development and assumes that the military may be the only group capable of laying the groundwork for further economic breakthroughs because the civilian sector seems less able to do so than the military in driving the development process (Onwioduokit 1999). According to Odetola, between 1964 and 1966 (civil government), the GNP level was 4.2; in 1969–71 (junta), it rose to 9.1, and in 1972 to 12.1 (Ndiomu 2000). Since the military is unstatic, dynamic, and not impervious to the international climate (Janowitz 1964), the qualities that make it unique such as strict discipline, are considered suitable for political leadership, especially as the propensity for military intervention is likely to increase with increased social mobilization. This empirical evidence, coupled with current realities, confirms this and draws from the happenings in countries like Taiwan, North Korea, China, Hong Kong, and Singapore. Thus, the social mobilization that demanded the military dictatorship's withdrawal from power in 1999 and led to a transition to democracy can now be a tool for a return to military rule.

11. Conclusion

In conclusion, if an economic system undermines social justice, the creation of a democratic political culture is delayed, and an enabling environment is created for undemocratic political orientations and practices that lead to the regression of military rule and hinder democratic consolidation. As such, the "loser class" may find a military rule better suited to ensure their economic well-being in the hope that the military government will be influenced by a democratic international climate or wave to respect and promote personal liberties. After all, a Nigerian adage states that a woman who has married two men is better positioned to know which is manlier. The policy implications are clear: If policymakers and other stakeholders can envision and implement a framework that resolves the income polarization in Nigeria, then all democratic actors will shift into institutionalizing what has been achieved. Consequently, the country will move to a point where there is democratic continuity, entrenchment, survival, and irreversibility.

Notes

1. The benchmark study, "Democratic Values" was conducted in September 2000 in 11 post-communist countries and was part of a major international initiative of the same name funded by the European Commission's 5th Framework Programme (HPSE-CT-2001-00062). The project was led by Professor Dr. Detlef Pollack of the European University Viadrina in Frankfurt (O), Germany. The survey data can be found on the following websites: www.democraticvalues.eu.tf and http://amcd.hit.bg.
2. The author notes that the percentage of African states classified as 'not free' by Freedom House has declined significantly from 59% in 1983 to 35% in 2003. A transcript of a speech by President Muhammadu Buhari at Chatham House is available online at: https://www.chathamhouse.org/sites/default/files/field/field_document/20150226BuhariSpeechQA.pdf.

References

Abazie Humphreym, I. M. (2014). Engaging the Nigerian Niger Delta Ex-Agitators: The Impacts of the Presidential Amnesty Program to Economic Development. Retrieved from, https://web.archive.org/web/20160125115453/http://www.eadi.org/typo3/fileadmin/Documents/Events/General_Conference/2014/gc2014-abazie-humphrey-41.pdf.

Acemoglu, D. and Robinson, J. A. (2006). *Economic Origins of Dictatorship and Democracy*. Cambridge and New York: Cambridge University Press.

Aigbokhan, B. E. (1999). The impact of adjustment policies and income distribution in Nigeria: An Empirical Study. Research Report No. 5, Development Policy Centre, Ibadan, Nigeria.

Bachrach, P. and Botwinick, A. (1992). *Power and Empowerment: A Radical Theory of Participatory Democracy*. Philadelphia: Temple University Press.

Bakare, A. S. (2012). Measuring the income inequality in Nigeria: The Lorenz Curve and Gini co-efficient approach. *Am. J. Econ.*, 2: 47–52.

Bartels, L. M. (2008). *Unequal Democracy: The Political Economy of the New Gilded Age*. Princeton: Princeton University Press.

Boix, C. (2003). *Democracy and Redistribution*. Cambridge and New York: Cambridge University Press.

Bratton, M. and Van de Walle, N. (1997). Democratic experiments in Africa: Regime transitions in comparative perspective. New York: Cambridge University Press.

Buhari, M. (2015). Prospects for Democratic Consolidation in Africa: Nigeria's Transition, Speech delivered at CHATHAM HOUSE, The Royal Institute of International Affairs, Africa Program Transcript https://www.chathamhouse.org/.../chathamhouse/.../2015.

Clementi, F., Dabalen, A. L., Molini, V. and Schettino, F. (2017). Review of Income and Wealth Series 63, Number 4, December 2017 DOI: 10.1111/roiw.12212.

De Maio, F. G. (2007). Income inequality measures. *J. Epidemiol. Community Health.*, 61(10): 849–52. DOI: 10.1136/jech.2006.052969.

Esebonu, E. N. (2012). The mass media and the struggle for democracy in Africa: The nigerian perspective. *Nordic Journal of African Studies*, 21(4): 183–198.

Eshikena, J. (2012). *Nigerian Government and Politics*. Lagos: Fortran Press Ltd.

European Commission. (2005). Democratic Values: Values Systems of the Citizens and Socio-Economic Conditions—Challenges from Democratisation for the EU-Enlargement, Brussels. Retrieved from https://cordis.europa.eu/docs/projects/files/HPSE/HPSE-CT-2001–00062/100124301–6_en.pdf.

Gilens, M. (2012). *Affluence and Influence: Economic Inequality and Political Power in America*. Princeton and New York: Princeton University Press/Russell Sage Foundation.

Hacker, J. S. and Pierson, P. (2010). *Winner-Take-All Politics: How Washington Made the Rich Richer—And Turned Its Back on the Middle Class*. New York: Simon and Schuster.

Haveman, R., Gary, S., Barbara, W. and Andrea, V. (2004). Trends in children's attainments and their determinants as family income inequality has increased. *In*: Kathryn Neckerman (ed.). Social Inequality. New York: Russell Sage Foundation.

Higgins, K. (2009). Regional Inequality and the Niger Delta, Policy Brief No. 5 Prepared for the World Development Report, Overseas Development Institute.

Huntington, S. P. (1991). The third wave: Democratization in the late twentieth century. University of Oklahoma Press.

Isma'ila, Y. and Madu, A. Y. (2016). Democratic transition in nigeria: Towards understanding the disguise of international dynamics challenging the fourth republic consolidation. pp. 289–295. *In*: Mohamad, B. (ed.). *Challenge of Ensuring Research Rigor in Soft Sciences*, vol 14. European Proceedings of Social and Behavioural Sciences. Future Academy. https://doi.org/10.15405/epsbs.2016.08.41.

Janowitz, M. (1964). The Military in the Political Development of New Nations: An Essay in Comprehensive Analysis. Chicago: University of Chicago Press.

Kuznar, A. L. (2019). Nigeria Country Report-An NSI Aggrieved Populations Analysis, NSI, Cambridge, Retrieved from https://nsiteam.com/social/wp-content/uploads/2019/10/NSI-Aggrieved-Populations-Reports_Nigeria-Country-Report_Final.pdf.

Levin-Waldman, O. M. (2013). Income, civic participation and achieving greater democracy. *Journal of Socio-Economics*, 43,2: 83–92.

Linz, J. J. and Stepan, A. (1996). Problems of Democratic Transition and Consolidation: Southern Europe, South America, and Post-Communist Europe. Baltimore: Johns Hopkins University Press.

Lipset, S. M. (1959). Some Social requisites of democracy: Economic development and political legitimacy. *American Political Science Review*, 53(1): 69–105.

Mahler, V., Jesuit, D. and Paradowski, R. (2013). The Political Sources of Government Redistribution in the Developed World: A Focus on the Middle Class, Paper prepared for presentation at the Luxembourg Income Study, Walferdange.

Meltzer, A. H. and Scott, F. R. (1981). A rational theory of the size of government. *Journal of Political Economy*, 89,5: 914–927.

Ndiomu, C. (2000). Military government, and the Nigerian economy. *Bull. Central Bank Nig.*, 24: 35–35.

National Bureau of Statistics (2020). 2019 Poverty and Inequality in Nigeria: Executive Summary, Nigeria.

Obiagu, A. and Ajaps, S. (2022). Civic engagement in Nigeria: A new sociopolitical development perspective. *International Journal of Social Sciences and Educational Studies*, 9: 128–147. https://doi.org/10.23918/ijsses.v9i1p128.

Obiora, L. A. (2005). Nigeria: Political Transition and the Complexities of Gender Empowerment, United Nation Report.

O'Donnell, G. (1992). Transitions, Continuities, and Paradoxes. Issues in Democratic Consolidation: The New South American Democracies in Comparative Perspective, eds Scott Mainwaring, et al. Notre Dame: University of Notre Dame Press 1756.

Onwioduokit, E. A. (1999). Governance and economic performance in Nigeria. *Bull. Abuja Central Bank Nig.*, 23: 89–94.

Oxfam International. (2017). Inequality in Nigeria: Exploring the Drivers. Retrieved from visit www.oxfam.org.

Oyakhilomen, O. and Zibah, R. G. (2014). Agricultural production and economic growth in Nigeria: Implication for rural poverty alleviation. *Quarterly Journal of International Agriculture*, 53(3): 207–223.

Piketty, T. (2014). Capital in the Twenty-First Century. Cambridge, MA, and London: Belknap Press of Harvard University Press.

Przeworski, A., Alvarez, M., Cheibub, J. A. and Limongi, F. (1996). What makes democracies endure? *Journal of Democracy*, 7(1): 39–55.

Tonta, R. (2016). The role of economic factors in the processes of democratic consolidation—The case of post-revolutionary Tunisia. *Athenaeum Polish Political Studies*, 52: 95–116. https://doi.org/10.15804/athena.2016.52.05.

United Nations Department of Economic and Social Affairs. (2020). World Social Report: Inequality in a Rapidly Changing World (Publisher) Retrieved from https://www.un.org/development/desa/dspd/wp-content/uploads/sites/22/2020/02/World-Social-Report2020-FullReport.pdf May 31, 2022.

United Nations Development Programme. (2006). Niger Delta Human Development Report. United Nations Development Programme (2006a) 'Nigeria,' Human Development Report 2006 Retrieved on June 1, 2022.

United Nations Development Programme. (2007b). MDG Monitor. http://www.mdgmonitor.org/index.cfm (November 6, 2007). Retrieved on June 1, 2022.

Uslaner, E. (2008). The Foundations of trust: macro and micro. *Cambridge Journal of Economics*, 32: 289–294.

Uslaner, E. M. and Brown, M. (2005). Inequality, trust, and civic engagement. *American Politics Research*, 33(6): 868–894.

Volscho, T. W. and Nathan, J. K. (2012). The rise of the super-rich: Power resources, taxes, financial markets, and the dynamics of the top 1 Percent, 1949–2008. *American Sociological Review*, 77,5: 679–699.

World Development Indicators. (2014). Washington, DC: World Bank. DOI:10.1596/978- 1–4648–0163–1. License: Creative Commons Attribution CC BY 3.0 IGO Retrieved from https://openknowledge.worldbank.org/bitstream/handle/10986/18237/9781464801631.pdf.

World Economic Forum. (2015). How can Nigeria grow its middle class? Retrieved from https://www.weforum.org/agenda/2015/05/how-can-nigeria-grow-its-middle-class/#:~:text=Using%20the%20VAMC%20estimate%2C%20we,standards%2C%20even%20for%20developing%20countries.

Chapter 13

Bitcoin's Reaction to the Federal Reserve's Announcements and Stimulus Measures, the Equity Market, and Economic Policy Uncertainty

Evidence during the COVID-19 Pandemic

Murad A. Bein,[1] *Bright Akwasi Gyamfi,*[2]
Festus Victor Bekun[3,4,*] and *Amjad Taha*[5]

1. Introduction

The COVID-19 pandemic is still underway, which means that the crisis's full extent cannot yet be known. Nevertheless, investigations into the economic effects of the pandemic and its impact on the financial markets have begun. Since the onset of the COVID-19 crisis, the stock markets in both developing and developed economies and the international crude oil and currency markets have experienced the worst volatility on record. During periods of uncertainty, investors typically attempt to reduce their exposure or diversify their investments and search for safer assets. Recently, there has been an increase in the debate regarding whether Bitcoin can serve as a haven during periods of market turmoil. While regular currencies are guaranteed by the

[1] Cyprus International University Nicosia, North Cyprus Via Mersin 10, Turkey.
[2] School of Management, Sir Padampat Singhania University, Bhatewar, Udaipur, Rajasthan, India.
[3] Department of Logistics and Transportation, Istanbul Gelisim University, Istanbul, Turkey.
[4] Adnan Kassar School of Business, Department of Economics Lebanese American University, Beirut, Lebanon.
[5] Banking and Finance Department Eastern Mediterranean University 99628, Famagusta, North Cyprus Via Mersin 10, Turkey.
Emails: muradabdurahmanb@gmail.com; brightgaymfi1987@gmail.com; amjad200543@gmail.com
* Corresponding author: fbekun@gelisim.edu.tr

governments or organizations that issue them, Bitcoin, as a digital currency, is not guaranteed by anybody (Weber 2016).

Interestingly, due to its innovative "distributed ledger," Bitcoin is entirely decentralized, which allows it to be utilized in a decentralized transaction framework. Satoshi Nakamoto introduced Bitcoin during the 2008 global financial crisis, which saw invisible financial and economic uncertainties disrupt international financial institutions and capital markets in industrialized and unindustrialized economies. During that period, trust in the financial system's sustainability and the resultant economic security decreased exponentially, while market instability, as calculated using the implied volatility, erupted worldwide.

Undoubtedly, Bitcoin benefited from being introduced into a deeply unpredictable world, with that unpredictability persisting after the recession of 2008–2009 (Weber 2014). Bitcoin's divisive characteristics soon attracted the interest of practitioners, researchers, and the financial press. Later crises, including the 2010–2013 European sovereign debt crisis (ESDC) and the 2012–2013 Cypriot banking crisis, also exacerbated Bitcoin's success and its discourse. During such times of tension, Bitcoin increased in value. It acquired further momentum when people started to see it as a refuge from the confusion affecting the traditional financial systems. Several journal reports written around that time acknowledged the migration from paper (fiat) currencies and bank deposits to Bitcoin, particularly in geographical areas heavily hit by the ESDC and Cypriot crises, such as Greece, Cyprus, and Spain.[1] In line with this, several authors examined the relationship between Bitcoin returns and other financial assets with the aim of assessing the potential of Bitcoin as a haven and for hedging purposes (see, e.g., Bouri et al. 2017, Bouri et al. 2018, Jareño et al. 2020, Shahzad et al. 2019). The study by Bouri et al. (2017) used the dynamic conditional correlation to examine whether Bitcoin could act as a hedge and provide diversification benefits for global equity market general commodities, oil, gold, and bonds. The authors found that Bitcoin was useful concerning diversification, although it proved to be poor concerning hedging.

The onset of the COVID-19 pandemic has led to many variations in global social reactions, economic limitations, and observational studies. A similar difference has been identified in both the financial markets and investors' reactions when trying to determine the size of the problems they face. The Chinese financial markets served as the early provocative epicenter (Corbet et al. 2020a), although the pandemic's foreign contagion consequences rapidly accelerated. In the middle of all this chaos, the conventional oil flight-to-safety networks deteriorated, leading to a worsening of inflationary pressures primarily facilitated by the United States (US), Russia, and Saudi Arabia, with investors scrambling to identify reliable, safe havens (Corbet et al. 2020b, Goodell 2020).

[1] http://www.CNBC.com/id/100597242;http://money.cnn.com/2013/03/28/investing/bitcoin-cyprus/; http://www.forbes.com/sites/petercohan/2013/04/02/are-bitcoins-safer-than-cyprus/#42f9ceef3c7a; http://abcnews.go.com/Business/cyprus-crisis-boosting-unique-currency-bitcoin/story?id=18792763.

For example, prior to the current crisis, Bloom et al. (2018) highlighted the economic vulnerabilities likely to stem from future pandemics, including the health-sector and social-isolating costs, the decline in work production, the decline in market demand, the effect on hospitality and related sectors, and the effect on foreign direct investment. Fan et al. (2018) echoed the same worries. They noted the significance of the unmet need for increased investments in readiness for significant outbreaks of infectious diseases. Moreover, they evaluated the anticipated annual pandemic threat liabilities to be approximately US$500 billion, or 0.6% of global revenue. Yet, concerning the anticipated cost of the COVID-19 pandemic, this seemingly substantial figure appears to be an underestimate.

Other studies that have addressed the need for economic risk activities concerning possible future epidemics include the work by Bloom and Canning (2004), Cavallo et al. (2013), Madhav et al. (2017), and Tam et al. (2016). Further, Lagoarde-Segot and Leoni (2013) investigated the probable burden of major pandemics on the financial industries of emerging countries. Additionally, the COVID-19 pandemic might also contribute to very long-term changes in the cost of equity. For instance, Goodell (2020) and Lee and McKibbin (2004) identified a 200-point rise in the national default risk for China and Hong Kong following the severe acute respiratory syndrome (SARS) pandemic. However, while China and Hong Kong were unique SARS threat zones, the COVID-19 pandemic is a worldwide phenomenon. A substantial rise in the potential equity risk is, therefore, relatively strongly anticipated.

Data also show that the implicit volatility index (VIX), including the US VIX, is adversely associated with equity (e.g., Jubinski and Lipton 2012). The VIX is the primary market risk tracker, representing stock prices and investor aspirations. As such, it is commonly used in market participants' risk-control schemes. Higher VIX values suggest greater investor volatility and vice versa. Notably, a rise in the VIX usually leads to so-called "safety flights" (Sarwar 2017). For instance, Jubinski and Lipton (2012) illustrated the negative association of the treasury- and investment-grade VIX bond yields. This implies that rates decline about implied volatility rises, suggesting that the bond price rises as the stock price declines. Appropriately, during such times, investors strategically switch from volatile investments and the equity markets to safe-haven investments as well as gold and treasury bonds.

However, the exact relation between market instability and Bitcoin remains mostly unexplored, although Bouri et al. (2016) included the US VIX within their generalized autoregressive conditional heteroskedasticity (GARCH)-based system and determined that Bitcoin volatility is negatively associated with US instability. As the evidence supplied by the VIX serves as a helpful guide for investors, it is imperative to include the VIX in every analysis of Bitcoin's potential to hedge and of its relationships with other properties. In line with this, researchers have investigated the relationship between changes in the level of economic policy uncertainty (EPU) and Bitcoin returns. For example, Demir et al. (2018) explored the effect of EPU on Bitcoin returns and documented a negative relationship during regular times. Moreover, they found the relationship positive in the case of bull and bear markets and concluded that Bitcoin could be used to hedge against EPU in extreme quantiles. Fang et al. (2019) used monthly data to study whether global EPU affects Bitcoin volatility, global equities, commodities, and bonds in the long run. They found that EPU affects the volatility of all those assets. Several other authors have also studied the effect of EPU on Bitcoin returns (Mokni et al. 2020, Wang et al. 2019, Wang et al. 2020).

Determining the existence and the symbol of the relationship between uncertainty measures and Bitcoin returns is critical for those market participants seeking to discover more about Bitcoin's elusive financial characteristics. In particular, the use of a global indicator of ambiguity would render such an analysis more vigorous. In this regard, prior studies have analyzed Bitcoin return uncertainty (Katsiampa 2017), aware trading (Feng et al. 2017), market clustering (Urquhart 2017), asset bubbles (Cheah and Fry 2015, Corbet et al. 2018), and exchange costs (Kim 2017).

Several authors have also investigated how the Bitcoin price is determined. For instance, Kristoufek (2013) explored Google and Wikipedia's effect on the Bitcoin price and identified both a strong correlation and a causal relationship between the Bitcoin price and both internet engines. Similarly, Claser et al. (2014) found that Google searches affect the price of Bitcoin. Corbet et al. (2020) are among the few authors to have investigated the effect of federal fund target changes and quantitative easing on fifty-eight digital assets. Their study classified digital assets into three groups, namely currencies, protocols, and decentralized applications. They then documented how the groups of digital assets responded differently, with the currencies group reacting immediately to US monetary policy announcements. Shaikh (2020) studied the impacts of EPU in several countries (US, United Kingdom, Japan, China, and Hong Kong), announcements following Federal Open Market Committee (FOMC) meetings, and gross domestic product (GDP) and macroeconomic data on Bitcoin returns and found a negative relationship with EPU in the US and Japan as well as with FOMC meeting announcement, GDP data, and macroeconomic data uncertainty. In line with this, Pay and Lee (2018) studied the effect of FOMC and macroeconomic announcements on the Bitcoin price and documented that the price declined by 1% following FOMC announcements. Moreover, when there were no announcements, the Bitcoin price increased by 0.26%, while on the day preceding such reports, the price increased by 0.96%.

However, the studies mentioned above did not consider the effect of FOMC announcements during the COVID-19 pandemic, although it is noteworthy that the US Federal Reserve (the Fed) has implemented several actions to save the economy and financial market from total collapse. Therefore, the present study aims to investigate whether the FOMC announcements (both scheduled and unscheduled meetings) and proposals by Congress regarding stimulus measures impact Bitcoin's abnormal returns. More specifically, the study will determine whether such announcements and proposals positively or negatively impact Bitcoin returns. Further, it will explore the effect of the VIX and EPU on Bitcoin's abnormal returns. Changes in the Federal Reserve's policies and stimulus proposals directly reflect on the EPU and VIX.

Furthermore, prior studies have proved inconclusive regarding the impacts of equity market uncertainty and EPU on Bitcoin returns, especially those studies that have explored the effects of such uncertainty during the COVID-19 pandemic. Therefore, using both ordinary least squares (OLS) and quantile regression (QR) methods, the present study explores the impact of the US VIX and EPU on Bitcoin returns. QR is a useful tool for investigating the effects of bull and bear markets. In terms of previous investigations concerning the impact of uncertainty on Bitcoin returns, Bouri et al. (2016) considered the VIX, while Demir et al. (2018) studied EPU.

Lastly, the present study also examines the time-varying relations between Bitcoin returns and the two uncertainty measures (VIX and EPU) using the GARCH-DCC (dynamic condition correlation) model. Determining the time-varying relations will assist in understanding the correlation during both tranquil and turbulent periods. This study has several significant findings. We have documented how the Fed's announcement days have higher return- and risk-to-reward ratios when compared with non-announcement days. Moreover, the Fed's announcements had a more significant positive impact on the abnormal returns when the Fed announced interventions and simultaneous news regarding stimulus measures. Further, the empirical results concerning the effects of monetary policy changes, the VIX, and EPU on abnormal returns demonstrated negative associations. The results also revealed how a Senate proposal regarding a second stimulus package led to an increase in the abnormal returns. In addition, the QR results indicated that the VIX at the median upper and lower quantiles had a negative and significant effect on Bitcoin returns.

In contrast, EPU positively impacted at 0.10 and 0.20 before turning negative until the upper quantile. There was also a consistently negative relationship between the VIX and EPU and Bitcoin returns. There was some similarity in terms of the trend and pattern of the time-varying relationships between those variables.

The remainder of this article is organized as follows. Section 2 summarizes our data and methodology, while Section 3 presents the empirical results. Finally, Section 4 sets out the conclusions of the study.

2. Data and Methodology

2.1 Data and graphical presentation

The present study used the daily Bitcoin price, the EPU, and the VIX from 2 January to 7 October 2020. We also obtained a three-month treasury bill to construct a reward for bearing risk, computed as the risk premium over the standard deviation. The data concerning the EPU were obtained from Baker et al. (2016), while the rest of the data were obtained from yahoofinance.com.[3] Figure 1 offers a graphical representation of Bitcoin (price and return) and the uncertainty measures (EPU and VIX) at different levels and changes. The Bitcoin price was at its lowest during March 2020 due to the COVID-19 outbreak and resulting event. It is important to note that there were several unprecedented meetings held by the Federal Reserve and, further, that the US announced a national lockdown and travel ban in the same month that the World

[3] data for EUP was obtain from Baker et al. (2016) whiles the rest were obtain from www.yahoofinance.com.

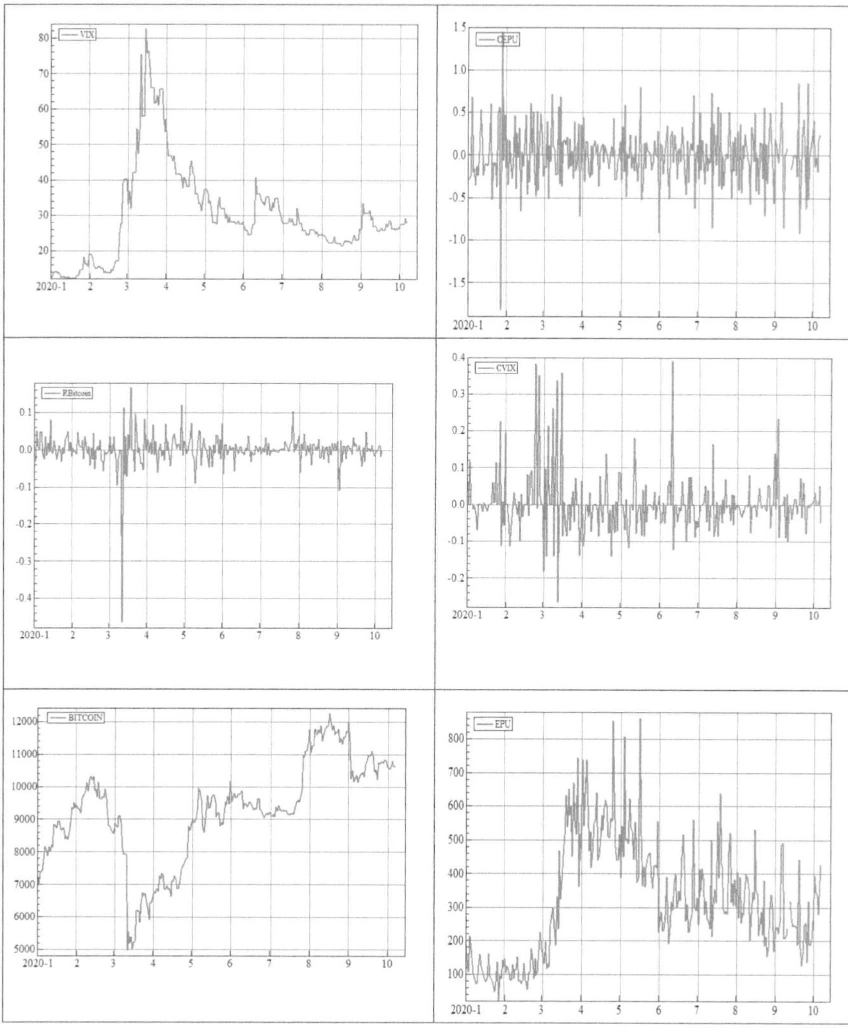

Figure 1: Bitcoin price and returns and uncertainty measures (at level and changes).
Note: all FOMC announcement are obtained from the official web site of the Federal Reserve Board, and news regarding stimulus is obtained from credible sources.[4]

Health Organization (WHO) declared the COVID-19 outbreak to be a pandemic. In addition, several other nations announced travel bans and lockdowns. The Bitcoin price was at its peak during August, which was right after the Republican Party proposed a second stimulus package. The highest (lowest) Bitcoin price during the

[4] https://www.cnbc.com/2020/04/13/coronavirus-update-here-is-everything-the-fed-has-done-to-save-the-economy.html
https://www.cnet.com/personal-finance/stimulus-package-how-the-new-heroes-act-compares-to-the-cares-act-from-march/
https://www.federalreserve.gov/newsevents/pressreleases/monetary20200916a1.htm.

Table 1: FOMC announcements and stimulus package proposals.

Date	Type	Change(s) Made/proposed
Jan 29	Scheduled	Decided to maintain the federal fund target rate
Mar 3	Unscheduled	0.5% rate cut in the federal fund target rate
Mar 15	Unscheduled	1% rate cut in the federal fund target rate
Mar 19	Intervention/ stimulus proposal	Senate passed the Cares Act/Fed announced a new operation focused on currency swaps
Mar 23	Intervention	Fed announced unlimited asset purchases and a $300 billion credit program for businesses and consumers
Mar 31	Intervention	Fed announced a temporary repurchase agreement facility for foreign and international monetary authorities
Apr 29	Scheduled	Decided to maintain the federal fund target rate
Jun 10	Scheduled	Decided to maintain the federal fund target rate
Jul 29	Scheduled	Decided to maintain the federal fund target rate
Aug 27	Intervention	Unanimous approval of updates to its Statement on Longer-Run Goals and Monetary Policy Strategy
Sep 16	Scheduled	Decided to maintain the federal fund target rate
May 15	Stimulus proposal	House passed the Hero Act
Jul 27	Stimulus proposal	HEALS introduced by Senate
Sep 2	Stimulus proposal	Senate minority leader criticized the narrower coronavirus stimulus bill Republicans wanted a vote on next week
Sep 11	Stimulus proposal	Senate Republicans' COVID-19 relief bill blocked

months mentioned above was US$12,254 (US$5,200). In terms of the uncertainty indicators, for both the EPU and VIX, the highest levels were observed during March, while the lowest levels were seen in August. This further indicates that there exists a negative relationship between the two uncertainty measures and the Bitcoin price. Looking at the changes in the EPU and VIX levels and the Bitcoin returns, it can be observed that the difference in the EPU was more volatile than the changes in the VIX and the Bitcoin returns. Table 1 details the Federal Reserve's announcements and the stimulus proposals and describes related events.

2.2 Methodology

2.2.1 Measuring bitcoin's reaction to the FOMC announcements and stimulus proposals

Researchers have widely used the event study to investigate market reactions to policy changes, earnings and dividend announcements, mergers and acquisitions, and other announcements that may affect the asset price. The process was first developed by Fama et al. (1969). There are three ways to calculate abnormal returns in terms of event studies, namely the Mean Adj, the Market Adj, and the Mood Adj methods. To examine Bitcoin's returns reaction to the FOMC announcements and stimulus proposals, this study used the Mean Adj method because it can overcome biases within the model (Brown and Warner 1985). The abnormal returns during the estimation windows can be calculated using the following equation:

$$AR_t = R_t - \overline{R}_t \tag{1}$$

Here, AR_t is the abnormal return, R_t is the actual return of the Bitcoin price on date t, and $\overline{R}_t$ is the average Bitcoin return over the estimation window. The actual returns R_t and the changes (VIX and EPU) are obtained by first transferring the data into the logarithmic form and then taking the first difference. Using the above formula, Pacicco et al. (2019) recently explored the impact of European Central Bank policy innovation on European equity. Further, the effects of changes in the federal fund target rate (FFTR), VIX, and EPU were studied using the OLS method. To examine the Bitcoin price reaction to the FFTR and the uncertainty measures, this study considered ±5 and 15 days around the announcements. We considered some days before the event because investors may expect an announcement or news may leak before the public announcement. In contrast, we considered some days after the event because it takes some time for the full effects of an event to influence the Bitcoin price. The specific model estimated is given below:

$$ARj_{(-15,\,15)} = a + \beta_1\,\Delta VIX_j + \beta_2\,\Delta EPU_j + \beta_3\,\Delta FFTR_j + \varepsilon_j \tag{2}$$

where AR is the abnormal return during the FOMC and stimulus announcements, while ΔVIX and ΔEPU are the changes in the VIX and EPU, respectively, estimating the above regression is useful because the EPU is based on newspaper coverage and captures when the Federal Reserve adjusted its policy rate and other economic policies in the US (Baker et al. 2016). In contrast, the VIX is computed based on the S&P 500 call and put options and measures investors' sentiments concerning the US equity return.

2.2.2 Quantile regression

The QR approach was developed by Koenker and Bassett (1978), and it has since become trendy due to the advantages it offers over the conventional linear regression approach. For example, using the QR approach, it is possible to capture the regular phase (median) of the dependent variable and the phases in different quantiles (lower and upper tails). The QR can be computed using the equation below:

$$Q\{\tau \backslash U_{it}, \beta(\tau)\} = U'_{it}, \beta(\tau) \tag{3}$$

where $\beta(\tau)$ is the vector of the coefficient with a given quantile. The τ-quantile estimates can be obtained by minimizing the equation:

$$\widehat{\beta}_n(\tau) = min_{\beta(\beta(\tau))}\left\{\Sigma_i \rho T\left(R_t - U_{it}, \beta(\tau)\right)\right\} \tag{4}$$

2.2.3 Multivariate DCC-GARCH

Multivariate DCC-GARCH model of Engle (2002) overcome heteroscedasticity, endogeneity, and omitted variable bias. The estimation consists of two steps. First, the univariate GARCH model is estimated for the Bitcoin returns and uncertainty measures (VIX, EPU). Second, the estimation of the conditional correlations that vary over time. Engle's (2002) model can be expressed as:

$$K_t = F_t L_t F_t \quad (5)$$

where K_t is the conditional covariance matrix, which is decomposed into dependent standard deviations, $F_t = diag(y_{1,1,t}^{1/2},\ldots\ldots, y_{N,N,t}^{1/2})$ which $y_{i,\,i,\,t}$ is any univariate GARCH process, while L_t, the time-dependent co nditional correlations matrix, can be specified as:

$$L_t = diag(z_{11,t}^{-1/2},\ldots., z_{NN,t}^{-1/2})Z_t(z_{11,t}^{-1/2},\ldots\ldots, z_{NN,t}^{-1/2}) \quad (6)$$

Here Z_t is a symmetric positive definite matrix that defines the dynamic correlation structure as:

$$Z_t = (1 - a - b)\overline{Z} + au_{t-1}u_{t-1}^{'} + Zb_{t-1} \quad (7)$$

u_i A vector of the standardized residuals z is the unconditional variance matrix of u_i, and a and b are non-negative autoregressive and variance coefficients satisfying $a + b < 1$. Subsequently, DCC between the Bitcoin returns and uncertainty measures can be expressed as:

$$p_{ij,t} = \frac{q_{ij,t}}{\sqrt{q_{ii,t}q_{jj,t}}} \quad (8)$$

Here ρ_{ij} is the element on the jth line and jth column of the matrix z.

3. Empirical Results

Panel A in Table 2 presents summary statistics concerning the daily Bitcoin returns for all the days, the non-announcement days, the announcement days, and the Federal Reserve's announcement and stimulus proposal days. In general, the average returns are higher during the announcement days than all the days and the non-announcement days. The Federal Reserve's announcement days have more massive average returns. Similarly, the Sharpe ratios are better during the Federal Reserve's announcement days, with an excess return of 0.36. The highest reward-to-risk ratio is also observed during the Federal Reserve's announcement days (0.64). However, the reward for bearing the risk (Sharpe ratio) is negative for all the days, the non-announcement days, and as well as during the stimulus proposals. This indicates that investing in Bitcoin is associated with some risk during the COVID-19 pandemic. Besides, a positive association between risk and return is evident during the announcement days. The VIX exhibits a higher negative correlation with the Bitcoin returns. Panel A in Table 1 also presents summary statistics concerning the uncertainty measures (VIX and EPU). Considering the volatility indicator, the change in the EPU is more significant than the difference in the VIX. Panel B of Table 1 displays the correlation using both pairwise correlation and two non-parametric correlation analyses, namely Spearman's and Kendall's. The three methods reveal a negative correlation between Bitcoin and uncertainty measures (VIX and EPU).

Table 2: Summary statistics of daily Bitcoin return and changes in uncertainty.

Table 2 Panel A	**N**	**Mean**	**Std**	**Skewness**	**Excess Kurtosis**	**Jarque-Bera**	**Sharp ratio**
All days 2, Jan 2020-7 October	281	0.0014	0.0422	-4.474	52.41	32977	–0.0985
Non-announcement days	267	0.0015	0.0430	–4.358	51.41	30030	–0.0990
Announcement days	14	0.0278	0.0719	0.2592	–0.288	0.2053	–0.0075
Fed announcements	10	0.0432	0.0624	0.950	–0.524	1.619	0.6411
Stimulus	4	–0.0107	0.0892	0.3043	–1.144	0.2797	–0.1316
Change EPU	281	0.003	0.349	-0.238	3.035	110.08	
Change VIX	281	0.003	0.079	1.884	7.624	843.77	
	ARCH (5)	Q(20)	Q220)	ADF (5)			
BITCOIN	0.279	22.43	6.562	–10.30			
Change EPU	11.33	62.74	48.38	–14.45			
Change VIX	8.791	23.19	98.31	–8.620			
Table 2 Panel B		**Correlation**					
	Pairwise	Spearman's	Kendall's				
Bit and CEPU	0.011	–0.016	–0.010				
Bit and CVIX	–0.396	–0.255	–W0.174				
VIX and EPU	–0.174	–0.069	–0.044				

Table 3: FOMC announcements and Bitcoin return.

Date	**E .w**	**-2**	**-1**	**0**	**1**	**2**	**CAR (-1,0)**	**CAR (0,+1)**
29-Jan-20	MAR	0.038	0.0519	-0.0029	0.0222	-0.0151		
	CAR	0.065	0.117	0.114	0.137	0.122	0.049	0.019
3-Mar-20	MAR	0.0355	-0.0096	-0.0041	0.0366	0.0366		
	CAR	0.009	-0.001	-0.005	0.032	0.068	0.032	-0.014
15-Mar-20	MAR	0.119a	-0.065b	0.0367	-0.070b	0.0418		
	CAR	-0.255	-0.321	-0.284	-0.354	-0.312	-0.034	-0.029
19-Mar-20	MAR	0.0422	0.0026	0.182a	0.0013	-0.0021		
	CAR	-0.056	-0.053	0.129	0.13	0.128	0.183a	0.185a
23-Mar-20	MAJ	-0.0016	-0.057b	0.1011a	0.0502a	-0.0074		
	CAR	0.186	0.129	0.23	0.28	0.273	0.151a	0.044
31-Mar-20	MAR	-0.0342	-0.050c	0.0867a	0.0023	0.0271		
	CAR	-0.071	-0.121	-0.034	-0.032	-0.005	0.089a	0.036
29-Apr-20	MAR	0.0146	0.001	0.1269a	-0.0166	0.0234		
	CAR	0.047	0.048	0.175	0.158	0.181	0.11c	0.128b
10-Jun-20	MAR	-0.0013	-0.0002	0.005	-0.0582	0.0144		
	CAR	-0.013	-0.013	-0.009	-0.067	-0.052	-0.053	0.005
29-Jul-20	MAR	0.1077b	-0.009	0.0153	-0.001	0.0172		
	CAR	0.136	0.127	0.142	0.141	0.158	0.014	0.006
27-Aug-20	MAR	-0.0368	0.0086	-0.0165	0.0172	-0.0052		
	CAR	-0.028	-0.019	-0.035	-0.018	-0.023	0.001	-0.008
16-Sep-20	MAR	0.0319	0.0082	0.0138	-0.005	-0.0031		
	CAR	0.02	0.028	0.042	0.037	0.034	0.009	0.022

Note: (MAR) stands for Mean Adj, AR. (CAR) cumulative abnormal returns. a, b and c significant 1%, 5%, and 10% respectively.

The present study considers all the FOMC announcements (both scheduled and unscheduled) and press releases to examine the Bitcoin abnormal returns. Table 3 presents the results of this examination. It can be observed that the Bitcoin returns usually are (mostly) positive during the announcement period. Highly positive and statistically significant abnormal returns can be observed on March 19, March 23, March 31, and April 30. Interestingly, the impact on the abnormal return is high when the Fed's announcement is related to emergency lending facilities.[5] And when there is simultaneous news released regarding stimulus measures. The most massive positive and statistically significant abnormal return can be seen on March 19 (reaching as high as 18%), followed by March 31 (reaching 10%). It should be remembered that

[5] Federal Reserve's crisis-fighting tools during the corona crisis include setting up corporate credit facilities, municipal liquidity, main street facilities, paycheck protection program liquidity, money market mutual fund liquidity, commercial paper funding, primary dealer credit, and term asset-backed securities loan.

on 19 March, the Republicans proposed \$1,200 stimulus packages for individuals, while on March 27, President Trump agreed to institute the highest stimulus package on record, which was worth \$2 trillion, while several other countries announced stimulus package on March 30. Our event window considers the five days before and after each announcement to capture investors' expectations regarding the asset returns and their reactions to the announcements. The most volatile abnormal return is evident during the event window of the March 15 announcement, where one day before and after the announcement, there were significant negative abnormal returns. However, on the announcement day itself (March 15), there was a positive abnormal return, which might be because the FOMC announcement sent a positive signal to Bitcoin investors. Table 3 also presents the cumulative abnormal returns (–2, 2), which are mostly negative, indicating a fall in investors' risk appetite. On the June 10, March 15, and March 31 announcements days, there were positive abnormal returns even though the cumulative abnormal returns were negative.

Table 4 displays the Bitcoin returns' reactions to the stimulus packages for the period events window (–2, +2). The empirical results show that the Republicans' proposal for the second round of \$1,200 stimulus checks led to positive and significant Bitcoin abnormal returns. During the event day (July 27), the abnormal return reached 10.7%, while the cumulative average return was positive for the events window period (–2, +2). In addition, the cumulative return (–1, 0) was positive and significant. This indicates that the new stimulus check proposal positively affected the abnormal returns. However, looking at the event window of the September 3 announcement, there was an adverse reaction to the stimulus package announcements. There was a press release (see Table 1) concerning disagreement among the political parties regarding dispute among the stimulus checks. During this period, a negative and significant abnormal return was noticed on September 3, reaching as high as 10.4%. The cumulative abnormal return was negative before, on, and after the news release, and the cumulative return (–1, 0) was negatively significant. On May 15, the House passed the Hero Act, which had a negative impact on the abnormal return, albeit an insignificant one, which may be because the market participants were waiting on the Senate decision before reacting to it. Next, we investigated the reaction on September

Table 4: Stimulus packages and Bitcoin abnormal returns.

	E.w	**-2**	**-1**	**0**	**1**	**2**	**CAR(-1,0)**	**CAR(0,+1)**
15-May	MAR	0.0531	0.0503	-0.0414	0.0055	0.0316		
	CAR	-0.0276	0.0227	-0.0188	-0.0133	0.0182	-0.0360	0.0088
27-Jul	MAR	0.012	0.021	0.107b	-0.01	0.015		
	CAR	0.022	0.043	0.15	0.141	0.155	0.098	0.128b
3-Sep	MAR	0.024	-0.048	-0.104b	0.025	-0.034		
	CAR	0.032	-0.016	-0.119	-0.095	0.128	-0.079	-0.151b
11-Sep	MAR	0.0097	0.0105	0.0024	0.0027	-0.0126		
	CAR	0.0025	0.0131	0.0155	0.0182	0.0056	0.0051	0.0129

Note: (MAR) stand for Mean Adj, AR. (CAR) cumulative abnormal returns. a, b and C significant 1%, 5% and 10% respectively.

11 when the Senate failed to pass the second stimulus package and found that there was a positive abnormal return; however, it was not significant because investors had already reacted on September 3.

Table 5 presents the effects of the FFTR and uncertainty measures on the abnormal returns using conventional OLS methods. Since the start of the COVID-19 pandemic, the Fed adjusted the policy rate two (on March 3 and 15), and the OLS regression result reveals the negative impact of the FFTR on the abnormal returns. However, the effect is insignificant, which shows the impact of monetary policy changes to be negligible about the Bitcoin price, which is in line with the findings of Tomása and Ibañez (2018). Further, looking at the effects of the VIX and EPU on the abnormal returns, it can be observed that during the Fed announcement days, uncertainly had a negative effect (VIX and EPU) and, further, that the negative association is higher with the VIX, implying that stock market fear is negatively correlated with Bitcoin returns.

Table 5 shows the OLS and quantile regressions' estimation results for the Bitcoin returns and the uncertainty measures (VIX and EPU). The OLS regression results reveal a negative relation between the uncertainty measures and the Bitcoin returns. The VIX has a higher negative and statistically significant impact on the Bitcoin returns, whereas the EPU has a negative but insignificant impact. The QR results also reveal a negative and significant association between the VIX and Bitcoin during the upper and lower quantile regressions, which implies that Bitcoin responds negatively to uncertainty in the stock market during the bear bull market. This result is in line with the findings of Shaikh (2020). However, considering the impact of the EPU on the Bitcoin returns, it can be observed at the lowest quantile that there is a positive relationship with the Bitcoin returns, although, after 0.20, the effect turns negative up to the upper quantile. Therefore, the Bitcoin returns are asymmetrical at the two extreme quantiles, making it difficult to predict the Bitcoin returns. Our results partially contradict Demir's (2020) findings, who noted the positive impact of

Table 5: Ordinary least squares and quantile regression.

		Intercept	VIX	EPU	$R_{Btc(-1)}$
OLS		0.0024	-0.2122[a]	-0.0087	-0.1306[b]
LQR	τ quantile				
	0.10	-0.0292[a]	-0.1767[a]	0.0042	-0.0966
	0.20	-0.0149[a]	-0.1480[a]	0.0039	-0.1851[a]
	0.30	-0.0065[a]	-0.1231[a]	-0.0041	-0.1698[a]
	0.40	-0.0029[b]	-0.1245[a]	-0.0038	-0.1598[a]
	0.50	0.0026[c]	-0.0894[a]	-0.0055	-0.1351[a]
	0.60	0.0066[a]	-0.0852[a]	-0.0035	-0.1417[a]
	0.70	0.0123[a]	-0.1003[a]	-0.0048	-0.1643[a]
	0.80	0.0198[a]	-0.1062[a]	-0.0117	-0.1533[b]
	0.90	0.0365[a]	-0.1434[b]	-0.0203	-0.2014

Note: a, b and C significant 1%, 5%, and 10%, respectively.

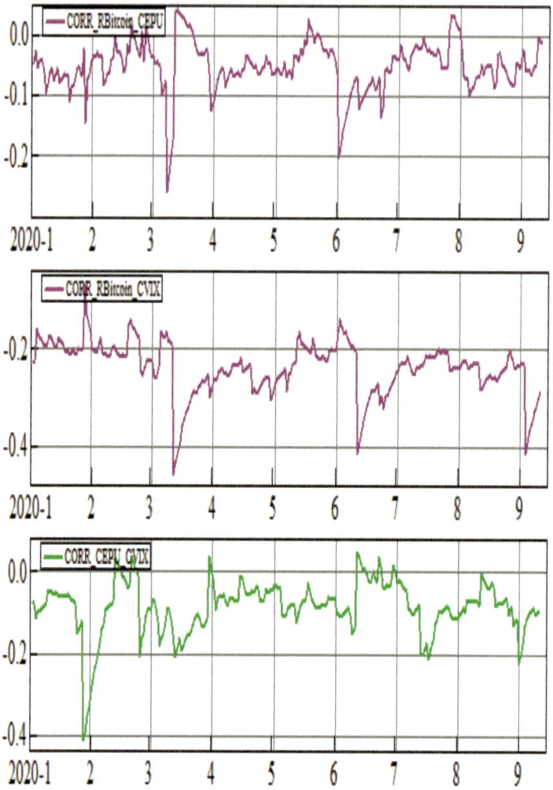

Figure 2: DCC between uncertainty measures and bitcoin returns.

EPU during the extreme quantile (bull market). The OLS and QR estimation results also show that the one-period lag is negatively related to the Bitcoin returns.

Figure 2 displays the time-varying relationships between the Bitcoin returns and the uncertainty measures (VIX and EPU) during the ongoing COVID-19 crisis. This study used the GARCH (1, 1) estimate to generate the DCC between the Bitcoin returns and the uncertainty. Subsequently, the multivariate DCC equation results revealed that a and b are non-negative and, further, the condition $a + b < 1$ is met. Looking at the time-varying relation between the VIX and EPU and the Bitcoin returns, it can be observed that there is a negative time-varying relationship throughout the period. More specifically, during the turbulent period, this relation becomes stronger. For example, in the bull market, the negative association increased. Interestingly, the DCC pattern and trend between the uncertainty measures (VIX and EPU) and the Bitcoin returns are the same throughout the period. Finally, the time-varying relation between the VIX and EPU is negative throughout the sample period.

4. Conclusion and Policy Direction

The present study examined the effects of US equity market uncertainty, US economic policy uncertainty, the Fed's announcements, and US stimulus packages on the Bitcoin returns during the COVID-19 crisis. The study also investigated the time-varying relations between the Bitcoin returns and the VIX and EPU, in addition to the impacts of the uncertainty measures (VIX and EPU) on the Bitcoin returns in bull and bear markets.

The study used the event studies, GARCH-DCC, QR, and conventional OLS methods to achieve the aims mentioned above. The empirical findings of the study are as follows. First, the Bitcoin price reacts to FOMC announcements. The reaction is stronger when the Fed announcements are combined with announcements regarding stimulus packages by Congress and when the Fed announcements are related to crisis-fighting tools (emergency lending facilities). For example, the highest positive abnormal return was noticed on March 19 when the Senate passed the Cares Act stimulus package. Simultaneously, the Fed announced a new operation focused on currency swaps. A highly positive abnormal return was also observed when the Fed announced unlimited asset purchases and a $300 billion credit program for businesses and consumers on March 23. These findings are in line with those of Corbet et al. (2020), who documented that the currency group of digital assets responded to the FOMC announcements.

The proposal of the HEALS stimulus package by the Senate on July 27 led to a positive and significant abnormal return. In contrast, news about disagreements among Senate members regarding the HEALS stimulus package negatively affected the Bitcoin returns. The empirical results also show that during the Fed announcements, investor sentiments, as measured using the VIX, had a negative and significant impact on the abnormal returns. In contrast, changes in the FFTR and EPU had negative effects on the abnormal return.

Second, the study provided evidence at the upper and lower quantiles that the VIX had negative and significant effects on the Bitcoin returns. However, the EPU had positive outcomes at the lower quantile (0.10, 0.20), which then turned negative but insignificant. According to our results, the VIX has a higher predictive power when it comes to Bitcoin returns. Our results partially contradict the findings of Demir et al. (2018), who noted the positive impact of the EPU on Bitcoin during the extreme quantile (bull market) and the negative effects during the standard period. Nonetheless, our findings are in line with those of Shiakh (2020), who found the VIX to affect both quantities negatively.

Next, this study noted a consistent negative time-varying relationship between the Bitcoin returns and the VIX and EPU. There are some similarities in terms of the trends and pattern of the time-varying relations among the Bitcoin returns and the VIX and EPU.

This study has several important implications for investors who wish to benefit from asset allocation and hedging uncertainty. It highlights how an increase in equity market uncertainty is associated with the Bitcoin market. Therefore, investors must take into account equity uncertainty when investing in Bitcoin. Also, this study answers the question as to whether Bitcoin is an efficient market by examining its reaction to the FOMC announcements and stimulus package proposals and determining that the Bitcoin market is efficient. Furthermore, due to the COVID-19 pandemic, governments worldwide are increasing their bailouts and stimulus packages for business, while corporations are increasingly relying on debt finance.

Thus, the growing size of debt financing and the considerable cost of the stimulus packages (mainly funded by printing money) will create a fragile financial market shortly, leading to great uncertainty for investors in the equity markets. Our results indicate that Bitcoin can provide a haven because investors will not be subjected to either currency depreciation or bankruptcy.

References

Baker, S. R., Bloom, N. and Davis, S. J. (2016). Measuring economic policy uncertainty. *The Quarterly Journal of Economics*, 131(4): 1593–1636.

Bein, M. (2017). Time-varying co-movement and volatility transmission between the oil price and stock markets in the Baltics and four European countries. *Engineering Economics*, 28(5): 482–493.

Bernanke, B. S. and Kuttner, K. N. (2005). What explains the stock market's reaction to Federal Reserve policy? *The Journal of Finance*, 60(3): 1221–1257.

Bloom, D. E. and Canning, D. (2004). Epidemics and economics. *Interactions between Global Change and Human Health (Scripta Varia)*, 106: 304–331.

Bloom, D. E., Daniel, C. and Sevilla, J. P. (2018). Epidemics and economics: New and resurgent infectious diseases can have far-reaching economic repercussions. *Finance and Development*, 55(2): 46–49.

Bouri, E., Azzi, G. and Dyhrberg, A. H. (2016). On the return-volatility relationship in the Bitcoin market around the price crash of 2013. Available at SSRN 2869855.

Bouri, E., Das, M., Gupta, R. and Roubaud, D. (2018). Spillovers between Bitcoin and other assets during bear and bull markets. *Applied Economics*, 50(55): 5935–5949.

Bouri, E., Molnár, P., Azzi, G., Roubaud, D. and Hagfors, L. I. (2017). On the hedge and haven properties of Bitcoin: Is it more than a diversifier? *Finance Research Letters*, 20: 192–198.

Brown, S. J. and Warner, J. B. (1985). Using daily stock returns: the case of event studies. *J. Fin. Econ.*, 14(1): 3–31.

Cavallo, E., Galiani, S., Noy, I. and Pantano, J. (2013). Catastrophic natural disasters and economic growth. *Review of Economics and Statistics*, 95(5): 1549–1561.

Cheah, E. T. and Fry, J. (2015). Speculative bubbles in Bitcoin markets? An empirical investigation into the fundamental value of Bitcoin. *Economics Letters*, 130: 32–36.

Claser, A., Fazioli, R. and Vecchia, P. (2014). Progettare la sostenibilità. libreriauniversitaria. it Edizioni.

Corbet, S., Hou, G., Hu, Y. and Oxley, L. (2020a). The influence of the COVID-19 pandemic on asset-price discovery: Testing the case of Chinese informational asymmetry. Available at SSRN 3623274.

Corbet, S., Hou, G., Hu, Y., Larkin, C. J. and Oxley, L. (2020). Any Port in a Storm: Cryptocurrency Safe-Havens during the COVID-19 Pandemic. Available at SSRN 3610461.

Corbet, S., Hou, Y., Hu, Y., Lucey, B. and Oxley, L. (2020b). Aye Corona! The contagion effects of being named Corona during the COVID-19 pandemic. *Finance Research Letters*, 101591.

Corbet, S., Larkin, C., Lucey, B., Meegan, A. and Yarovaya, L. (2020). Cryptocurrency reaction to fomc announcements: Evidence of heterogeneity based on blockchain stack position. *Journal of Financial Stability*, 46: 100706.

Corbet, S., Lucey, B. and Yarovaya, L. (2018). Datestamping the Bitcoin and Ethereum bubbles. *Finance Research Letters*, 26: 81–88.

Demir, E., Gozgor, G., Lau, C. K. M. and Vigne, S. A. (2018). Does economic policy uncertainty predict the Bitcoin returns? An empirical investigation. *Finance Research Letters*, 26: 145–149.

Engle, R. F. (2002). Dynamic conditional correlation—A simple class of multivariate GARCH.

Fama, E. F., Fisher, L., Jensen, M. C. and Roll, R. (1969). The adjustment of stock prices to new information. *International Economic Review*, 10(1): 1–21.

Fan, V. Y., Jamison, D. T. and Summers, L. H. (2018). Pandemic risk: how large are the expected losses? *Bulletin of the World Health Organization*, 96(2): 129.

Fang, L., Bouri, E., Gupta, R. and Roubaud, D. (2019). Does global economic uncertainty matter for the volatility and hedging effectiveness of Bitcoin? *International Review of Financial Analysis*, 61: 29–36.

Feng, W., Wang, Y. and Zhang, Z. (2017). Informed trading in the Bitcoin market. *Finance Research Letters*, 26: 63–70.

Glaser, F., Zimmermann, K., Haferkorn, M., Weber, M. C. and Siering, M. (2014). Bitcoin-asset or currency? Revealing users' hidden intentions. Revealing Users' Hidden Intentions ECIS. Available at SSRN: 2425247.

Global Economic Policy Uncertainty Index. Economic Policy Uncertainty, <http://www.policyuncertainty.com/global_monthly.html>.

Goodell, J. W. (2020). COVID-19 and finance: Agendas for future research. *Finance Research Letters*, 101512.

Goodell, J. W. and Goutte, S. (2020). Co-movement of COVID-19 and Bitcoin: Evidence from wavelet coherence analysis. *Finance Research Letters*, 101625.

Gozgor, G., Tiwari, A. K., Demir, E. and Akron, S. (2019). The relationship between Bitcoin returns and trade policy uncertainty. *Finance Research Letters*, 29: 75–82.

Jareño, F., de la O González, M., Tolentino, M. and Sierra, K. (2020). Bitcoin and gold price returns A quantile regression and NARDL analysis. *Resources Policy*, 67: 101666.

Jubinski, D. and Lipton, A. F. (2012). Equity volatility, bond yields, and yield spreads. *Journal of Futures Markets*, 32(5): 480–503.

Katsiampa, P. (2017). Volatility estimation for Bitcoin: A comparison of GARCH models. *Economics Letters*, 158: 3–6.

Kim, T. (2017). On the transaction cost of Bitcoin. *Finance Research Letters*, 23: 300–305.

Koenker, R. and Bassett Jr, G. (1978). Regression quantiles. *Econometrica: Journal of the Econometric Society*, 33–50.

Kristoufek, L. (2013). BitCoin meets Google Trends and Wikipedia: Quantifying the relationship between phenomena of the Internet era. *Scientific Reports*, 3(1): 1–7.

Lagoarde-Segot, T. and Leoni, P. L. (2013). Pandemics of the poor and banking stability. *Journal of Banking and Finance*, 37(11): 4574–4583.

Lee, J. W. and McKibbin, W. J. (2004, April). Estimating the global economic costs of SARS. In *Learning from SARS: Preparing for the Next Disease Outbreak: Workshop Summary* (pp. 92–109). Washington, DC: National Academies Press.

López-Cabarcos, M. Á., Pérez-Pico, A. M., Piñeiro-Chousa, J. and Šević, A. (2019). Bitcoin volatility, stock market and investor sentiment. Are they connected? *Finance Research Letters*, 101399.

Madhav, N., Oppenheim, B., Gallivan, M., Mulembakani, P., Rubin, E. and Wolfe, N. (2017). Pandemics: Risks, impacts, and mitigation.

Mokni, K., Ajmi, A. N., Bouri, E. and Vo, X. V. (2020). Economic policy uncertainty and the Bitcoin-US stock nexus. *Journal of Multinational Financial Management*, 100656.

Pacicco, F., Vena, L. and Venegoni, A. (2019). Market reactions to ECB policy innovations: A cross-country analysis. *Journal of International Money and Finance*, 91: 126–137.

Pyo, S. and Lee, J. (2019). Do FOMC and macroeconomic announcements affect Bitcoin prices? *Finance Research Letters*, 101386.

Sarwar, G. (2017). Examining the flight-to-safety with the implied volatilities. *Finance Research Letters*, 20: 118–124.

Shahzad, S. J. H., Bouri, E., Roubaud, D., Kristoufek, L. and Lucey, B. (2019). Is Bitcoin a better safe-haven investment than gold and commodities? *International Review of Financial Analysis*, 63: 322–330.

Shaikh, I. (2020). Policy uncertainty and bitcoin returns. Borsa Istanbul Review.

Tam, C. C., Khan, M. S. and Legido-Quigley, H. (2016). Where economics and epidemics collide: Migrant workers and emerging infections. *The Lancet*, 388(10052): 1374–1376.

Urquhart, A. (2017). Price clustering in Bitcoin. *Economics Letters*, 159: 145–148.

Vidal-Tomás, D. and Ibañez, A. (2018). Semi-strong efficiency of Bitcoin. *Finance Research Letters*, 27: 259–265.

Wang, G. J., Xie, C., Wen, D. and Zhao, L. (2019). When Bitcoin meets economic policy uncertainty (EPU): Measuring risk spillover effect from EPU to Bitcoin. *Finance Research Letters*, 31.

Wang, P., Li, X., Shen, D. and Zhang, W. (2020). How does economic policy uncertainty affect the bitcoin market? *Research in International Business and Finance*, 101234.

Weber, B. (2016). Bitcoin and the legitimacy crisis of money. *Cambridge Journal of Economics*, 40(1): 17–41.

Chapter 14

Socio-economic Effects and Evidence-based Households' Response to Covid-19 in Nigeria

Olubunmi O. Alawode[1,*] and *Yetunde O.M. Oladokun*[2]

1. Introduction

Coronavirus disease (COVID-19) was first reported to the World Health Organization (WHO) in December 2019 in Wuhan (Nigerian Centre for Disease Control 2020a). The outbreak of coronavirus in Wuhan had killed more than eighteen hundred and infected over seventy thousand individuals within the first fifty days of the epidemic; about 188 countries and territories have confirmed cases of about 20.7 million including more than 751,000 deaths (Muhammad et al. 2020).

African countries are not left out in the pandemic as numbers of confirmed cases keep rising rapidly with particularly severe situation in South Africa, Egypt, Algeria, Morocco, and Cameroon (Peterson 2020a). According to WHO, as of 11 August 2020, Africa had a cumulative total of 905,782 confirmed COVID-19 cases, having more than half 63% (566,109) of all the confirmed cases in the region. There are also large thousands of confirmed cases in Nigeria (47,290), Ghana (41,404), Algeria (36,204), Kenya (27,425), Ethiopia (24,175), Cameroon (18,263), Côte d'Ivoire (16,847), Madagascar (13,317), and Senegal (11,380).

Nigerian Centre for Disease Control (NCDC) reported the first (index) case of COVID-19 on 27 February 2020. According to NCDC (2020b), a second confirmed case was reported on the 9 March 2020, who was a contact of the index case. As of 11 August 2020, samples tested were 338,084 with 47,290 confirmed cases in 37 states including FCT, 33,609 discharged, and 956 deaths. The Nigerian government, in a way to reduce the spread of the disease, had to introduce containment measures

[1] Department of Agricultural Economics, University of Ibadan, Nigeria.
[2] Cocoa Research Institute of Nigeria, Ibadan, Nigeria.
* Corresponding author: busolaferanmi@gmail.com, oo.alawode@ui.edu.ng

such as restrictions on the movement of people, limitations to numbers of people in gatherings, closure of air borders, extension of its land border closures (to include human traffic), and putting a curfew in place (World Bank 2020). With voluntary behavioural changes by individuals and firms, the containment measures had a great impact on all sectors of the economy.

From the global perspective, COVID-19 has been the major driver of economic hardship in the past few months (Bhuiyan et al. 2020); the economic impact has been global in scale and magnitude, leading to a global recession (Fernandes 2020), which has been the worst since the World War II (World Bank 2020). International Labour Organization (ILO 2020) estimated that a global workforce of 3.3 billion has been massively affected due to containment measures against the coronavirus spread. According to United Nations Development Programme (UNDP 2020), reduction in air traffic with 98 percent fall in international tourist numbers made the tourism sector one of the worst hit by the COVID-19 outbreak with a forecasted loss of US$910 billion to US$1.2 trillion in export revenues, and an estimated job loss of around 100 and 120 million. An estimated 1.6 billion informal workers, which is about half the global workforce lost 60 percent of their income, with little to no savings and no access to social protection (ILO 2020).

Covid-19 impacted food security by disruption of food systems, which was consequent upon the negative effects of lockdowns on households' incomes and their physical access to basic services and food. COVID-19 and its underlying consequences disrupted food production, processing, and marketing (Devereux et al. 2020). Coronavirus pandemic is a real challenge to the UN Sustainable Development Goal One of ending poverty by 2030 as the number of people living in poverty is expected to increase by 420–580 million relative to the latest official figures for 2018 (Sumner et al. 2020).

In Nigeria, from mid-March 2020, the government put in place several measures such as lockdowns, movement restrictions, social and physical distancing measures, and other public health measures in order to curb and mitigate the impact of COVID-19. These measures, according to the Nigerian Bureau of Statistics (NBS 2020a), have disrupted the economy and the supply of employment, exacerbated poverty, and impacted food security negatively. Nigeria's GDP suffered a 34.1 percent loss due to COVID-19, amounting to USD 16 billion, with two-thirds of the losses coming from the service sector, and agriculture sector suffering a 13.1 percent loss in output (USD 1.2 billion). Also, NBS (2020a) specified that employment has been impacted with 42 percent reduction in the number of workforce. The economic impacts of Covid-19 include a 14 percent temporary increase in the poverty headcount rate for Nigeria, implying that 27 million additional people fell below the poverty line during lockdown. Key macroeconomic indicators are deeply impacted with reduction in tax, remittances, and employment (Idris and Oruonye 2020).

As noted by Human Rights Watch (HRW 2020), the Nigerian government, through the Central Bank of Nigeria (CBN), announced the release of 50 billion naira as a special relief package to cushion the effect of the pandemic on households. The government set up 1 trillion naira in loans to support local manufacturing and production sector. Other economic stimulus measures put in place by the Nigerian government include one year extension on principal repayments for CBN intervention facilities, reduction of interest rate on intervention loans from 9 percent to 5 percent, support of pharmaceutical companies with 100 billion naira intervention fund, credit assistance for health industry as well as extension of more credit to private sector (Chukwuka and Mma 2020).

Employment retention is also being promoted by the federal government to discourage private businesses from laying off workers. This was presented as an Emergency Economic Stimulus Bill 2020 to the lawmakers. According to the bill, any company that did not retrench staff between the 31st of March 2020 and the 31st of December 2020 (except for reasons related to a breach of Labour Act) would get a 50 percent tax refund. This bill was later passed by the National Assembly in late March, 2020. These are examples of the proactive steps by the government to protect jobs and take financial burden off the citizens.

Also, following the health and economic emergencies occasioned by the COVID-19 pandemic, the federal government established the Economic Sustainability Committee (ESC) on 30 March 2020, inaugurated by President Muhammadu Buhari. According to the Nigeria Economic Sustainability Plan document (NESP 2020), the ESC would develop a clear Economic Sustainability Plan in response to challenges posed by the COVID-19 Pandemic, identify fiscal and monetary policies, and articulate specific measures to support the thirty-six states and the Federal Capital Territory (FCT), and come up with clear-cut strategy to keep existing jobs and creating new ones. To achieve these, the ESC would identify required legislative support to deliver the Plan and secure sufficient resources to fund the Plan. The NESP is expected to be the post COVID-19 economic recovery plan (bouncing back), covering all the sectors of the economy, and supporting the economy diversification bid into the non-oil sectors such as agriculture, solid minerals and tourism.

The pandemic significantly affected farm labour, transportation and security, while food security was threatened by insufficient labour, poor transportation, low farmers' morale and poor farm co-ordination (Inegbedion 2020). Food insecurity during the pandemic was significantly occasioned by an increase in prices, fall in income, and adverse financial well-being (Akinleye et al. 2020). The Federal Government distributed palliatives in terms of cash transfers and in-kind assistance called 'palliative stimulus packages' using the socio-register of vulnerable households, who were referred to as 'the poorest of the poor'. It was found by Ojo (2020) that eight out of every ten respondents did not have access to palliative stimulus packages from the federal government, which is expected to serve as safety nets and social protection for individuals and households. On coping strategies by households in the Oyo State, Ojo (2020) found that households had to rely on taking less preferred food, borrow or beg for food, buy food on credit, allow household members to eat elsewhere, and limit or restrict adult portions of food or meals.

The spread of COVID-19 into communities and the restrictive measures to contain it disrupted interventions. For example, the COVID-19 outbreak in Bornu, Anambra, and Yobe impacts the livelihoods of 7 million people, thus increasing the number of food insecure individuals by 3.4 million (UNDP 2020). Abdul (2020) identified the impact of the pandemic on rural households; increase in vulnerability, shortage of farm labour, increase in transport cost and postharvest loss, low purchasing power, domestic violence, poor market linkages, and worsening healthcare system. However, as the lockdown postponement continued, Nigerians found ways to cope with it. This paper examines the socio-economic effects and evidence based on the households response to the COVID-19 shock in Nigeria.

1.1 Research questions

This paper seeks to provide answers to the following research questions:

i. How did households react to the COVID-19 pandemic?
ii. Did the COVID-19 pandemic affect households' access to basic services and food security?
iii. What were the shocks that affected households' livelihood during COVID-19?
iv. How did households survive the pandemic?

1.2 Research objectives

This paper broadly examines the socio-economic effects of the COVID-19 pandemic and households' response to it in Nigeria. Specifically, this paper achieves the following objectives:

i. Examines the knowledge and attitude of households to COVID-19.
ii. Analyses the effects of COVID-19 on households' access to basic services and food security.
iii. Identifies the adverse effects on the households' livelihoods during COVID-19.
iv. Assesses coping strategies to the adverse impact of the pandemic and social safety nets provided for the survival of the households.

2. Empirical Review

Peterson (2020a) examined the socio-economic impact of COVID-19 and the policy response in African countries. Discourse analysis was conducted using information obtained from several reputable sources such as the World Health Organization (WHO), United Nations Educational, Scientific and Cultural Organization (UNESCO), and information obtained from the media and other public sources. He showed that the coronavirus outbreak affected the social and economic well-being of most citizens in many African countries during the period even though the social distancing measures were temporary. According to him, the coronavirus outbreak unveiled the fact that a biological crisis can transform into a sociological subject. He noted that the most important sociological consequence of the outbreak for African citizens was the creation of social anxiety among families and households in the region. Peterson recommended that policymakers should enforce social policies that will unite communities in bad times with a view to reducing social anxiety. He concluded that the right mix of monetary and fiscal tools will be needed to revive the economy of African countries during the crisis and also when the coronavirus crisis ends.

Asongu and Adesoji (2020) studied the economic consequences of the COVID-19 pandemic in Nigeria and presented an early review of the macroeconomic impact of it. From their findings, while the number of infected cases reflects significant correlations with economic activity from the perspective of a trend analysis, the estimates from dynamic ordinary least squares (DOLS) show that connections between the number of confirmed cases and attendant macroeconomic outcomes are largely insignificant. The study therefore showed that the COVID-19 pandemic has insignificant negative impacts on basic macroeconomic variables in Nigeria such as inflation, employment, exchange rate, and GDP growth, among others. This means that there is a gap between the impact of Covid-19 and its economic causality. They concluded that since the government has engaged in Nigeria Economic Sustainable Plan (NESP 2020), a post COVID-19 recovery plan, it is hoped that the policies would be effectively implemented so as to provide the critical measure in repositioning the country's economy on the path towards inclusive and sustained economic development.

Omaka-Amari et al. (2020) examined the coronavirus Covid-19 pandemic in Nigeria; specifically, the preventive and control challenges within the first two months of outbreak. Data from the daily updates of Nigeria Centre for Disease Control (NCDC) were graphically used to describe the trend of spread while facts from both verified online and print media reports on COVID-19 were used to assess the challenges. The outcome of the study showed a steady increase in COVID-19 from one case on the 27th of February 2020 to 1,932 confirmed cases, 58 fatalities, and 319 discharged cases by 30th of April 2020. They found that COVID-19 preventive measures in the country included hand washing, use of sanitizers, wearing of face masks, and implementing lockdowns and social distancing. The factors that made government preventive efforts less effective were a poor compliance attitude, selective lockdown, social media interference, misconceptions and myths, stigmatization, fear, inadequate health facilities, and distrust of the government. They recommended that there is need to amend the adverse effects of COVID-19 misconceptions and myths through evidence-based campaigns using all sources of information.

Shabir and Aijaz (2020) carried out a research on COVID-19 pandemic; specifically from an African perspective. They aimed at providing a comprehensive update of the COVID-19 pandemic in Africa and highlighted the main topics which include aetiology, epidemiology, vulnerability, preparedness, and economic impact of the pandemic on the African continent. They searched and reviewed published work on COVID-19 from the 1st of January 2020 to 18th of April 2020. They found the African continent to be the last and the least to be affected by COVID-19 pandemic as of when the research was carried out but could not conclude the true epidemiology of COVID-19 in the continent due to inadequate testing capacity for COVID-19, which makes the true number of cases undetected.

In another study, Peterson (2020b) assessed the relationship between COVID-19 pandemic and the economic crisis caused by it—the Nigerian experience and structural causes. He found that COVID-19 outbreak affects Nigeria's economy mainly in five aspects; reduction in the borrowers' capacity to service loans, which gave rise to non-performing loans that depressed banks' earnings and eventually impaired bank soundness and stability, oil demand shocks, which was reflected in the sharp decline in oil price, and supply shocks in the global supply chain as many importers shut down their factories and closed their borders, particularly to China. Also, the national budget was affected, and the COVID-19 pandemic affected the Nigerian stock market. According to him, the government made efforts to reduce the spread of the virus but proved futile due to some structural factors like poor public health infrastructure, weak and underdeveloped digital economy, and lack of social welfare program. He concluded that the crisis period is a time for the nation to enrich its physical, spiritual, and emotional health, and not just focusing on avoiding the coronavirus.

3. Research Methodology

3.1 Study area

The study area is Nigeria and the report in this paper is based on the Nigeria COVID-19 National Longitudinal Phone Survey (NLPS Baseline) conducted in 2020 by the National Bureau of Statistics (NBS 2020b). Nigeria is divided into thirty-six states and the Federal Capital Territory (FCT, Abuja). It has rural and urban sectors, as well as six geopolitical zones (North Central, North East, North West, South East, South South and South West). In sub-Saharan Africa, Nigeria was one of the first few countries to report cases of COVID-19 and has experienced increasing cases and fatalities (NLPS 2020). As a result of this, strict measures were taken by the Nigerian Government to contain the spread of the virus.

Data sources

The secondary data used were the Nigeria COVID-19 National Longitudinal Phone Survey (NLPS Baseline), which was released in June 2020 as the first round of the Covid-19 NLPS. The survey has national coverage. The data were collected based on the households involved in the Wave 4 of the General Household Survey (GHS) Panel in 2018/2019. The GHS-Panel sample included 4,976 households that were interviewed in the post-harvest event of the fourth wave in January/February 2019. The GHS sample households are representative nationally since they cut across the six geopolitical zones. The NLPS baseline survey was leveraged on the fact that the GHS sample households were interviewed a year before the COVID-19 pandemic. For the NLPS survey, 3,000 households were selected from the GHS Wave 4 frame, and the final successful sample comprised 1,950 households. The NLPS data covered interview responses from households on a monthly basis to identify and reflect on the key areas where individuals and households were affected by COVID-19.

This paper presents report on the knowledge and attitude of individuals and households on COVID-19, impact of COVID-19 on access to basic services and food security, shocks to households' livelihood during COVID-19, the coping strategies by individuals and households, as well as the social safety nets.

Data analysis

This paper uses the NLPS baseline data to address the research questions and achieve the objectives of this paper. Frequency counts and percentages were used to analyze the responses of households to the key variables covered by the data; knowledge and attitude of households to COVID-19, access to basic services, food security, adverse effects during COVID-19, as well as coping strategies and social safety nets. The descriptive results are presented in frequency distribution tables and charts. Cross-tabs were done to establish the relationships between variables to account for the magnitude of the effects of COVID-19 on food security of households by sectors and geopolitical zones in Nigeria. The results are also presented in tables and charts.

4. Results and Discussions

Results of the analysis of Nigeria COVID-19 National Longitudinal Phone Survey to examine the socio-economic effects of Covid-19 pandemic and households' response to Covid-19 shock in Nigeria are presented in this section.

4.1 Knowledge and attitude of households to COVID-19

The control of the coronavirus pandemic, according to World Health Organization (WHO) and Nigerian Centre for Disease Control (NCDC), requires that people, all over the world, understand the way the disease is spread and the measures needed to contain it. This underscores the importance of the knowledge and attitude of the people of every country to the pandemic. The results on the knowledge and attitude of Nigerian households to the coronavirus disease are presented in Table 1 and Figure 1. Results show that almost all households (99.9%) have heard about coronavirus disease. This is expected since it is a pandemic and the containment measures in Nigeria reached all nooks and crannies, with no movement except for basic services of health and food. Awareness is the first critical stage to overcoming the spread of the disease.

The second important step is the attitude of the people to the disease in terms of the rules and guidelines to curtail the spread of the virus. Results also show that majority of the households observed the guidelines of hygiene and social distancing;

Table 1: Knowledge and attitude of Nigerian households to COVID-19.

Knowledge/Attitude	**Number of Households (N = 1950)**	**Percentage**
Have you heard about coronavirus?	1948	99.9
Did you wash your hands with soap more often than you used to?	1837	94.2
Did you avoid handshakes/physical greetings?	1849	94.8
Did you avoid groups of more than 10 people?	1833	94.0

Source: Computed from NBS COVID-19 NLPS Data (2020).

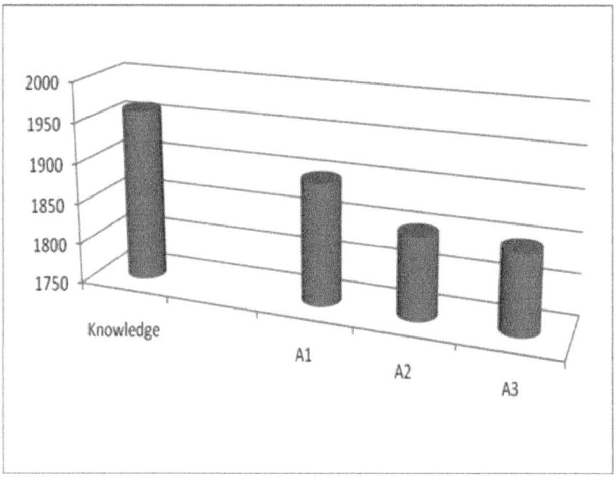

Figure 1: Knowledge and attitude of households to COVID-19.
Source: Computed from NBS COVID-19 NLPS Data (2020).
Note: Knowledge - Have you heard about the coronavirus?
A1 - Did you wash your hands with soap more often than you used to?
A2 - Did you avoid handshakes/physical greetings?
A3 - Did you avoid groups of more than 10 people?

94.2% washed their hands more often with soap and water, 94.8% avoided handshake and physical greetings, and 94.0% avoided groups of more than 10 people (observing social distancing and use of nose masks). As good as these are, it is still important that all people observe these rules because a single infected person can still infect many others.

Figure 1 also shows that despite the fact that almost all the households are aware of the disease and have experienced the restrictions associated with the containment of the pandemic, reduced number of households reported observing all the basic rules of hand washing, avoiding handshakes and physical greetings, and avoiding groups of more than 10 people.

4.2 Effects of COVID-19 on households' access to basic services and food security

One of the ways by which COVID-19 has impacted households in Nigeria is in their access to basic services and food security. This is due to the fact that there were strong restrictions on movement of people, shops, and markets to contain the spread of the virus. Although there were exceptions for movement of people that perform health and food services.

Households' access to basic services

Figure 2 clearly shows that not all households required the listed basic necessities during the lockdown period probably because the households that could afford to get full stock of supplies have done so at the time of announcement of lockdowns, leading to panic buying. Also the figure shows that not all households that required access actually had access to basic services. The figure shows the visual representations of the households that needed access to basic necessities of food and other households' needs as compared to those that actually had access. The differences between the two blocks of blue and red colours show the impact of COVID-19 on households' access to basic necessities. The basic necessities include medicine, soap, cleaning supplies, rice, beans, cassava, yam, guinea corn/sorghum, medical treatment, and financial services in the use of bank/money agent/Automated teller machine (ATM). Figure 2 also shows that the most required necessity was soap while the least required was medical treatment. High soap requirements could be due to the higher requirements for cleanliness and hygiene to contain the spread of the virus. The least requirement for medical treatment could be because households were constrained to be in their homes and having compulsory rest at the time of lockdown.

In quantitative terms, Table 2 shows that more than half of the households required access to soap (74.5%), rice (58.8%), and cleaning supplies (58.1%). Also, households actually got access mostly to soap (67.4%), cleaning supplies (47.7%) and medicine (40.4%). Access was possible because the government eased lockdowns for two or three days for households to re-stock their homes with whatever they have. Lack of access to these basic needs has been adduced to the basic necessities being out of stock, too high prices, limited or no transportation, and households having no money to buy. Results also show the effect of COVID-19 on the access to these basic needs (difference between the proportion that needed the services and those

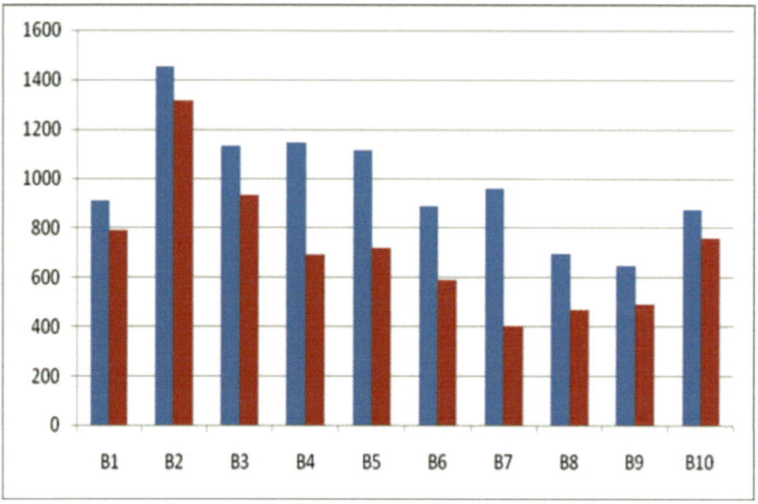

Figure 2: Effect of COVID-19 on households' access to basic services.
Source: Computed from Nigeria COVID-19 NLPS Data (2020).
Note: Colour Blue: Households needed the basic services.
Colour Red: Households were able to get the basic services.
B1–B10: Basic services: medicine, soap, cleaning supplies, rice, beans, cassava, yam, guinea corn/ sorghum, medical treatment, bank/money agent/atm).

Table 2: Effect of COVID-19 on households' access to basic services.

Need for basic services in the past seven days	Number of households (N = 1950)	Access to basic services in the past seven days	Number of households (N = 1950)	Difference (Effect of COVID-19)
Household needed to buy medicine	908 (46.5)	Household was able to buy medicine	789 (40.4)	119 (6.1)
Household needed to buy soap	1454 (74.5)	Household was able to buy Soap	1314 (67.4)	137 (7.0)
Household needed to buy cleaning supplies	1133 (58.1)	Household able to buy cleaning supplies	931 (47.7)	202 (10.4)
Household needed to buy rice	1146 (58.8)	Household was able to buy rice	690 (35.4)	456 (23.4)
Household needed to buy beans	1113 (57.1)	Household was able to buy bean	719 (36.9)	394 (20.2)
Household needed to buy cassava	889 (45.6)	Household was able to buy cassava	588 (30.2)	301 (15.4)
Household needed to buy yam	959 (49.2)	Household was able to buy yam	401 (20.6)	558 (28.6)
Household needed to buy guinea corn/sorghum	695 (35.6)	Household was able to buy guinea corn/sorghum	469 (24.1)	226 (11.6)
Household needed medical treatment	648 (33.2)	Household able to access medical treatment	492 (25.2)	156 (8.0)
Household needed to go to bank, money agent, or use the ATM	875 (44.9)	Household was able to successfully access bank, money agent, or use the ATM	759 (38.9)	116 (5.9)

Source: Computed from NBS COVID-19 NLPS Data (2020).
Note: Figures in parentheses are percentages.

that actually got). It is observed that the greatest proportions of the effect were on the basic staple foods: yam (28.6%), rice (23.4%), beans (20.2%), and cassava (15.4%). These show that households went through hunger due to COVID-19 pandemic and the inability to access other basic necessities.

Effect of COVID-19 on households' food security in Nigeria

The NLPS data contained four food security questions which are presented in Figure 3 and Table 3. From results in Table 2, it was seen that households' access to staple foods were negatively impacted. This is confirmed in the results on food security as Figure 3 shows that at least one adult in the household had to skip a meal to ensure food availability to younger persons (minors).

Table 3 shows that households were not food secure as 49.7% of the households had to go into panic buying and stocked on more food than normal. At least, one adult ran out of food in more than half (57.2%) of households, one adult had to skip a meal in majority (73.4%) of the households, and one adult could eat once in a day

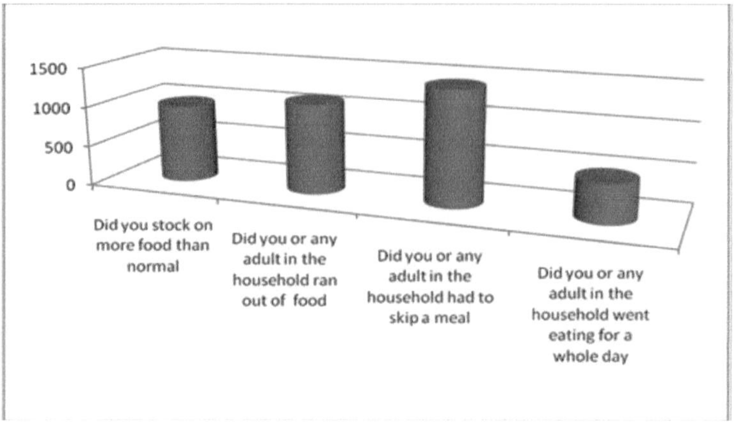

Figure 3: Effect of COVID-19 on households' food security.
Source: Computed from Nigeria COVID-19 NLPS Data (2020).

Table 3: Effect of COVID-19 on households' food security in Nigeria.

Food Security Question	Frequency N = 1950	Percentage
Did you stock on more food than normal?	970	49.7
Did you or any adult in the household run out of food?	1117	57.2
Did you or any adult in the household have to skip a meal?	1431	73.4
Did you or any adult in the household have food for a whole day?	482	24.7

Source: Computed from NBS COVID-19 NLPS Data (2020).

in about one-quarter (24.7%) of the households. Akinleye et al. (2020) found that food insecurity of households during the pandemic was significantly occasioned by increase in prices, fall in income, and adverse financial well-being. Also, Ojo (2020) found that Nigerian households had to rely on taking less preferred food, borrow or beg for food, buy food on credit, allow household members to eat elsewhere, and limit or restrict adult portions of food or meals.

Effect of COVID-19 on food security across sectors

There are different levels of impact of COVID-19 on households' food security across sectors in Nigeria. Figure 4 shows that there are greater negative effects on food security in rural than urban area, except that there were more adults that had food for the whole day in rural sector. Households in the rural sector stocked more food than normal, but their adults still ran out of food and skipped meal more than households in the urban sector. This cannot be disconnected with the fact that they saved all the available food they had just before mid-March and the period was not a time of harvest of food crops in most parts of Nigeria. This accounts for greater increases in prices of staples during the full lockdown and even as lockdowns were eased phase by phase. It should be noted that though adults in households in the rural

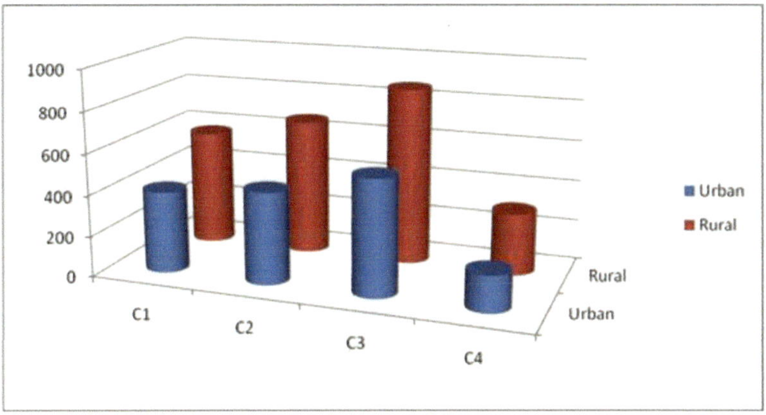

Figure 4: Effect of COVID-19 on food security across sectors.
Source: Computed from NBS COVID-19 NLPS Data (2020).
C1 - Did you stock on more food than normal?
C2 - Did you or any adult in the household run out of food?
C3 - Did you or any adult in the household have to skip a meal?
C4 - Did you or any adult in the household have food for a whole day?

Table 4: Effect of COVID-19 on food security across sectors.

Food Security Question	**Urban Number of Households**	**Rural Number of Households**	**Total Number of Households**
Did you stock on more food than normal?	400 (41.2)	570 (58.8)	970 (100)
Did you or any adult in the household run out of food?	449 (40.2)	668 (59.8)	1117 (100)
Did you or any adult in the household have to skip a meal?	564 (39.4)	867 (60.5)	1431 (100)
Did you or any adult in the household have food for a whole day?	176 (36.5)	306 (63.5)	482 (100)

Source: Computed from NBS COVID-19 NLPS Data (2020).
Figures in parentheses are percentages.

sector ran out of food and skipped meals, most of the adults did not go hungry for a whole day. Households in the rural sector can find some reprieve since they are closer to the farms, rather than go hungry for a whole day.

From Table 4, more than half (58.8%) of rural households but less than half (41.2%) of urban households stocked more food than normal. This is an indication of panic buying among urban households and keeping of available foods among rural households. Also, more than half, 59.8% and 60.5 of rural households had at least one adult who ran out of food or had to skip a meal, respectively, while less than half, 40.2% and 39.4 of urban households had at least one adult who ran out of food or had to skip a meal, respectively. These emphasize greater levels of negative impacts of COVID-19 on food security of rural households. However, more than three-fifths

(63.5%) of rural households had adults that were able to have a meal per day while less than two-fifths (36.5%) of urban households fell in the same category.

Effect of COVID-19 on food security across geopolitical zones

Across the geopolitical zones in Nigeria, the negative impacts of COVID-19 followed the same pattern as shown in Figure 5. The figure shows that highest proportions of households in each geopolitical zone had at least one adult that skipped a meal, followed by households with at least one adult that ran out of food and households that stocked more food more than normal. Least in all geopolitical zones are the proportions of households where at least 1 adult had food for a day.

From Table 5, the highest proportion of households that stocked more food than normal were in the South West (20.6%), those having at least one adult that ran out of food were found in the South East (19.0%), and those having at least one adult that had to skip a meal were found in the South West (20.8%) to make food available for their minors based on their societal value. However, the highest proportion of households that had at least one adult that had food for a day were found in the North East (22.4%).

4.3 Shocks to household livelihood during COVID-19

There were different types of shocks experienced by households as occasioned by the Covid-19 pandemic. There were strict lockdown rules which made people to stay in their homes. Businesses were shut down, all schools—primary, secondary and

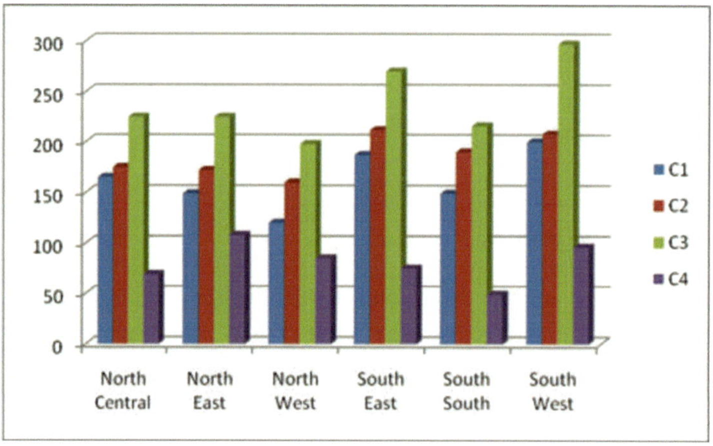

Figure 5: Effect of COVID-19 on food security across geopolitical zones.
Source: Computed from Nigeria COVID-19 NLPS Data (2020).
C1 - Did you stock on more food than normal?
C2 - Did you or any adult in the household run out of food?
C3 - Did you or any adult in the household have to skip a meal?
C4 - Did you or any adult in the household have food for a whole day?

Table 5: Effect of COVID-19 on food security across geopolitical zones.

Variable	North Central	North East	North West	South East	South South	South West	Total
Did you stock on more food than normal	165 (17.0)	149 (15.3)	120 (12.4)	187 (19.3)	149 (15.4)	200 (20.6)	970 (100)
Did you or any adult in the household run out of food	175 (15.7)	172 (15.4)	160 (14.3)	212 (19.0)	190 (17.0)	208 (18.6)	1117 (100)
Did you or any adult in the household had to skip a meal	225 (15.7)	225 (15.7)	198 (13.8)	270 (18.9)	216 (15.1)	297 (20.8)	1431 (100)
Did you or any adult in the household have food for a whole day?	69 (14.3)	108 (22.4)	85 (17.6)	75 (15.6)	49 (10.2)	96 (19.9)	482 (100)

Source: Computed from NBS COVID-19 NLPS Data (2020).
Figures in parentheses are percentages.

tertiary institutions were shut down (both government and private), and all businesses came to a standstill. This led to various types of shocks as enumerated in Table 6. Some households suffered illness, injuries, or even death due to the risks imposed by coronavirus lockdowns, while some died due to the virus infection, especially households' income earners or breadwinners, making such households much more vulnerable. There were shutdowns of non-farm businesses due to lockdown restrictions where both business owners and clients/customers could no longer meet. Also, due to business shutdowns, especially in the private sector, there were many job losses, and crippling of private businesses, especially the small scale businesses and start-ups. The lockdowns promoted sharp practices and theft/looting of private stores and properties since owners were forced to completely stay away from their stores. There were daylight theft of cash and properties by street/area boys and motor garage touts in the urban centres, who depend on their daily activities and other people for daily subsistence. Due to lockdowns, they could not find daily bread.

For farming households, there were disruptions of crop farming, livestock, and fishing activities. This led to increase in prices of farming or business activities for

Table 6: Shocks to household livelihood during COVID-19.

Type of Shocks
Illness, injury, or death of income earner
Job loss
Non-farm business closure
Theft/looting of cash and other property by thieves and street boys
Disruption of crop farming, livestock, and fishing activities
Increase in price of farming/business for owners
Fall in the price of farming/business for hired hands
Increase in price of major food items

Source: Extracted from Nigeria COVID-19 NLPS Data (2020).

owners and fall in price of farming or business activities for hired hands. In all, supply became less than demand, so that even in the face of little or no cash in the hands of household members, there was an upward surge in the prices of major food items which persists till this August 2020.

4.4 Coping strategies to shocks and social safety nets for survival of households

Coping strategies to shocks by individual household members

As a result of the shocks experienced by households as highlighted in Table 6, individual household members adopted coping mechanisms or strategies for survival during the pandemic. In this section, individual coping strategies were recorded. Figure 6 shows strategies adopted by individual members of households to cope with the various shocks induced by the COVID-19 pandemic. The prominent strategies include reducing food consumption, relying on meagre savings, reducing non-food consumption, and other strategies such as eating from stored produce, planting of other crops and doing farm activities by self, among others. Also, it is prominent that some did nothing about the adverse impact caused by the pandemic (risk neutral).

Results in Table 7 emphasize the coping strategies adopted by individuals against COVID-19 induced shocks. Reduced consumption by individuals (32.0%)

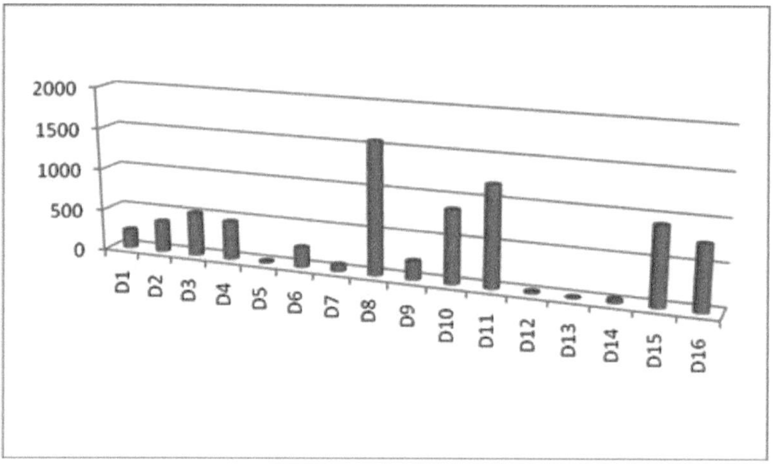

Figure 6: Coping strategies to shocks by individual household members.

Source: Computed from NBS COVID-19 NLPS Data (2020).

D1-Sales of assets, D2-Engaged in additional income generating, D3-Received assistance from friends and family, D4-Borrowed from friends and Family, D5-Took a loan from a Financial Institution, D6-Credited Purchases, D7-Delayed payment obligations, D8-Reduced Food Consumption, D9-Sold harvest in advance, D10-Reduced Non Food Consumption, D11-Relied on savings, D12-Received assistance from NGO, D13-Took advanced payment from Employer, D14-Received assistance from government, D15-Did nothing, D16-Others (eating from stored produce, planted other crops, doing farm activities by self, among others).

Table 7: Coping strategies to shocks by individual household member.

Coping Strategies	Number (N = 4937)	Percentage
Sales of assets	225	4.6
Engaged in additional income generating	376	7.6
Received assistance from friends and family	523	10.5
Borrowed from friends and Family	460	9.3
Took loan from a Financial Institution	20	0.4
Credited Purchases	236	4.8
Delayed payment obligations	75	1.5
Reduced food consumption	1579	32.0
Sold harvest in advance	219	4.4
Reduced non food consumption	872	17.7
Relied on savings	1201	24.3
Received assistance from NGO	21	0.4
Took advanced payment from Employer	6	0.1
Received assistance from government 38	38	0.8
Did nothing	924	18.7
Others (eating from stored produce, planted other crops, doing farm activities by self, among others)	762	15.4

Source: Computed from Nigeria COVID-19 NLPS Data (2020).

corroborates the findings on food security situation in Table 4 where at least one adult ran out of food or skipped meals. Relying on savings by almost one-quarter (24.3%) of the individuals shows the sign of reduced investments by such individuals and their households in the post COVID-19 period.

Reduced non-food consumption (17.7%) underscores the fact that food is the most basic requirements for survival. Some other coping strategies by other individuals which affect small-scale business owners, especially those that deal in food and other basic necessities are purchasing on credit, delay in payments, borrowing, and having to give assistance to others. This loss of business capital are quite likely to affect such small-scale business owners and start-ups at post COVID-19 period. Negligible proportion (0.4%) received assistance from governments, 0.1% took advanced payment from employers, while 18.7% of the members of households did nothing about the shocks. Specifically, farming households started eating from stored produce, corroborating the fact that rural households who had at least one adult having food daily were more than urban households (Table 5).

Social safety nets for survival by individual household members

As a result of the enforcement of lockdown across the thirty-six states of Nigeria and the Federal Capital Territory Abuja, many individuals and households became disadvantaged in procuring basic necessities for survival. Many individuals and households were not prepared for the sudden lockdowns to contain the spread of

COVID-19, therefore those that rely on wages and daily activities for survival became greatly handicapped to get daily needs. There was need for social protection for such people. Some were tagged as the 'poorest of the poor' while some became poor because they worked with private business owners who could not pay wages or salaries without daily business activities. Payments were cut off for workers in such private businesses. Forms of assistance received include food, direct cash transfer, and other in-kind transfers that were called 'palliatives' in government circles.

From Table 8, only 6.9% of the individuals received assistance. Figure 7 visually shows that these assistances were received from Nigerian Governments; Federal, State, and Local, as well as Non-Government Organizations (NGOs), Community-Based Organisations (CBOs) and International Organisations, Religious Bodies, and others such as individuals at home and abroad. Figure 7 shows that these beneficiaries received assistance mainly from governments of their states, religious bodies, and others including individuals at home and abroad.

According to results on Table 8, the most prominent of the social safety nets were the state governments who gave in-kind transfers (palliatives) to their citizens (32.4% of the beneficiaries). This was followed by religious bodies who gave assistance to about one-quarter (25.6%) of the beneficiaries. Surprisingly, only 7.9% of the beneficiaries received assistance from the Federal Government of Nigeria, who according to reports, gave direct cash and cash transfers as palliatives to the 'poorest of the poor' using much disputed 'socio-register' of vulnerable households. This is in consonance with the findings of Ojo (2020) that 8 out of every 10 respondents in Oyo state did not have access to palliative stimulus packages from the federal government. Also, the federal government distributed in-kind palliatives through School Feeding Programme to the parents of school children during the lockdown period. The impact of such palliatives may not be easily felt as many individuals and households depend on daily wage pay for daily needs.

Table 8: Social safety nets for survival by individual household members.

Received Assistance	**Number of individuals (N = 4,937)**	**Percentage**
Yes	340	6.9
No	4,597	93.1
Social safety nets	**(n = 340)**	**Percentage**
Federal Government	27	7.9
State Government	110	32.4
Local Government	34	10.0
Community Organization/Cooperative	27	7.9
Non-Governmental Organization (NGO)	12	3.5
International Organization	1	0.3
Religious Bodies	87	25.6
Others	42	12.4

Source: Computed from NBS COVID-19 NLPS Data (2020).

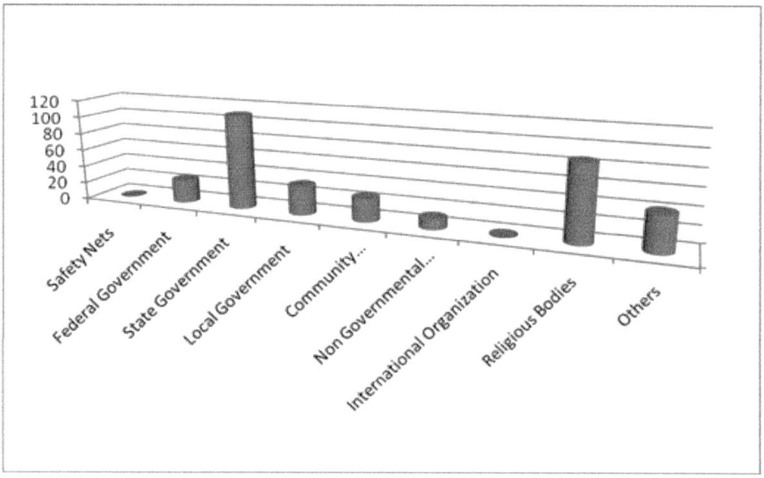

Figure 7: Social safety nets for survival by individual household members.
Source: Computed from Nigeria COVID-19 NLPS Data (2020).

5. Conclusion

Nigerian households have good knowledge of COVID-19 pandemic and have good attitude towards the containment of its spread. The lockdowns led to the shutdowns of businesses which led to job loss and other shocks such as illness/injuries/death, job losses, and crippling of private businesses, sharp practices, and theft/looting of private stores. For farming households, there were disruptions of crop farming, livestock, and fishing activities. COVID-19 has great negative impact on households' access to basic services and households' food security. There are greater negative effects on food security of households in the rural sector than the urban sector. However, households adopted coping strategies such as reducing food consumption and skipping meals by adults. Notwithstanding, rural/farming households were able to fall back on stored produce and their farms to ensure that adults have access to a meal in a day. This underscores the importance of households' farms and gardens to relieve the stress that Covid-19 has imposed on individuals and households.

The most prominent social safety net was the assistance received from the state governments and religious bodies. The palliatives distributed by the federal government using the socio-register was not felt by households as only few individuals got it. The Nigerian governments should embark on updating socio-register to include wide range of vulnerable households across all the geopolitical zones in the country, as well as create comprehensive database for the records of all Nigerian citizens to serve as safety nets for individuals and households. This will ensure social protection for all Nigerians. Also, the Nigeria Economic Sustainable Plan (NESP 2020), a post COVID-19 recovery plan of the Federal Government should be effectively implemented to generate new jobs to ameliorate the sufferings of individuals and households due to job loss and crippling of businesses.

References

Abdul, I. M. (2020). Covid-19, lockdown and transitory Food Insecurity in Nigeria. *Food and Agribusiness Management (FABM)*, 1(1): 31–35.

Akinleye, O., Dauda, S. and Oladoyin, R. (2020). Impact of COVID-19 Pandemic on Financial Health and Food Security: A Survey-Based Analysis. Retrieved from: papers.ssrn.com/sol3/papers.cfm?abstractid=3619245.

Asongu, S. A. and Adesoji, F. (2020). The Economic Consequences of the COVID-19 Pandemic in Nigeria. Excas Working Paper, pp. 1–24.

Bhuiyan, A. I., Sakib, N., Pakpour, A. H., Griffiths, M. D. and Mamun, M.A. (2020). COVID-19-related suicides in Bangladesh due to lockdown and economic factors: Case study evidence from media reports. *International Journal of Mental Health and Addiction*. doi.org/10.1007/s11469–020–00307, Letter to the Editor.

Chukwuka, O. and Mma, A.E. (2020). Africa in Focus: Understanding the Impact of the Covid-19 outbreak on the Nigerian economy. https://www.brookings.edu/blog/africa-infocus/2020/04/08/understanding-the-impact-of-the-covid-19-outbreak-on-the-nigerian-economy/ (Accessed 11/8/2020).

Devereux, S., Béné, C. and Hoddinott, J. (2020). Conceptualising COVID-19's impacts on household food security. *Food Security*, 12: 769–772. https://doi.org/10.1007/s12571–020–01085–0 Economic effects of coronavirus outbreak (COVID-19) on the world economy. Available at SSRN 3557504.

Human Rights Watch. (2020). Nigeria: Protect most vulnerable in Covid 19 Response. https://www.hrw.org/news/2020/04/14/nigeria-protect-most-vulnerable-covid-19-response. April 14.

Idris, M. and Oruonye, E. D. (2020). Socioeconomic impact of COVID-19 in oil exporting countries: An analytical review of the macroeconomic indicators in Nigeria. *International Journal of World Policy and Development Studies*, 6(5): 44–50.

Inegbedion, H. E. (2020). COVID-19 lockdown: Implication for food security. *Journal of Agribusiness in Developing and Emerging Economies*, Vol. ahead-of-print No. ahead-of-print. https://doi.org/10.1108/JADEE-06–2020–0130.

International Labour Organization. (2020). ILO: As job losses escalate, nearly half of global workforce at risk of losing livelihoods. https://www.ilo.org/global/about-the-ilo/newsroom/news/WCMS_743036/lang--en/index.htm (Accessed 11/8/2020).

Muhammad, A. S., Abeer, K., Suliman, K. and Rabeea, S. (2020). COVID-19 infection: Origin, transmission, and characteristics of human coronaviruses. *Journal of Advanced Research*, 24: 91–98.

National Bureau of Statistics NBS. (2020a). Impact Monitoring Report, https://nigeria-coronavirus-response-data-hub-nbs-nigeria.hub.arcgis.com (Accessed 11/8/2020).

National Bureau of Statistics (NBS). (2020b). Nigeria Covid-19 National Longitudinal Phone Survey (NLPS Baseline). Data catalog at: https://microdata.worldbank.org/index.php.

NESP. (2020). Bouncing Back: Nigeria Economic Sustainability Plan, 76 pp.

Nigerian Centre for Disease Control NCDC. (2020a). Frequently asked questions, https://covid19.ncdc.gov.ng/faq/ (Accessed 11/8/2020).

Nigerian Centre for Disease Control NCDC. (2020b). COVID-19 Situation Report. https://covid19.ncdc.gov.ng (Accessed 11/8/2020).

Ojo, I. O. (2020). COVID-19 and Idiosyncratic Food Dearth Risk Resilience in Nigeria. ZBW-Leibniz Information Centre for Economics, Kiel, Hamburg http://hdl.handle.net/10419/222240.

Omaka-Amari, L. N., Aleke, C. O., Obande-Ogbuinya, N. E., Ngwakwe, P. C., Nwankwo, O. and Afoke, E. N. (2020). Coronavirus (COVID-19) Pandemic in Nigeria: Preventive and Control Challenges within the First Two Months of Outbreak. *African Journal of Reproductive Health*, 24(2): 87–97.

Peterson, K. O. (2020a). COVID-19 in Africa: Socioeconomic impact, policy response and opportunities. *International Journal of Sociology and Social Policy*, pp. 1–24.

Peterson, K. O. (2020b). COVID-19 pandemic and economic crisis: The nigeria experience and structural causes. *SSRN Electronic Journal*.

Shabir, A. L. and Aijaz, A. (2020). Covid-19 pandemic: An African perspective. *Emerging Microbes & Infections*, 9, 1: 1300–1308.

Sumner, A., Hoy, C. and Ortiz-Juarez, E. (2020). Estimates of the Impact of COVID-19 on Global Poverty. ISSN 1798–7237 ISBN 978–92–9256–800–9 https://doi.org/10.35188/UNU-WIDER/2020/800–9.

United Nations Development Programme UNDP. (2020). Covid-19 and the SDGs https://feature.undp.org/covid-19-and-the-sdgs/ (Accessed 11/8/2020).

World Bank. (2020). Covid-19 to plunge Economy into worst recession since World War II. https://www.worldbank.org/en/news/press-release/2020/06/08/covid-19-to-plunge-global-economy-into-worst-recession-since-world-war-ii (Accessed 8/6/2020).

World Health Organization (WHO) African region. (2020). https://www.afro.who.int/health-topics/coronavirus-covid-19 (Accessed 11/8/2020).

Chapter 15

Mapping the Post-Pandemic Tourism Research

Insights from Bibliometric Analysis

Zobayer Ahmed[1] and *Stephen Taiwo Onifade*[2,*]

1. Introduction

COVID-19 has made a paradigm shift in the tourism sector throughout the world as the tourism industry remained one of the worst-hit sectors (WTO 2021, Co-operation and Development 2020, Onifade et al. 2022, Zenker and Kock 2020, Göktuğ Kaya et al. 2022). However, from the last quarter of the year 2021 till the first quarter of 2022, the effect of COVID-19 on international tourism started to witness some levels of recovery. Nearly three times as many tourists from abroad arrived in various destinations during the same period in 2021, and Europe alongside the Americas (North and South) majorly led the recovery trends with about 19% and 17% corresponding growth in the levels of international arrivals (WTO 2021).

A significant number of researchers have thereafter dedicated themselves to assessing the economic losses incurred by nations due to the pandemic, especially in the tourism sector. For instance, Qiu et al. (2020) studied the factors that determine travelers' willingness to pay for tourism. Ataguba (2020) also examined the economic consequences of the virus outbreak in Africa as a continent mostly notable for fragile economies with high levels of socio-economic challenges including poverty, unemployment, and vulnerabilities to other public health problems (Adebisi et al. 2021, Onifade 2022, Hakan et al. 2022, Erdoğan et al. 2022, Onifade et al. 2020,

[1] Assistant Professor, Department of Economics & Banking, International Islamic University Chittagong, Bangladesh.
[2] School of Finance and Accounting, Department of Economics, University of Vaasa, FI-65200 Vaasa, Finland.
Email: ecozobayer@gmail.com
* Corresponding author: estephentaiwo.onifade@gmail.com

Çevik et al. 2020). Some other studies focused only on a partial quantitative analysis of the impact of the COVID-19 pandemic on tourism-related employments in different countries or across regions like the European Union (Vărzaru et al. 2021, Sun et al. 2021, Shah et al. 2021, Bajrami et al. 2021).

On the other hand, some people have tried to trace the major research aims and objectives in the tourism industry during the pandemic (Zenker and Kock 2020), while some others have attempted to review the research conducted during and after the pandemic. Some of these reviews also include studies on the consequences of the pandemic on other aspects of the society not directly related to the human facets of the discussion. For instance, with regards to the latter category, Usui et al. conducted a systematic review of ninety-two reports that are related to the effects of the pandemic on mammals in tourism destinations. Another article (Bhuiyan et al. 2021) reviewed 100 articles on the tourism industry, economic crisis, and mitigation process of the loss in pre, during, and post-pandemic situations. (Persson-Fischer and Liu 2021) conducted a content analysis of eighty-seven articles to find the research aims and objectives in tourism during the pandemic.

In a nutshell, there have been several review attempts during the pandemic with a special focus on a systematic review, content analysis, and conventional reviews. However, the present study is unique as it aims to analyze the articles published on tourism in the post-pandemic era using quantitative bibliometric analysis. Bibliometric analysis is the quantitative analysis of the meta-data of the research documents published on a certain topic (Patra et al. 2006). Among the various types of reviews, bibliometric analysis involves large meta-data that requires extensive use of computer software. Thus, this quantitative analysis here focuses on the analysis of the sources, documents, and keywords to identify major research focus and gaps in the existing research. As such, the outcome of our review analysis in this chapter will help to determine future research agenda towards tackling the menace created by the COVID-19 pandemic in the global economy and most especially in the hard-hit tourism industry.

The chapter is outlined beginning with an introduction followed by detailed information about the data and methodology used. The next section focuses on the main findings of the bibliometric analysis, and it is partitioned into three sub-sections, namely, analysis of the sources, documents analysis, and keywords analysis. The chapter finally ends with a conclusion.

2. Data and Methods

2.1 Data collection process

The data for the bibliometric analysis have been collected from the SCOPUS database in seven consecutive steps. According to Table 1, the initial step involves searching the documents using post-covid OR post-corona OR post-pandemic OR corona recovery keywords in the title-abstract-keywords of the documents. The keyword 'touris*' is used to include all the keywords having suffixes after 'touris' (e.g., tourism, tourist). Through this search, a total of 624 documents were detected as of 28 July 2022.

Table 1: Stages of data collection.

Steps	Search and Screening	No. of Documents
Step 1	Documents that include keywords= post-covid OR post-corona OR post-pandemic OR corona recovery	**11,374**
Step 2	Documents that include keyword = touris*	**624**
Step 3	Limit-to (publication year, 2021-2022)	**549**
Step 4	Limit to subject area = Social sciences, business studies, economics, earth science, multidisciplinary	**474**
Step 5	Limit to (Source Type, 'Journal')	**410**
Step 6	(Limit-to (language, "English")	**394**
Step 7	(LIMIT-TO (PUBSTAGE , "final")	**309**

In the next step, we limited the publication year only to 2021 and 2022 up to date. Although COVID-19 started in China at the end of the year 2019, the pandemic spread all over the world in 2020 thus affecting the tourism sector. However, in the middle of 2020 and towards the end of the year, in most countries, the tourism sector started to signal the possibility of gradual recovery. The WTO (2021) noted that there were signs of a moderate rebound in the level of international tourism in 2021. Thus, since the present study focuses on post-pandemic tourism, we essentially consider only the publications of the years 2021 and 2022. At this stage, a total of 549 documents were visible. Regarding the subjects of study, we limited the documents published to the field of social sciences, economics, business studies, earth science, and multidisciplinary fields. After this step, the number of documents became 474. The next steps were sources (journal), language (English), and publication stage (final), thus providing the number of documents (410, 394, and 309, respectively).

2.2 Analytical tools

We analyze bibliometric data using three different tools: Rshiny for bibliometrics, VOSviewer, and Microsoft Excel. Rshiny is a shiny application in R-Studio, a popular tool for generating bibliographic data for datasets (Ahmi 2022). The networking and contenting analysis were done with VOSviewer. It is possible to create and visualize bibliometric networks using the VOSviewer application (Van Eck and Waltman 2010, 2013). These visual representations are frequently referred to as maps, and they are used to undertake a variety of network analyses, for example, co-occurrence, co-citations, and co-authorship. This network of relationships can be built for sources, authors, keywords, and countries. Additionally, VOSviewer was utilized to perform bibliographic coupling. We utilized Microsoft excel to supplement R-Studio, as the latter's graphics were not easily modifiable (Alshater et al. 2022).

3. Analysis of the Data

In this section, first, we present an overview of the data. By then, the entire analysis has been divided into three major categories, namely, sources analysis, documents analysis, and keyword analysis.

3.1 An overview of the data

Under the specific search criteria, we collected 309 documents for the bibliometric analysis. These 309 documents come from 125 different sources. The author's keywords and the keywords plus of the documents are 1176 and 539, respectively. A total of 886 authors appeared 967 times in the documents. Eventually, the documents per author became 0.349 while the authors per document was 2.87 (see Table 2).

Table 2: Summary of the data.

Description	Results	Description	Results
Sources (Journals, Books, etc)	125	Single-authored documents	46
Documents	309	Documents per Author	0.349
Keywords Plus (ID)	539	Authors per Document	2.87
Author's Keywords (DE)	1176	Co-Authors per Documents	3.13
Authors	886	Collaboration Index	3.19
Author Appearances	967		

3.2 Sources analysis

In 1934, S. C. Bradford first explained Bradford's law which is basically a pattern that shows the core journals on a particular topic (Brookes 1977). In terms of percentage of the total publications, as shown in the Bradford's Law, in Figure 1, one-third of the publications (core sources) on the topic come from four sources, as shown by a blue shaded area. The majority of the publications were in the Sustainability (Switzerland), followed by the *Worldwide Hospitality and Tourism Themes, Current Issues in Tourism, and the African Journal of Hospitality, Tourism and Leisure.*

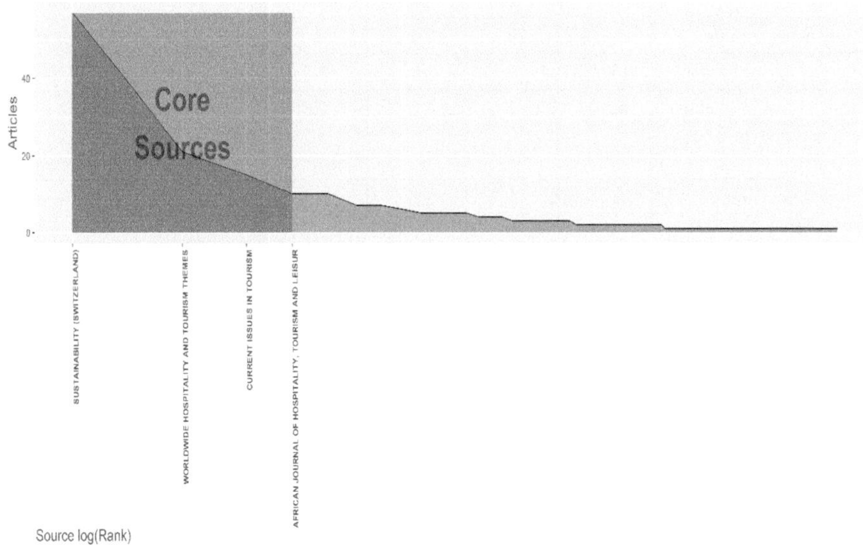

Figure 1: Impactful sources (Bradford's Law).

Table 3 gives a more detailed analysis of the sources. The top 15 contributing journals are listed here according to the number of publications. Other than the top four journals listed in Figure 1 (Bradford's Law), eleven others journals and their performance have been presented in this table (Table 3). The analysis here includes the impact factor in general of the journals and the h-index, g-index, m-index, total citations (TC), number of publications (Chanpermpoonpol and Kirawanich, 2021), and the per article citation (TC/NP) of the sample articles (309) considered for this study.

From the impact factor analysis in general, the journal of *Tourism Management* is the top journal with an impact factor of 13.78. The second and the third journals among the sample are *Annals of Tourism Research and Research* and the *Research in Globalization*, respectively. As per the total number of publications, although *Sustainability* (*Switzerland*) is at the top with a total of 57 publications, the average

Table 3: Sources' impact (According to the number of publications).

Source	Impact Factor	h_index	g_index	m_index	TC	NP	TC/NP
Sustainability (switzerland)	4.166	7	13	3.5	241	57	4.22807018
Worldwide hospitality and tourism themes	1.509	4	5	2	55	21	2.61904762
Current issues in tourism	7.578	9	15	4.5	416	15	27.7333333
African journal of hospitality, tourism and leisure	0.824	3	4	1.5	23	10	2.3
Annals of tourism research	12.853	5	10	2.5	221	10	22.1
Geojournal of tourism and geosites	2.384	2	2	1	8	7	1.14285714
Tourism economics	5.588	2	5	1	32	7	4.57142857
Tourism management	13.788	5	6	2.5	303	6	50.5
International journal of religious tourism and pilgrimage	0.906	1	1	0.5	1	5	0.2
International journal of tourism policy	0.943	1	1	0.5	2	5	0.4
Journal of destination marketing and management	7.748	3	5	1.5	47	5	9.4
Journal of travel and tourism marketing	8.623	3	5	1.5	34	5	6.8
Anatolia	3.264	4	4	2	96	4	24
Journal of environmental management and tourism	1.032	1	1	0.5	3	4	0.75
Research in globalization	9.259	2	4	1	22	4	5.5

citation per article is highest for the *Current Issues in Tourism*. This journal is also at the top position, considering the h-index.

3.3 Documents analysis

Document analysis concentrates on the impact of research documents collected in the sample. Table 4 shows the top ten impactful documents according to the local citations. Local citation here refers to the citations within the sample of 309 articles. However, the global citation includes citations both in and outside the sample (Chung 2007, Nielsen and Andersen 2021).

Among the most local cited articles, the article written by Zheng et al. (2021) is in the first position with fourteen local citations so far. Moreover, this article has received 194 global citations, which is also the highest among the sample articles. But as the percentage of local citation to global citation, the LC/GC score of this article becomes 7.22% only. At the same time, the highest LC/GC ratio is for the article written by Xie

Table 4: Most impactful documents.

Title	Authors and Year	Local Citations	Global Citations	LC/GC Ratio (%)
Afraid to travel after COVID-19? Self-protection, coping and resilience against pandemic 'travel fear'	Zheng et al. (2021)	14	194	7.22
Reviving tourism industry post-COVID-19: A resilience-based framework	(Sharma et al. 2021)	12	170	7.06
Coronavirus impacts on post-pandemic planned travel behaviors	(Li et al. 2021)	6	116	5.17
Understanding the new post-COVID-19 risk scenario: Outlooks and challenges for a new era of tourism	(Villacé-Molinero et al. 2021)	5	44	11.36
Destination image during the COVID-19 pandemic and future travel behavior: The moderating role of past experience	(Rasoolimanesh et al. 2021)	5	34	14.71
Proximal and distal post-COVID travel behavior	(Miao et al.2021)	5	47	10.64
The effects of risk message frames on post-pandemic travel intentions: the moderation of empathy and perceived waiting time	(Xie et al. 2021)	5	16	31.25
The effect of coronavirus disease-19 (COVID-19) risk perception on behavioral intention towards 'untact' tourism in South Korea during the first wave of the pandemic (March 2020)	(Bae and Chang 2021)	5	190	2.63
COVID-19 and a justice framework to guide tourism recovery	(Rastegar et al. 2021)	4	31	12.90
The Effects of COVID-19 on Future Domestic Travel Intentions in South Africa: A Stakeholder Perspective	(Bama and Nyikana 2021)	4	13	30.77

et al. (2021). The top cited articles here give an idea about the most focusing topics of the post-pandemic tourism industry. The major area of concentration is the behavioral aspect of the tourists and the factors affecting travel planning.

3.4 Word analysis

In a bibliometric analysis, the analysis of the keywords is important to understand the major topics covered by sample articles. In this section, we analyzed the keywords in several steps. First, the tree map analysis presents the most frequently used twenty-five keywords. Second, we present the relative importance of keywords in a word cloud. Third, the three-field plot presents the relationship among the affiliated countries, keywords, and keywords plus. Finally, from the Vosviewer software output, keywords are analyzed with the help of a co-occurrence network.

In Figure 2, a tree map of the keywords has been shown. This map is a data-visualization technique used for large, hierarchical data sets in many other studies as well (Jadeja and Shah 2015). The map highlights two data characteristics: the value of particular data positions; and the order composition (Sharif and Abdullah 2013). Of the top twenty-five keywords in the tree map, COVID-19 and tourism cover almost 30% of the keyword frequency. The 'tourist destinations', 'tourism development', and the 'sustainability' are placed in the next column of the hierarchy. Some other frequently used keywords are 'epidemic', 'tourism management', 'tourism market', and the 'tourist behavior'.

A more specific visualization of the frequently used keywords in the sample articles is presented in Figure 3 with the help of the word cloud. This time, we used keyword plus only. The font size of each keyword shows the relative frequency of the word used in the sample articles.

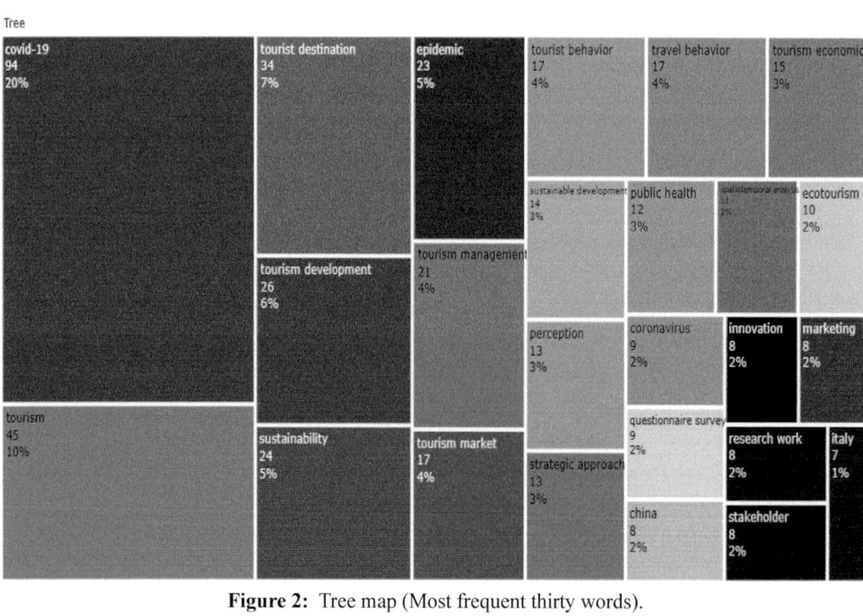

Figure 2: Tree map (Most frequent thirty words).

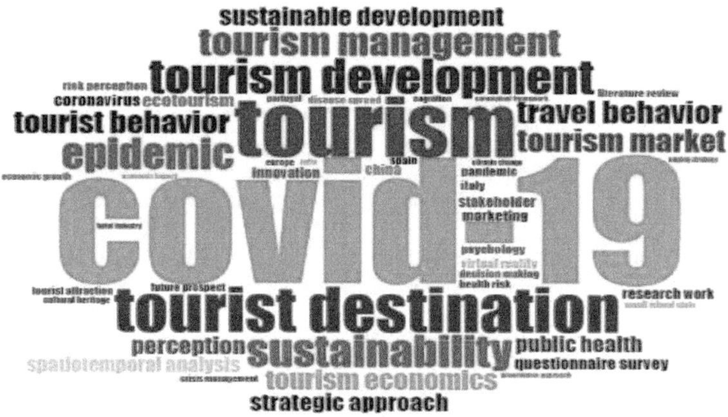

Figure 3: Word cloud (Keyword plus).

Both the tree map and word cloud show that the major focus areas of the articles in post-pandemic tourism research are tourist destinations, tourism development, sustainability, tourist and travel behaviors, ecotourism, sustainable development, tourism marketing, and tourism management. All these issues are somehow connected to the tourism economy and the sustainability of the environment. Throughout the world, the COVID-19 pandemic was primarily a public health concern as millions of people died after the pandemic started (Douglas et al. 2020). The second-largest damage was the economic loss faced by all the key sectors of every economy (Haleem et al. 2020). Countries that have a major economic structure based on tourism suffered from the pandemic significantly.

Another important analysis in this context is the three-field plot analysis. It is a Sankey diagram that is a visual representation to show how values flow from one set to another. In Figure 4, the left column shows the top fifteen affiliated countries in the sample articles. The middle column and the right column present the most frequently used fifteen keywords and keyword plus, respectively. The graph shows how the different items in each column are connected to other items in the rest of the columns. The size of a rectangle is proportional to its frequency in the sample articles.

Figure 4 shows that almost all the countries have focused on COVID-19 and tourism. The flows from left to the right depict that the majority of the topics connected to COVID-19 and tourism are tourism management, tourist destination, tourism development, sustainability, and sustainable development. Moreover, flows from right to left shows COVID-19 and tourism are mostly connected to tourism recovery, sustainable tourism, sustainability, sustainable development, etc. Thus, the three-field plot gives an idea of the major focusing areas of the present research articles. It is clear that the sample articles mostly focused on the development and management aspects of tourism along with sustainability in the post-pandemic world.

In order to understand further, the co-occurrence network analysis of the keywords by using Vosviewer is considered in this study. Co-occurrence analysis is basically the counting of matched data inside a collection unit.

In order to gain an understanding of the knowledge structures and temporal dynamics of an emerging study subject, keyword co-occurrence networks are used

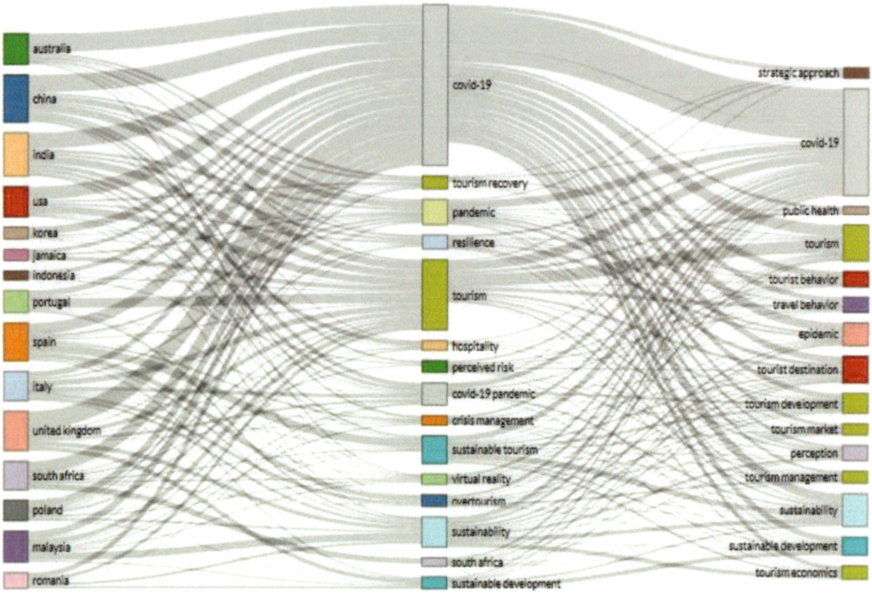

Figure 4: Three field plot (Countries-Keyword-Keywords Plus).

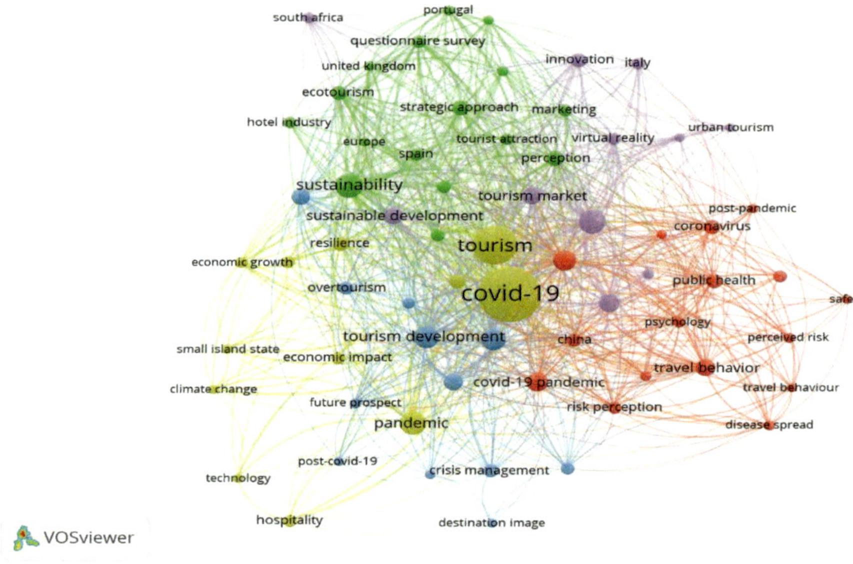

Figure 5: Co-occurance network.

(Radhakrishnan et al. 2017). Moreover, the keywords co-occurrence Network displays the links among the keywords and the relative link strengths. The Vosviewer-generated co-occurrence network divides the keywords into five clusters with five separate colors, as shown in Figure 5. A list of the items in each cluster is given in

Table 5: Clusters of the co-occurrence network.

Cluster	Major Focused Area	Items (Alphabetic Sequence)	Number of Items
Cluster 1	Travel behavior	China, Cognition, Coronavirus, Covid-19 pandemic, Disease spread, Epidemic, Health risk, Perceived risk, Post-pandemic, Psychology, Public health, Risk perception, Safety, Tourism recovery, Travel behavior, Travel behavior.	(16 items)
Cluster 2	Sustainability in tourism	Cultural heritage, Ecotourism, Ecuador, Europe, Hotel industry, Marketing, Perception, Portugal, Questionnaire survey, Research work, Spain, Stakeholder, Strategic approach, Sustainability, Tourist attraction, United Kingdom.	(16 items)
Cluster 3	Tourism management	Crisis management, Destination image, Destination Management, Future prospect, India, Literature review, Over tourism, Post-covid-19, Sustainable tourism, Tourism Development, Tourism economics, Tourism management.	(12 items)
Cluster 4	Economic and development	Climate change, Covid-19, Decision making, Economic growth, Economic impact, Hospitality, Pandemic, Resilience, Small Island state, Spatiotemporal analysis, Technology, Tourism.	(12 items)
Cluster 5	Innovation and technology	Indonesia, Innovation, Italy, South Africa, Sustainable developer, Tourism market, Tourist behavior, Tourist destination, Urban tourism, Virtual reality, Virtual tourism.	(11 items)

Table 5. The first cluster is colored red, and it contains sixteen items. The major focusing areas under this cluster are travel behaviors connected to safety and risk perception derived from the COVID-19 pandemic that originated in China. The second cluster (green) concentrated on European tourism and sustainability. However, the third cluster (blue) is mostly related to the management aspect tourism industry in the post-pandemic world. The next cluster of the keywords (yellow) is tourism's economic and development aspects. The final cluster (purple) is related to the innovation and technological issues in the tourism industry.

4. Conclusion

The tourism industry has been recovering from the adverse shock faced during the COVID-19 pandemic years. Academic studies from the perspective of social sciences, business studies, and economics have focused on various aspects of the tourism sector during, and in post-pandemic periods. Moreover, some systematic reviews and content analyses of the existing research that cover a small sample size are available. However, the present chapter focuses on analyzing the articles published on tourism in the post-pandemic era using quantitative bibliometric analysis. A total of 309 articles were collected from the SCOPUS database under specific search criteria to meet the objective of the study. Later, the results generated from bibliometric Rshiny, Vosviewer, and Microsoft Excel have been presented in three major categories, namely, sources analysis, documents analysis, and keywords analysis. The majority of the publications were made in the Sustainability (Switzerland), followed by the Worldwide Hospitality and Tourism Themes, Current Issues in Tourism, and the African Journal of Hospitality, Tourism and Leisure.

The analysis of the document revealed that the major concentration areas are the behavioral aspect of the tourists and the factors affecting travel planning. More specifically, the analysis of the keyword, which covers the lion's share of the entire bibliometric analysis in the study, reveals some specific aspects of the sample articles. Of the most frequently used keywords, the 'tourism, 'tourist destinations', 'tourism development' and the 'sustainability' are significant. Some other frequently used keywords are 'epidemic', 'tourism management', 'tourism market', and the 'tourist behavior'. From the three-field plot analysis, the sample articles have mostly focused on the development and management aspects of tourism along with sustainability in the post-pandemic world.

Furthermore, co-occurrence network analysis clustered the keyword in five different clusters show the linkages among the keywords. The major clusters are travel behaviors, sustainability in tourism, tourism management, tourism economics and development, innovation, and technology. The analysis in this chapter thus clearly reveals that there are some less focused areas of study. For instance, hospitality in the tourism industry in the post-pandemic era may be a new area of interest for future research and policy-making. Moreover, due to pandemics and other adverse shocks in the global economy in 2022, travel costs and their dynamics have changed significantly. Some other potential research areas could be ecotourism, crisis management in the tourism industry, and tourism economics. These areas need to be explored and expanded to design tourism marketing and other related issues towards tackling the menace created by the COVID-19 crisis for better tourism management and to foster sustainable tourism development in the post-pandemic era.

References

Adebisi, Y. A., Ekpenyong, A., Ntacyabukura, B., Lowe, M., Jimoh, N. D., Abdulkareem, T. O. and Lucero-Prisno III, D. E. (2021). COVID-19 highlights the need for inclusive responses to public health emergencies in Africa. *The American Journal of Tropical Medicine and Hygiene*, 104(2): 449.

Ahmi, A. (2022). *Bibliometric Analysis using R for Non-Coders: A Practical Handbook in Conducting Bibliometric Analysis Studies using Biblioshiny for Bibliometrix R Package*.

Alshater, M. M., Hassan, M. K., Rashid, M. and Hasan, R. (2022). A bibliometric review of the Waqf literature. *Eurasian Economic Review*, 12(2): 213–239. doi:10.1007/s40822–021–00183–4.

Ataguba, J. E. (2020). COVID-19 pandemic, a war to be won: Understanding its economic implications for Africa. *Applied Health Economics and Health Policy*, 18(3): 325–328.

Bae, S. Y. and Chang, P. J. (2021). The effect of coronavirus disease-19 (COVID-19) risk perception on behavioural intention towards 'untact' tourism in South Korea during the first wave of the pandemic (March 2020). *Current Issues in Tourism*, 24(7): 1017–1035. doi:10.1080/13683500.2020.1798895.

Bama, H. K. N. and Nyikana, S. (2021). The effects of COVID-19 on future domestic travel intentions in south africa: A stakeholder perspective. *African Journal of Hospitality, Tourism and Leisure*, 10(1): 179–193. doi:10.46222/AJHTL.19770720–94.

Bajrami, D. D., Terzić, A., Petrović, M. D., Radovanović, M., Tretiakova, T. N. and Hadoud, A. (2021). Will we have the same employees in hospitality after all? The impact of COVID-19 on employees' work attitudes and turnover intentions. *International Journal of Hospitality Management*, 94: 102754.

Bhuiyan, M. A., Crovella, T., Paiano, A. and Alves, H. (2021). A review of research on tourism industry, economic crisis and mitigation process of the loss: analysis on pre, during and post pandemic situation. *Sustainability*, 13(18): 10314. Retrieved from https://www.mdpi.com/2071–1050/13/18/10314.

Brookes, B. C. (1977). Theory of the bradford law. *Journal of Documentation*, 33(3): 180–209. doi:10.1108/eb026641.

Chanpermpoonpol, S. and Kirawanich, P. (2021). An analysis of cost improvement schemes for electromagnetic safety-related activities during railway project implementation. *IEEE Transactions on Engineering Management*. doi:10.1109/tem.2021.3072919.

Chung, H. -K. (2007). Evaluating academic journals using impact factor and local citation score. *The Journal of Academic Librarianship*, 33(3): 393–402.

Çevik, S., Onifade, S. T., Erdoğan, S., Asongu, S. and Bekun, F. V. (2020). An empirical retrospect of the impacts of government expenditures on economic growth: New evidence from the Nigerian economy. *Journal of Economic Structures*, 9(1): 6. https://doi.org/10.1186/s40008–020–0186–7.

Co-operation, O. F. E. and Development. (2020). Rebuilding tourism for the future: COVID-19 policy responses and recovery: OECD Publishing.

Douglas, M., Katikireddi, S. V., Taulbut, M., McKee, M. and McCartney, G. (2020). Mitigating the wider health effects of covid-19 pandemic response. *BMJ*, 369.

Erdoğan, S., Onifade, S. T., Altuntaş, M. and Bekun, F. V. (2022). Synthesizing urbanization and carbon emissions in Africa: how viable is environmental sustainability amid the quest for economic growth in a globalized world? *Environmental Science and Pollution Research*, 1–14. https://doi.org/10.1007/s11356–022–18829–4.

Göktuğ Kaya, M., Onifade, S. T. and Akpinar, A. (2022). Terrorism and Tourism: An empirical exemplification of consequences of terrorist attacks on tourism revenues in Turkey. *Tourism: An International Interdisciplinary Journal*, 70(1): 28–42. https://hrcak.srce.hr/267971.

Haleem, A., Javaid, M. and Vaishya, R. (2020). Effects of COVID-19 pandemic in daily life. *Current Medicine Research and Practice*, 10(2): 78.

Hakan, A., Onifade, S. T. and Savaş, Ç. (2022). Modeling the impacts of MSMEs' contributions to GDP and their constraints on unemployment: The case of African's most populous Country. *Studies in Business and Economics*, 17(1): 154–170. https://doi.org/10.2478/sbe-2022–0011.

Jadeja, M. and Shah, K. (2015). Tree-Map: A Visualization Tool for Large Data. Paper presented at the GSB@ SIGIR.

Li, J., Nguyen, T. H. H. and Coca-Stefaniak, J. A. (2021). Coronavirus impacts on post-pandemic planned travel behaviours. *Annals of Tourism Research*, 86. doi:10.1016/j.annals.2020.102964.

Miao, L., Im, J., Fu, X., Kim, H. and Zhang, Y. E. (2021). Proximal and distal post-COVID travel behavior. *Annals of Tourism Research*, 88. doi:10.1016/j.annals.2021.103159.

Nielsen, M. W. and Andersen, J. P. (2021). Global citation inequality is on the rise. *Proceedings of the National Academy of Sciences*, 118(7): e2012208118.

Onifade, S. T. (2022). Retrospecting on resource abundance in leading oil-producing African countries: How valid is the environmental Kuznets curve (EKC) hypothesis in a sectoral composition framework? *Environmental Science and Pollution Research*. http://dx.doi.org/10.1007/s11356–022–19575–3.

Onifade, S. T., Ay, A., Asongu, S. and Bekun, F. V. (2020). Revisiting the trade and unemployment nexus: Empirical evidence from the Nigerian economy. *Journal of Public Affairs*, 20(3): e2053. https://doi.org/10.1002/pa.2053.

Onifade, S. T., Gyamfi, B. A., Bekun, F. V. and Altuntaş, M. (2022). Significance of air transport to tourism-induced growth hypothesis in E7 Economies: Exploring the implications for environmental quality. *Tourism: An International Interdisciplinary Journal*, 70(3): 339–353. https://hrcak.srce.hr/279096.

Patra, S. K., Bhattacharya, P. and Verma, N. (2006). Bibliometric study of literature on bibliometrics. *DESIDOC Bulletin of Information Technology*, 26(1): 27–32.

Persson-Fischer, U. and Liu, S. (2021). The Impact of a global crisis on areas and topics of tourism research. *Sustainability*, 13(2): 906. Retrieved from https://www.mdpi.com/2071-1050/13/2/906.

Qiu, R. T. R., Park, J., Li, S. and Song, H. (2020). Social costs of tourism during the COVID-19 pandemic. *Annals of Tourism Research*, 84:, 102994. doi:https://doi.org/10.1016/j.annals.2020.102994.

Radhakrishnan, S., Erbis, S., Isaacs, J. A. and Kamarthi, S. (2017). Novel keyword co-occurrence network-based methods to foster systematic reviews of scientific literature. *Plos One*, 12(3): e0172778.

Rasoolimanesh, S. M., Seyfi, S., Rastegar, R. and Hall, C. M. (2021). Destination image during the COVID-19 pandemic and future travel behavior: The moderating role of past experience. *Journal of Destination Marketing and Management*, 21. doi:10.1016/j.jdmm.2021.100620.

Rastegar, R., Higgins-Desbiolles, F. and Ruhanen, L. (2021). COVID-19 and a justice framework to guide tourism recovery. *Annals of Tourism Research*, 91. doi:10.1016/j.annals.2021.103161.

Sharif, T. and Abdullah, J. A. S. (2013). Human resource development and economic growth in Bangladesh: An econometric analysis. *European Journal of Business and Management*, 5(7): 133–144.

Sharma, G. D., Thomas, A. and Paul, J. (2021). Reviving tourism industry post-COVID-19: A resilience-based framework. *Tourism Management Perspectives*, 37. doi:10.1016/j.tmp.2020.100786.

Shah, C., Chowdhury, A. and Gupta, V. (2021). Impact of COVID-19 on tourism and hospitality students' perceptions of career opportunities and future prospects in India. *Journal of Teaching in Travel and Tourism*, 21(4): 359–379.

Sun, Y. Y., Sie, L., Faturay, F., Auwalin, I. and Wang, J. (2021). Who are vulnerable in a tourism crisis? A tourism employment vulnerability analysis for the COVID-19 management. *Journal of Hospitality and Tourism Management*, 49: 304–308.

Usui, R., Sheeran, L. K., Asbury, A. M. and Blackson, M. (2021). Impacts of the COVID-19 pandemic on mammals at tourism destinations: A systematic review. *Mammal Review*, 51(4): 492–507. doi:https://doi.org/10.1111/mam.12245.

Van Eck, N. and Waltman, L. (2010). Software survey: VOSviewer, a computer program for bibliometric mapping. *Scientometrics*, 84(2): 523–538.

Van Eck, N. J. and Waltman, L. (2013). VOSviewer manual. *Leiden: Univeristeit Leiden*, 1(1): 1–53.

Vărzaru, A. A., Bocean, C. G. and Cazacu, M. (2021). Rethinking tourismindustry in pandemic COVID-19 period. *Sustainability*, 13(12): 6956. Retrieved from https://www.mdpi.com/2071-1050/13/12/6956.

Villacé-Molinero, T., Fernández-Muñoz, J. J., Orea-Giner, A. and Fuentes-Moraleda, L. (2021). Understanding the new post-COVID-19 risk scenario: Outlooks and challenges for a new era of tourism. *Tourism Management*, 86. doi:10.1016/j.tourman.2021.104324.

WTO. (2020). World Tourism Organization: World Tourism Barometer. 2020 Edition, Volume 18 Issue 7 December 2020 https://doi.org/10.18111/wtobarometereng.

Xie, C., Zhang, J., Morrison, A. M. and Coca-Stefaniak, J. A. (2021). The effects of risk message frames on post-pandemic travel intentions: The moderation of empathy and perceived waiting time. *Current Issues in Tourism*, 24(23): 3387–3406. doi:10.1080/13683500.2021.1881052.

Zenker, S. and Kock, F. (2020). The coronavirus pandemic—A critical discussion of a tourism research agenda. *Tourism Management*, 81: 104164. doi:https://doi.org/10.1016/j.tourman.2020.104164.

Zheng, D., Luo, Q. and Ritchie, B. W. (2021). Afraid to travel after COVID-19? Self-protection, coping and resilience against pandemic 'travel fear'. *Tourism Management*, 83. doi:10.1016/j.tourman.2020.104261.

Chapter 16

Medical Tourism

An Evaluation of its Global Situation

Omar Jamal Adam,[1] *Matina Ghasemi*[1,*] and
Mazyar Ghadiri Nejad[2]

1. Introduction

Medical tourism, or people who leave their country of residence outside established border arrangements, with the goal of accessing medical, reproductive, cosmetic, dental, transplant, and elective surgical care abroad, has been described as a phenomenon of the twentieth century, which has grown significantly in recent years (Crooks et al. 2010). This phenomenon has become an industry that attracts many domestic and international travelers seeking medical care (Cormany and Baloglu 2011). This has led to advances in technology, particularly in transportation, that offer more affordable air travel costs (Golabi and Nejad 2022). In addition, global marketing communications through the Internet have enabled medical tourism destinations to promote themselves to medical tourists (Lunt et al. 2010). Moreover, the presence of expert health personnel in developing countries for desk training means that doctors in these countries can provide the same or higher quality of medical care than developed countries such as Western Europe and North America (Edelheit 2008).

In Europe, the acceptance of medical care mobility is growing and more medical patients are traveling abroad to receive medical care (Legido-Quigley et al. 2011). In addition, the increasing demand for health care and the high health costs in many high-income countries increased the demand for medical tourism industry. As a result, significant growth in investment in medical destination countries by various healthcare providers has been recorded (Smith et al. 2009).

[1] Business and Economic Department, Girne American University, 99428 Kyrenia, TRNC, Turkey.
[2] Industrial Engineering Department, Cyprus International University, 99258 Nicosia, TRNC, Turkey.
* Corresponding author: matinaghasemi@gau.edu.tr

However, reliable information on medical travel is scarce (Alvarez et al. 2011). In addition, there is little information about the number of patients and their disease characteristics (Smith et al. 2009). Therefore, the medical tourism market needs more research. For example, there is a lack of information on the importance of medical tourism providers and facilitators and clinical outcomes for specific treatments. In this context, the effects of medical tourism on clients and local systems, the role of marketing, and ethical and legal issues are still not fully understood. Also, the patient's motivations for choosing to travel and why, when others do not, and their experiences are still unclear.

The main purpose of this study is to evaluate and present the level of knowledge in the field of medical tourism by examining the main elements of medical tourism such as medical tourists, medical facilitators, and domestic and foreign medical countries. Meanwhile, this study aims to present the current level of knowledge on medical tourism through a systematic review of relevant literature in the field of medical tourism. In particular, it seeks to better understand (1) the concept of medical tourism, (2) types of medical travel, (3) patient motivation, (4) the role of the Internet and third parties, and (5) the effects of medical travel on countries of origin and destination. The obtained results not only provide a better understanding of medical tourism but also help us to identify knowledge gaps and provide a basis for future studies.

2. Literature Review

Medical tourism has created a new form of patient mobility in which people travel abroad to receive treatments such as fertility, cosmetic and surgical while obtaining a better quality of care and access to facilities available in their home country (Carrera and Lunt 2010). In other words, medical tourism occurs when people travel long distances with the aim of maintaining or restoring health through medical intervention. This practice includes a full range of medical services such as dental care, cosmetic surgery, and fertility treatment while patients are simultaneously on vacation (Gray and Poland 2008). Traveling for medical care has been around for a long time, it is nothing new. Since time immemorial, people have been used to visiting health care centers and spas, such as Harley Street, which in the past decades was London's famous location for medical care in Europe (Connell 2006). This phenomenon is referred to as medical tourism, which is known as travelling to obtain health treatments (Burkett 2007, Balaban and Marano 2010, Connell 2013). Medical tourism is one of the international sectors in the tourism industry and is associated with rapid growth, especially in developing countries. Advances in information and technology, especially in medical transportation and treatment, are facilitating the rapid global growth of the medical care industry (Heung et al. 2011).

Medical tourism conceptually has different definitions (Yu and Ko 2012) which is due to the existence of various stakeholders involved in the industry, and each of them defines medical tourism from their field and interest (Alvarez et al. 2011). The concept of health tourism has always been confused with the concept of medical tourism and the scope of these two concepts has not been fully clarified (Connell

2013). The doubt about the structure of these two concepts has two sides. On the one hand, most researchers argue that the two terms have the same structure and can be used interchangeably (Kmietowicz 2004, Weber 2002).

On the other hand, some researchers state that these two terms are not exactly the same but are related concepts and there are certain factors that can distinguish them.Medical tourism is related to the broader concept of health tourism (Pocock and Phua 2011). In other words, health tourism is the broadest possible category of health-related activities that include medical tourism. Proponents such as Muller and Kaufmann (2001) argue that health tourism and medical tourism are the main components of health tourism. They also believe that patients travel to destinations that offer medical services for the express benefit of improving health through medical or wellness tourism.

Carrera and Bridges (2006), believe in the argument that health tourism can be defined as the organized travel of an individual outside their local environment that seeks to maintain, enhance, or restore the individual's well-being in mind and body through medical intervention. In this regard, Whittaker (2008) defines the difference between medical tourism and health tourism in the way that medical tourism is the correct term that is used in cases where there are medical, surgical, or dental interventions and anything else outside of this area is in line with health tourism. In addition, Smith and Puczko (2009) suggest that health tourism and medical tourism are components of health tourism, and the concept of medical tourism includes the application of medical procedures to patients, while health tourism refers to spas, resorts, and traditional travel.

However, Kajzar (2015) stated that health tourism is a term used to refer to the activities of medical tourism and health tourism, and the expected goal of both is to achieve the health or well-being of an individual through the use of medical products or related services. Chantrapornchai and Choksuchat (2016) classified health tourism into two types. First, there is the health treatment that patients travel abroad to restore their health, which includes cosmetic and dental interventions. Second, health promotion is when patients travel abroad to improve their health ability. In short, restoring health through medical intervention means that medical tourism falls under the umbrella of health tourism.

3. Research Design

The research strategy in this study, considering the research method provided by Smith, seeks to examine studies related to medical tourism and related services, in order to organize and present the level of knowledge in this field (Alvarez et al. 2011, Smith et al. 2011).

3.1 Data collection

Primary literature was obtained from the Academic Search Premier and Science Direct databases by searching in the Google Scholar and Sciencedirect.com websites, using the matched search terms in (Table 1) for each database. According to the objectives

Table 1: Key search terms.

No.	Search Terms
1.	Brokers and medical tourism
2.	Cost of medical treatment
3.	Decision making cycle for medical tourist
4.	Economic benefits of medical tourism
5.	Health tourism
6.	Health travel
7.	Inbound and outbound medical tourism
8.	Internet and medical tourism
9.	Inward and outward medical destination
10.	Marketing mix for a medical tourism destination
11.	Medical patient
12.	Medical patient experiences
13.	Medical tourism conceptualization
14.	Medical tourism opportunities
15.	National health service (NHS)
16.	Third-party involvement in medical tourism
17.	Medical treatment
18.	Medical tourism SWOT analysis
19.	Medical tourism evolution
20.	Medical tourism motivation.

of the study, articles published in English and focusing on medical tourism, its types, motivations, and consequences were used. It should be noted that in this study, articles focusing on transplant tourism, ethical issues, laws and guidelines, and news (such as Coward's analysis) were excluded.

3.2 Medical tourism conceptualization

The concept of health tourism is often confused with the concept of medical tourism; hence the limits of these two concepts are not fully clarified (Connell 2013). Through deep analysis, it can be said that both of these terms are related but not the same. To understand these two concepts, we need to consider three important factors (1) geographic boundaries, (2) the type of useful medical intervention, and (3) the medical center where the intervention is performed. Figure 1 shows the three mentioned factors.

As shown in Figure 1, if a person travels from one country to another to buy cheap medicine, it is classified as health tourism. On the other hand, if the purchased medicines are accompanied by medical interventions such as surgery, this action is classified as medical tourism (Fetscherin and Stephano 2016). For example, in cases like the United States of America as well as Canada, where people have crossed local

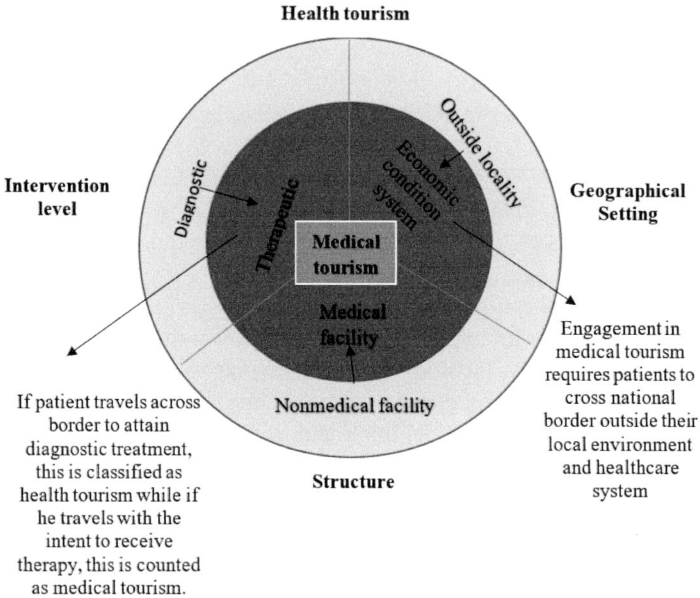

Figure 1: Scope of health and medical tourism (Carrera and Bridges 2006).

borders or states just to receive health services without medical intervention, this would not be considered medical tourism.

4. Types of Medical Tourists

The key element of the medical tourism industry is medical tourists who are the primary customers of medical products and services and have a significant impact on the success of this industry (Ghasemi and Hasanpoor 2021). In the field of tourism, medical tourists can be defined as those who travel across borders with the intention of obtaining medical treatment or health-related procedures while resting or performing tourism activities (Connell 2013). From the point of view of the medical field, medical tourists are those who, contrary to the definition of traditional tourism, travel across borders without intending to participate in tourist or recreational activities (Green 2008). Medical tourists can be classified based on place of stay, length of stay, and type of medical treatment. Each of these class includes different types of medical tourists with distinct qualities for each (Ghasemi et al. 2022).

4.1 Tourist's place of residence

Outbound medical tourism includes patients who leave their country (exporting country) to consume health services in another country (importing country) (OECD 2011). The lack of trust in the medical services provided by the local health care system and the lack of doctor and nurse staff cause the emergence of medical tourism abroad. Cross-border medical tourism offers patients the opportunity to receive medical services that are affordable, available, or legally prohibited (such as organ transplantation in the Philippines) in the home country but legal and permitted in other countries (Cohen 2008).

In the past, only countries such as USA, Switzerland, and the then East Germany received wealthy class tourists from other countries to receive top-quality medical services not available in their homeland. However, today developing countries receive patients from rich countries who seek superior health services and places to relax (Horowitz et al. 2007). In addition, patients traveling from rich countries to developing countries are looking for cheaper and more accessible care to avoid the long access queues and expensive costs of medical services at home. For example, India has made significant improvements in its medical infrastructure, transport, and travel networks, security, facilities, and tourist spots to attract citizens of countries like the US, UK, etc. Long queues to access expensivee medical care in a local health care system makes patients think and decide to go abroad because they can get affordable medical services while getting equal or better quality care compared to their country (Waikar et al. 2011).

First, governments of most developing countries, realizing the huge potential of medical tourism, have actively invested in the medical tourism industry to upgrade medical infrastructure, develop, and increase medical expertise. These measures have introduced countries in the field of medical tourism and gain international recognition for attracting medical professionals from abroad. One of the main examples in this field is the country of Malaysia, realizing the great potential that medical tourism can create, its government has taken major measures through the creation of new medical centers, improving the transportation system, and developing the sector. Hospitality provides the initiative to improve medicine in this country. The investment of the private sector in the medical sector and other related sectors in this country has made Malaysia known as a medical tourism destination at the global level (Chee 2010). However, Malaysia is not alone in this area and there are other countries in Asia such as Thailand and Singapore that are actively developing and promoting medical tourism.

Second, inbound medical tourism refers to the travel of domestic patients to another city, region, or state for medical tourism purposes (OECD 2011). Medical tourism creates competition that challenges local governments to improve and differentiate their local medical services (Sobo 2009). Medical tourism abroad is putting pressure on local health sectors such as hospitals and clinics to redesign the way they operate their business, as significant income is lost from local people (Malekzadeh et al. 2021). Forgione and Smith (2007) stated that the local medical tourism system must be developed to reach a wider segment of the community, reduce costs, improve operational efficiency, and provide appropriate prices and quality information to the consumer to compete effectively in today's world. As a result, some hospitals are starting to offer attractive domestic medical travel options to promote and enjoy the benefits of medical tourism. For example, America is starting to specialize in certain treatments and offer cost reduction packages for certain surgeries to simulate national medical tourism and face the global challenges that arise from medical tourism (Rhea 2009). In this regard, the states of Oklahoma and Missouri are attracting local medical patients from other states who are interested in oncology care and medical interventions (HealthLeaders-InterStudy 2011).

Third, intra-border medical tourism/diaspora refers to a group of migrant tourists in a country that return to their citizen countries for medical treatment.

Immigrants from countries such as China, India, Mexico, and Korea refer to their countries for these primary reasons. First, none of the desired treatments are available in their country of residence. Second, medical services are expensive in the host country, such as for Mexican immigrants in the United States. Third, there is a lack of trust in the country's medical care system. Fourth, cultural similarity. Immigrants sometimes suffer from a lack of shared culture in their host country, such as language barriers for Korean immigrants in Australia (Alsharif et al. 2010, Lee et al. 2010, Alvarez et al. 2011).

4.2 Period of stay

Medical tourists can be classified based on the length of stay at the medical destination. According to Carrera and Lunt (2010), medical tourists can be divided into three types based on the length of the medical trip. First, temporary visitors from abroad, who often choose to stay for a few days. Second, long-term residents, whose trip is likely to last more than a month. Third, common borders, countries which can be frequented due to a common border such as the European Union.

4.3 Medical tourist's motivations

The decision making of medical tourists is often influenced by certain motivations that shape it. Motivations can be defined as perceived needs that drive decision-making processes (Yolal et al. 2012). Tourist motivations are defined as a framework of awareness that allows a person to move while providing a rationale for this travel decision (Yolal et al. 2012). Attitudes play a key role in patient decision making. They become aware of the perceived needs of medical tourists based on the patient's value system and situational factors. For example, people's intention to perform a behavior is influenced by their attitude. If patients have a favorable attitude towards that behavior, the likelihood of conflict decreases. In addition, if the medical tourist is satisfied with the result of the health care activity, he will have a positive attitude towards medical tourism and will have a favorable desire to participate.

Choosing a medical tourism destination includes two parts: the first is the decision to choose a foreign country and the second is the choice of international facilities (Forgione and Smith 2007). Each of these decisions adapts and changes based on the customer's needs. Figure 2 shows the two stages of the patient decision-making model.

Forgione and Smith (2007) are of the opinion that the choice of a different country is influenced by several variables including the economic situation, political environment, and legal system of that country. Regarding the economic conditions, patients are more likely to choose countries that have a stable economic situation and growth, because they have the advantage of responding quickly and providing a

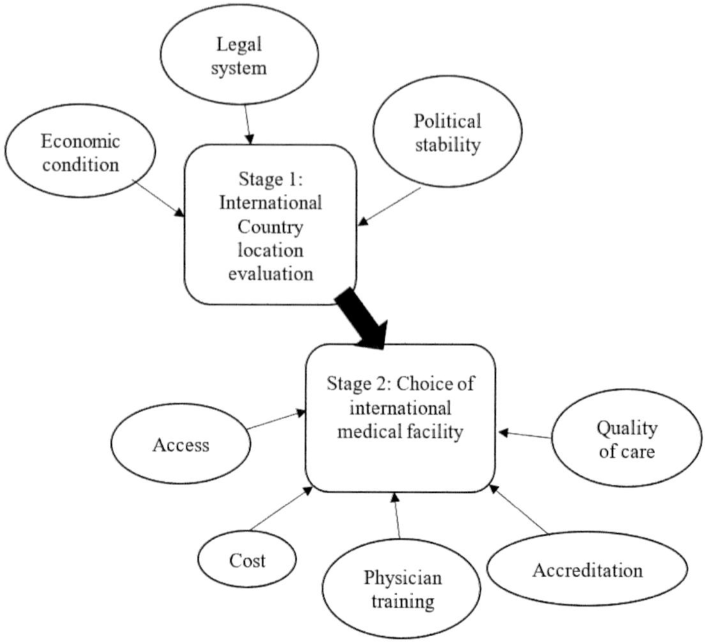

Figure 2: Factors affecting the choice of medical facility and international destination.

wider range of services to the patient sector. Regarding the political environment, it is said that the political situation of the chosen country plays a major role in attracting medical tourists. Travel safety and health are important priorities for the patient; they do not choose a destination that may harm their life. Patients will be attracted to countries where government power is well exercised and there is no danger of war. Finally, regarding the legal system, the legal and regulatory environment of the country is one of the major concerns of medical tourists. Therefore, patients are attracted to countries that have protective laws that protect their rights. For example, the US healthcare system created the Health Insurance Portability and Accountability Act of 1996 (HIPAA) to protect US patient information and restrict how patient health information can be shared or used (Mello 2006). The US healthcare system also establishes rules for compensating patients in the event of injury or unacceptable outcome standards.

In order for a medical tourist to determine the destination country, he begins by evaluating its international healthcare facilities. During this process, there are push and pull motivations that affect the evaluation of patients. Push motives are the factors that make people decide to travel outside their home country to seek treatment in another country (Weaver and Oppermann 2000). The pulling motives of tourists are classified into the following main categories:

First, personal escape. Patients sometimes want to get out of their normal lifestyle and environment and experience new things, the pressures of work life, and the desire to experience new adventures lead patients to seek medical tourism. The higher cost of medical services in one's own country compared to others and the long waiting time limits people's access to the desired medical procedures. For example, in Canada and the UK, people choose to travel abroad to avoid long waiting times (Gray and Poland 2008, Hall 2011, Hansen 2008, Hume and DeMico 2007). In some treatments, such as drug addiction treatment and gender reassignment, where confidentiality is a requirement, patients choose to receive treatment abroad, where they cannot be seen by anyone they know (Connell 2006). As a result, participants decide to leave the system and look for a better destination.

Second, interpersonal escape. This can be defined as avoiding stressful social environments or normal social environments (Snepenger 2006). Being sick forces them to leave the house to avoid causing trouble to their family members. These people feel ashamed that others have to take care of them. They want to leave to reduce the pressure on caregivers.

The third is interpersonal search, which is defined as attending events to gather family members and meet and seek new friends who share common interests. The opinions of family members and friends are very important in influencing patients' decisions to enroll in health tourism. Some people have the chance to combine their caregiving journey with a holiday with family at the same time (Connell 2006). Most people prefer travel destinations where they can meet people who can help them overcome their worries. In addition, medical tourists prefer to move to places where they can really enjoy the vacation and meet friends and relatives, enjoy support, and connect with members of the traveling group (Snepenger et al. 2006).

In addition to push motives, there are pull factors that influence the patient's decisionmaking. These factors are described as specific motivations that lead people to understand the desires for specific tourism experiences and attract them to travel to the destination country. They motivate medical tourism destinations based on the point of view of the pulling factor of the destination, patients, which include the following:

First, there is cost, which is undoubtedly the most important motivation considered by a medical tourist. Medical tourists feel that they can obtain significant benefits in a place other than the usual environment (Essier and Casken 2013, Lunt et al. 2013). The main attraction that drives patients to seek care in developed countries is reasonable cost (Connell 2006, Gray and Poland 2008, Horowitz and Rosensweig 2007, Hall 2011, Herrick 2007, Ormond 2011). The demand for medical tourism has increased due to the continuous increase in the cost of medical care in rich countries such as the United States and the continuous development and expansion of health facilities and services in emerging countries with better healthcare costs (Lunt et al. 2016, Yu and Ko 2012). In most cases, medical tourists tend to reduce the cost of elective medical treatments that can only be paid for out of their wallets (Crooks et al. 2010). Many medical services, such as dental care and gender reassignment surgery, are not covered by insurers or have limited discount rates, where people have to pay out of pocket for each required procedure, which is an opportunity for medical tourists to seek treatments not supported by local insurance policies (Österle et al. 2009). India is among the preferred destinations for American medical tourists due to major cost differences in its hospitals compared to the United States. For example, an acute knee operation in India costs only $1,500, while the same operation costs

$10,000 in American centers. These countries are able to offer such a low cost because of cheap labor and reduced costs in court cases (Herrick 2007). Cosmetic surgery is one of the most common health services in the health tourism market because it can be performed more cheaply in emerging countries (Marlowe and Sullivan 2007). In addition, Connell (2006) points out that travelers want to take advantage of cheap air transportation and attractive currency exchange.

Second, the quality of medical services and trust. Better quality treatment with high standards of care and hospitality, along with professional consistency in terms of procedures and operational expertise, promotes patient enrollment (Chang et al. 2013, Han and Hyun 2015). The emergence of world-class medical facilities and services, as well as the availability of modern medical technology in the destination area, play an important role in attracting customers and creating an image of the destination. In addition, the availability of well-trained physicians and nurses and the preference of physicians/specialists to perform a particular procedure or treatment are key considerations for choosing the location of high-quality health services. Some patients require first-class medical care. They are very sensitive to quality and seek higher standard medical care (Yap 2007). In addition, medical tourist group will focus on expert medical staff, post-operative care services, speed of service delivery, and privacy of international medical service standards. For example, Malaysia is becoming a desirable country for medical patients due to its highly effective medical personnel coupled with modern health and welfare infrastructure (Tourism Malaysia 2007).

The third is faster access to medical and health facilities. Another critical factor that encourages people to travel for medical tourism around the world is the long waiting time for health treatment in their home countries (Heung et al. 2010). Speed of access to medical services is one of the primary considerations for medical tourists (Turner 2010). For example, access to cancer treatment facilities in many countries requires a long waiting time and long treatment duration (Vijaya 2010). According to a study conducted in the United States, patients had to wait an average of 66 days for a medical examination in Boston, twenty-nine days for seeing the first doctor for a skin examination, and thirty-two days for a heart evaluation by a cardiologist in Washington.

Fourth, the availability of information. A medical patient is expected to travel to a destination with positive word of mouth and good reputation. In addition, patients are influenced by their experiences and motivated to travel when their friends enjoy medical travel and speak positively about the medical destination. Also, the presence of reliable knowledge on the Internet also plays a role in patients' decision making (Crooks et al. 2010). Medical tourists are more motivated when they can easily find the necessary information about the desired destination through the Internet and social networks.

Fifth, cultural similarities. Cultural similarity is another important driver in which people feel more comfortable going to a country with a common language, history, and geographical proximity (Jaapar et al. 2017). The more a country is similar to the patient's culture and language, the higher the potential for the patient to choose. For example, cultural similarities seem to be an important motivating factor for Indonesian patients to choose Kuala Lumpur as a medical destination, as Malaysia and Indonesia have similar cultures.

Sixth, legal provisions. Some procedures and treatments increase the demand for medical services because these treatments are either unavailable or legally prohibited in the countries where patients live. For example, the United Kingdom has banned abortion and families wishing to have an abortion must travel abroad (Connell 2006). The Philippine government has also banned organ transplants (Horowitz and Rosensweig 2007).

Seventh, hospital accreditation. Accreditation standards are an important factor that a medical tourist uses to evaluate the quality of services provided by an overseas medical facility. In foreign countries, medical centers should seek to meet or exceed the quality standard of facilities received by patients at home. Foreign patients prefer to choose medical centers that provide the same quality of treatment as or exceed the quality provided by their home facilities. These international accreditations indicate that medical centers employ only trained and professional nursing personnel (Ramanna 2006).

Internationally recognized accreditation, such as the Joint Commission International (JCI) initiative, seeks to establish standards that provide uniform expectations for structures, processes, and outcomes for hospitals with more than 120 beds. JCI health care facilities in Europe, India, Middle East, Asia, South America, and the Caribbean have eleven, and in Singapore and India have six approved medical centers, which the most favorable places are in US (JCI International 2007). Many other accredited bodies also operate in addition to JCI standards. For example, the Australian Council for Healthcare Standards (ACHS), the UK and Hong Kong Trent Accreditation Scheme (TAS), and the International Organization for Standardization (ISO).

Finally, training doctors. Many medical tourism countries are working on attracting patients by increasing the number of qualified doctors in their medical schools. For example, in Bangkok, Boomeranggrad Hospital advertises to American medical tourists that more than 200 board-certified physicians at that hospital are board-certified by US medical universities. The hospital advertises that its management team includes professional medical directors from the USA, Singapore, Thailand, and Australia. The doctor's willingness to speak the tourist language and specialized training in their countries of residence is another important factor for patients, which also increases the choice of hospitals. Figure 3 presents a framework of popular medical tourists' motivations.

5. Internet and Medical Tourism Facilitator Roles in Medical Tourism

5.1 Internet

Accurate information about medical care choices, medical centers, travel benefits, itineraries, and host countries for potential patients is a critical link to optimal medical tourism practices (Crooks et al. 2011). The Internet is the most important source of public travel data. This resource contains supplementary information for medical tourists seeking travel information (Rooney et al. 2004, Patterson 2007).

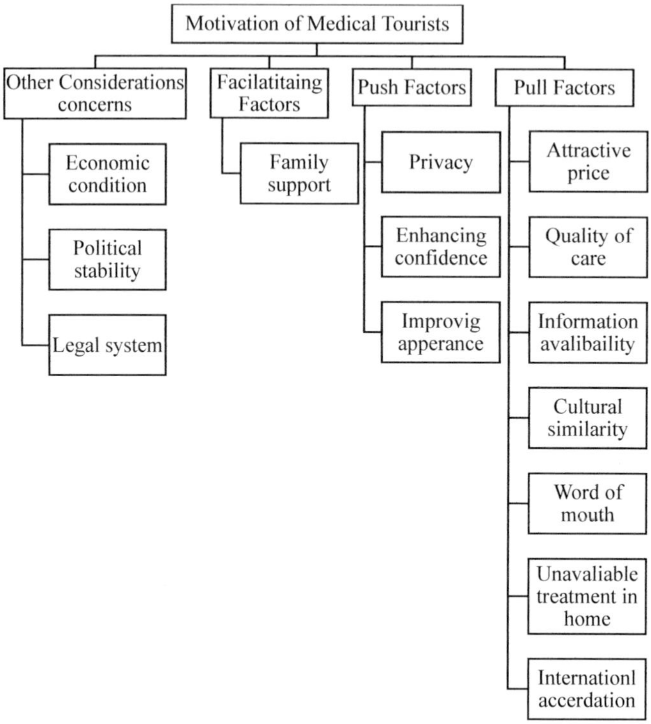

Figure 3: A framework of popular medical tourists' motivations.

The Internet plays an important role in promoting the demand for medical tourism. Every medical destination uses the Internet and its website to advertise its medical services (Connell 2006). The Internet provides a database for accessing information and advertising of medical treatment. This affects the patient's decision-making process and medical tourists are more likely to start exploring different destinations using the Internet to choose a destination (Lunt et al. 2010).

Internet websites are one of the main Internet products that have emerged to promote this industry (Lunt et al. 2010). Industry marketers create these websites to reach and assist people who use the Internet to search for medical care information. These websites can serve as a medium for medical travel providers and operators, and to introduce and create awareness about the various medical care treatments offered by hospitals and medical clinics to those seeking information about the services (Patterson 2007).

The Internet can also be used to advertise and promote medical treatment through active advertising and enrollment programs. According to Lunt et al. (2010), medical websites perform a number of functions. First, they create a gateway to medical information. The main goal of medical websites is to provide enough information to ensure the interest of medical tourists, in a coherent, simple, and reliable format to help. The purpose of a webpage is to help and ensure that patients make informed decisions and choices about clinical investigations, transportation, and medical care abroad. Second, connecting to related medical services. By providing relevant information and details to insurance providers, travel partners, and other websites, medical tourism sites connect patients with the necessary parties to complete the traveling matter. Third, evaluation and promotion of medical services. One of the important functions of medical tourism sites is to market their services to attract and maintain consumer interest. By creating awareness of medical tourism, medical websites aim to manipulate the emotional and behavioral aspects of customers to create a perceived desire to ensure that they seek medical tourism assistance for participation. Checking the consistency and validity of the quality of web data and information is a vital necessity. Moreover, sometimes the content may be vague, unreliable, and deceptive (Ibraghimova 2006). Medical tourism sites enable patients to access relevant health information and portals in times of uncertainty through analytical reviews of treatment, specialty, and health practices. Fourth, trade. The commerciality of medical websites depends on the processing of consumer information while browsing the websites. Important information is obtained in the form of cookies, surveys, posts in newsgroups, and web pages, and the data obtained from users to register on the website in order to access the entire content or sign up for newsletters can be used for future purposes. Finally, the opportunity to communicate. Medical websites are able to create a direct interaction between the doctor and the medical tourist as a customer in terms of communication and evaluation of a process involving the consumer's well-being. Such communication can provide an immediate response and facilitate the potential evaluation of the patient's health care after treatment by a medical professional (Nejad and Banar 2018).

5.2 Medical tourism facilitator

The global development of healthcare technologies and the increase in the number of health patients have led to the development of support services such as the emergence of medical facilitators in the tourism sector. Preparation for travel is very complex and intensive, especially for patients with no previous travel experience. For example, a prospective patient with no prior knowledge trying to directly interact with a doctor or hospital abroad is motivated to engage in extensive searching on hospital websites, along with searching and processing information.

Language and conversational difficulties are likely to present another degree of difficulty. Medical tourism facilitation companies have been established to facilitate the process and provide the required information (Ghasemi et al. 2021). These companies operate through the network. They sell medical services to a person seeking medical care abroad. Facilitators of medical tourism play an important role in marketing, promoting, and assisting in travel planning (Crooks et al. 2011, Khawand et al. 2022). The main purpose of such companies is to inform the public, create consumer demand, and ensure that the buyer is inspired to buy the service (Lunt et al. 2010).

These facilitators are intermediaries who connect a potential client to an international health professional and help them organize medical, dental, or cosmetic services outside the patient's country of residence. They are new companies that provide support to the medical tourism industry such as risk management companies, healthcare solutions companies, and aftercare recovery companies (such as Homewatch International). Traditionally, a travel agency has played an important role in arranging travel plans and coordinating overall customer service, and due to its impact on individuals and businesses, travel operators are considered a vital part of the tourism system (Budeanu 2005).

The optimal performance of medical tourism requires good coordination and cooperation with key players in the fields of medical care and tourism such as hospitals, travel companies, and resorts to meet the needs of customers and attract them (Heung et al. 2010, Anifowose et al. 2022, Ghasemi et al. 2022). Facilitators of medical tourism act as a pulling force to increase demand in the medical tourism market (Budeanu 2005). Medical travel agencies act as a one-stop solution that provides first-hand details with integrated knowledge of medical services and networking with leading medical care resources that can assist medical travelers (Jagyasi 2010).

Medical facilitators perform many tasks. (1) Facilitator of full-service medical travel, working on the assumption that international patients' access to doctors and medical services in destinations where medical travel is challenging due to language and technical differences between the two countries. (2) To overcome language gaps and social distance, full-service facilitators strive to provide fully organized programs, starting from travel permits, accommodation, doctors, various medical facilities, post-operative treatment, and translation services, information, and assistance. They facilitate the distribution of patient information and online browsing through popular Internet media to promote potential medical destinations and procedures that assist medical tourists in their selection (Ytterhus 2000). Jagyasi (2010) indicated that medical tourists, especially those who are not comfortable with the host country, require medical tourism facilitators to reduce waste by seeking out reliable providers, ensuring hassle-free itineraries, and providing integrated prior knowledge. These service operators minimize time: While a patient searches for a specific country, it matches the potential consumer with reliable foreign medical service providers, saving patients valuable time. This also helps them in arranging the desired medical treatment abroad for the domestic consumer. In addition, they assist patients in making logistical preparations and travel arrangements such as visa procedures, ticketing, using airports, providing interpreters, booking hotels, planning tours in the destination country, providing insurance services, currency exchange, etc. (Gan and Frederick 2011, Forgione and Smith 2007).

In addition to all this, they help medical patients with aftercare treatments in their country of residence. They transfer medical records when needed and connect patients with hospitals and clinics to assist them with follow-up treatment while the patients are in their home country (Horowitz and Rosensweig 2007). They evaluate and monitor the quality of their services during and after recovery to ensure customer satisfaction. This will reduce negative perceptions and positively contribute to the medical tourism industry (Law 2000, Shirkhani and Ghasemi 2020).

One such example is the Medtrava website (www. Medtrava.com), which sends patients to India, Thailand, Mexico, and Costa Rica, and North America. Medtrava strives to articulate medical care and convey to patients the same feeling they have in their local communities. For example, if a patient wants to go to Mexico and seek medical care, the patient first contacts Medtrava. Then the organization determines the date of the trip and after filling the medical history questionnaire, provides the total cost of the trip. As the day of departure approaches, the company organizes travel, accommodation, hospitalization, and necessary documents to make patients feel reassured. This means that the medical tourist leaves the terminal, is transported to the treatment facility, and then returned to the terminal (Whittaker 2008).

There is another type of medical travel facilitator called a referral service facilitator. Typically, referral service facilitators aim to connect medical tourists with care providers. They offer limited services such as arranging appointments, organizing part of the trip, and information about routine care. Referral service facilitators operate in one country and focus on connecting patients rather than dealing with logistical issues compared to full-service medical travel facilitators. One such example was a referral support provider along the US-Mexico border by a 50-year-old US citizen. He started by recommending dentists and doctors on a website that he knows personally. He helped American patients overcome the cultural differences associated with Mexican healthcare, such as communication, aesthetics, and patient-physician experiences, as these factors make the selection of healthcare professionals challenging. Involved patients can access this person's webpage, message him, and pay a fee. He then contacts the attending physician who is familiar with the patient's needs to schedule an appointment.

Medical travel agencies can be classified into four major groups: first, hotel organizations. For example, the ITC-WelcomGroup in India, which is expanding its services as patient provider intermediaries. Second, travel companies such as Commonwealth Travel in Singapore, which provides travel arrangements and organizes logistical requirements for medical tourists. MediTravel Malaysia helps connect potential patients with international medical care providers and help them organize health, dental, or cosmetic services outside their home country. Third, medical travel organizers such as Planet Hospital and Global Choice Healthcare, which act as advocates for patients seeking care abroad, or Cosmetic Surgery Travel, which is a US-based medical tourism organization specializing in cosmetic surgery from the US to Thailand.

This agency makes all arrangements for American patients' stay in Thailand and helps them with surgical evaluation and further planning. Also, Med de Tour, which is headquartered in the UK, has been developed to help UK patients get world-class medical treatment at lower prices compared to the UK at India's best medical centers with the advantage of reduced waiting times. Fourth, the service organization, such as Bumrungrad in Thailand has developed healthcare services exclusively for foreign medical tourists (Keckley and Underwood 2008).

The mediator role may also be played by health insurers or employers. In their policies, providers and insurance companies are considering outsourcing some treatments as a way to minimize costs and offer discounts on coverage to people who go abroad (Caffarini 2008, Reed 2008). According to the Centers for Medicare (2015), health insurance and employer healthcare costs in the United States increased from 5.3% to 17.5% of GDP in 2014. As a result, private and public employers were forced to minimize the provision of health benefits.

In addition, insurers and employers have listed several overseas medical centers as licensed health providers and offer discounts to clients for using medical tourism (Arueyingho et al. 2022). For example, Tehe Blue Cross Blue Shield of California has provided financial benefits to US citizens for further treatment in Mexico (Vequist and Valdez 2008). Insurance companies help consumers overcome concerns about the quality of care and patient safety by working with JCI-certified medical centers or centers with outstanding ratings.

6. Opportunities and Risks of Medical Tourism

To examine the opportunities and threats of tourism, the term "export" is used to describe medical systems where medical tourists are imported from other countries to receive health treatment. The term "imports" is also used to describe patients leaving their local medical system to seek healthcare elsewhere.

6.1 Exporting medical tourism countries

6.1.1 Exporting opportunities

The medical tourism market is expanding rapidly. It is one of the fastest-growing tourism sectors that enable countries to improve their economies (Han and Hyun 2015). The medical tourism industry offers great potential for developing countries, and since exporting countries are generally lower-income countries, medical tourism can benefit them greatly through the revenues it generates (De Arellano 2007, Timmermans 2004, Turner 2007). From an economic perspective, medical tourism offers economic opportunities with financial incentives for the healthcare system (Whittaker 2008). Trying to attract medical patients can be a means of raising additional money. Tax receipts collected from partnerships with the tourism, insurance, food, and hotel sectors may be used in hospitals and technological equipment, which can lead to cost savings in the domestic medical care system; some examples of which are Singapore and Cuba (Lee et al. 2010).

In this regard, medical tourism has the possibility to facilitate the growth of the local economy by bringing foreign currency from foreign patients to the host country, which increases income and motivates the private sector to invest more in the medical sector (Beladi et al. 2015). According to the Tourism Research and Marketing Group, about 37 million health-related trips are made annually, generating €33 billion in revenue (Smith et al. 2009). As a result, more jobs in health and related industries such as manufacturing, housing, transportation, and tourism will open up for local residents. In addition, it creates the potential to attract other high-quality medical vendors in the region such as medical device companies, technicians, medical systems technology, and other companies related to the medical sector. This movement will serve the existing medical facilities and promote the domestic economy and increase local taxes. In addition, it improves the economies of countries that are heavily dependent on a few limited sectors, such as destinations that rely on the tourism industry (such as Turkey) as a significant economic activity.

Some industries are associated with off-peak periods where the economy is likely to recover slowly (the tourism industry is likely to slow down in winter). Providing some elective surgery and other medical care that has a more appropriate schedule can help reduce periodic recessions or off-peak seasons in the country's economy. In addition, the pricing for the travel of a therapeutic traveler can be considered in this context. Lower-than-average off-peak pricing can help patients reduce costs and in turn attract more patients to the destination.

On the social side, international patients allow healthcare providers to learn more about different cultures, social norms, and regional beliefs that affect healthcare, and to improve communication skills and advanced management strategies on a global scale (Ghasemi et al. 2017). Private medical institutions can share their services and human capital with the public sector due to the high profit of the private sector from medical tourism. This encourages residential patients to enjoy high facilities and reduce the burden on the public sector (Chen and Flood 2013).

Therefore, the profits from medical tourism can be invested in the public healthcare system, which will benefit the public sector. Medical tourism enables the local community to avoid "brain drain" problems. That is, expanding medical tourism can help exporting countries retain domestic medical professionals and reverse the brain drain by offering competitive salaries and job opportunities (Connell 2008, Chinai and Goswami 2007, Dunn 2007). Medical tourism requires an educated and skilled workforce to work in this industry. It should also strive to provide job opportunities for graduate students in the national health industry instead of leaving the region.

On the medical side, medical tourism allows health providers to gain first-hand experience in dealing with diseases that are not commonly known in the world. In addition, all the revenues from medical tourism can be used to improve and upgrade the current medical services and treatment standards. Strengthening the destination brand is an important marketing objective to attract medical travelers seeking high-quality medical care (Ratner 2009). This creates an incentive for the public sector to invest and improve its medical facilities to diversify the tourism base and reduce destination marketing costs (Chen and Flood 2013).

6.1.2 Exporting risks

Despite the benefits of medical travel for developing countries, there are many concerns associated with medical tourism for exporting countries. For example, there is concern about the cost of medical services for local people (Chen and Flood 2013). Medical tourism may have adverse effects on local medical systems because when foreign patients' demand for medical services increases, prices are likely to increase and local citizens will incur higher costs. This phenomenon has been shown in Thailand where many people cannot access medical treatment in the private sector due to high prices for medical services.

Medical tourism can lead to overcrowding of local people and reduced access to healthcare services, leading to inappropriate relationships between local and foreign patients (Connell 2013). In addition, the unequal distribution of healthcare services between foreign patients and domestic patients, especially for those who are completely dependent on public healthcare, creates a dual market for health services in which only foreigners have access to highly qualified staff, and world-class private hospitals ,that use high-tech sanitary devices. On the other hand, the domestic citizen can only access basic medical facilities without resources. All this puts a lot of pressure on domestic patients who are completely dependent on their government's public medical services (Cohen 2010).

Another concern is the lack of government regulations and policies that guarantee the standard of patient treatment, safety and efficiency, and quality of the medical care sector (Burkett 2007). Ensuring the quality of medical care is a key factor in sustaining the medical tourism industry. In the absence of such regulations on the part of the governments, there will be a lack of trust on the part of the patient, which will negatively affect the attractiveness of the country. For example, most medical tourists returning from a healthcare trip often have difficulty obtaining follow-up services from local hospitals for the period after their treatment. In addition, a study confirms that the lack of government support to protect the quality of health care and the effectiveness of certain treatments is ranked as the number one issue for American patients in the medical tourism market (Herrick 2007).

While it is possible that medical tourism will reverse the mass brain drain, there is concern that medical tourism will lead to an internal brain drain. Doctors are expected to migrate to work in private clinics that attract medical visitors rather than public health system tourists, where job opportunities and higher salaries are offered. As a result, there will be a shortage of healthcare professionals in the public health system, which will affect the quality of public health services (Arunanondchai and Carsten 2007, Burkett 2007, Chinai and Goswami 2007).

Most of the concerns of medical tourism include ethical issues (Malekzadeh et al. 2020). The development of medical tourism, especially in developing countries, leads to many ethical problems that arise as a result of poor governance. These practices, like the development of profit-oriented medicine, focus on increasing revenue and ignore ethics (Gan and Frederick 2011, Han and Hwang 2013). The legal issues are considered a risk through the possibility of imposing legal restrictions that restrict people from moving from one destination to another, such as complex visa procedures for patients entering a particular country. Imposing complex visa procedures is detrimental to promoting the growth of medical tourism globally. At the same time, it provides opportunities for other local people and thus reduces the competitiveness of the destination (Herrick 2007, Whittaker 2008).

Patients are always exposed to infections after surgery or during medical procedures in hospitals, and the lack of blood in hospitals in destination countries, as well as traveling during recovery, obviously have adverse effects on the patient's condition and health. In addition, medical tourists may be vulnerable to unforeseen health threats that require medical treatment or facilities that are not in the patients' home country. For example, a kidney transplant, which is illegal in most countries, may cause a patient to experience tuberculosis, pneumonia, HIV, urinary tract infections, and wound infections. Such a risk threatens the patient's life because these treatments are not allowed in their countries and if post-operative care is needed, the home healthcare system is not ready to accept it.

Another aspect is language barriers and ineffective communication concerns. Successful information about medical care choices, treatment centers, tourism benefits, travel plans, and countries for prospective patients is a critical key to optimal medical tourism practices (Crooks et al. 2011). Medical tourism is a global industry with a large number of international patients, each of whom speaks a different language. This complicates the communication process and leads to not providing correct information to potential customers.

As part of the competition, the medical tourism industry faces risks. The medical industry is characterized by intense competition between different medical destinations, where each destination offers different treatments and services to meet the needs of patients. As a result, the lack of effective marketing and customer-oriented approach creates more pressure on tourism destinations to differentiate themselves and market effectively. In addition, the medical tourism industry faces significant risks regarding insurance companies and related services. There is limited insurance coverage, underdeveloped insurance industry, insurance fraud, and denial of compensation by international companies.

Among the other negative concerns of medical tourism in exporting countries, we can mention the use of investment in high and private urban hospitals to meet the demand of medical tourists. Ignoring the needs of rural primary care and health centers for local residents living in remote areas to access medical treatments and facilities is unfair. For example, during the preparation for the 2014 World Cup, the Brazilian government provided the best medical facilities with advanced medical equipment and well-trained doctors in urban areas, while while ignoring the rural public health care sector (Connell 2008, Garud 2005, Leahy 2008, De Arellano 2007).

Another issue that a tourist destination has to become a competent medical tourism industry is a weak economy. The unstable economic situation represents a push factor for further investment and growth of the medical market. Economic problems, including domestic currency fluctuations due to unstable conditions, cause economic stagnation and decrease access to credits. Also, lack of communication services such as lack of access, lack of coordination mechanism, lack of proper catering services, inadequate electricity supply, and poor water supply fuel this issue.

6.2 Importing countries

6.2.1 Importing opportunities

Medical tourists traveling abroad can benefit from imported destinations. When a country designates a certain number of patients to be cared for in certain medical centers, prices can be negotiated based on these facilities to save costs. These savings from patient care abroad will be integrated into the overall budget, which will allow financial resources to be properly prepared and organized. One report estimates that if 1 in 10 Americans suffering from 1 to 15 diseases moved abroad for medical care, the net annual savings would be $1.4 billion (Mattoo and Rathindran 2006).

In addition, medical tourism allows importing countries to reduce their waiting lists and average waiting times. By preparing the sent patients, how many, where, and with what treatment, it will be easier for the government to organize the public health system and increase the standard of care provided to the citizens (Malekzadeh et al. 2021). In addition, it provides the patient with a wider range of medical choices with higher levels of personalized care than a medical tourist can choose from it. Moreover, it allows the patient to experience a high level of privacy and confidentiality during medical treatment (Burkett 2007, Horowitz et al. 2007, Muzaffar and Hussain 2007).

6.2.2 Importing risks

For import destinations, there are some unfavorable factors associated with medical tourism. The main risk is the standard of treatment provided by the exporting countries. Importing countries are concerned that the quality of care their patients receive abroad is not equal to or inferior to that available at home. Exporting countries have tried to solve this problem by developing regional and global certifications for production and promoting the reputation of their medical industry. These accreditation bodies ensure that medical centers in exporting countries provide care in accordance with local and global requirements and that medical services are equivalent to or better than those provided by their native centers to prospective clients. The Joint Commission International (JCI), the Australian Healthcare Standards Council, the International Association for Healthcare Accreditation, and the Canadian Health Services Council are the main global accreditation bodies.

Another concern is the lack of appropriate legislation in exporting countries (Dunn 2007, Turner 2007). Errors by the medical staff will inevitably occur, and what countries can do to protect the rights of the patient may be limited and not exceed a small compensation. In addition, lack of post-operative care is another concern (Connell 2008, Dunn 2007). After the patient returns home after the operation, if complications occur, the internal medicine system must deal with them. In addition, the lack of data exchange between physicians in importing and exporting destinations will make continued treatment and reduced profits in the importing country's health sector increasingly challenging (Martinez et al. 2008, Merrell et al. 2008).

Another concern of medical tourism is the increase in demand for it. As the medical tourism market continues to grow and expand, while domestic healthcare systems suffer from limited resources to develop their medical facilities, governments, and health system leaders must make decisions to improve their medical services that address potential risks. Medical tourism should be eliminated.

7. Conclusion

The investigations carried out in this study show the lack of clarity between the conditions of medical and health tourism. Guided by this research, we defined medical tourism as organized travel outside of one's natural health care for the purpose of restoring health through medical intervention (Whittaker 2008). While health tourism is a broader concept of which medical tourism is a part and includes organized travel outside one's home environment to maintain, improve, or restore health in one's mind and body (Pocock and Phua 2011). We discussed the scale of health and medical tourism depending on the geographic context, implementation of a health-related intervention, and the structural facility in which the intervention is used, to better understand health and medical tourism. Through this review, we also identified types of medical tourists using different aspects based on the type of stay, length of stay, and medical product to provide a better understanding of medical tourists, their impact, and their needs. In addition, we discussed personal and country factors that drive patients to seek medical care beyond their local system, as well as medical destination pull factors that drive medical patient choice.

As discussed, the main push factors for medical tourism are high prices and long waiting times in the home country, which place a heavy burden on medical patients, while the low costs and high quality of medical facilities abroad. It is considered the strongest traction factor.

The Internet plays a vital and fundamental role as well as a mediator of medical tourism in the practice of medical tourism. Due to the obvious role of both in communication and influencing the selection of medical patients, they are considered a key success factor in the medical tourism sector. Facilitators come in many forms such as hotel groups, travel agencies, and medical travel planners. The purpose of all of them is to serve medical patients and connect them with medical facilities, but the scope and depth of services provided, such as translation, etc., depends on the patient's choice.

Considering the impact of medical tourism, countries should be aware of its advantages and disadvantages of medical tourism. It is not right for a country to be completely dependent on overseas medical tourism, as it will harm its economy and its overall system such as the reliability of that country. On the other hand, the cities that accept medical tourists should take care of the quality of their services to the citizens along with foreign patients. It is not acceptable to provide quality services to foreign patients while low-quality, expensive services are provided to their own citizens. Prioritization is necessary for the medical tourism sector. To develop the medical sector as a medical destination, a balance must be maintained to ensure success on a local and global scale. In addition, countries that export medical patients abroad should be aware of how far they can push their citizens out and what reforms they need to make to improve the local system.

7.1 Recommendations

This study recommends that further analysis is needed for the behavior of medical tourists and how the role of facilitators is formed as a key pull factor in the medical tourism sector and the impacts of medical tourism. It is hoped that by covering these sections and the concepts of medical tourism, this research can guide and help future researchers in conducting basic research.

7.2 Limitation of the study

The lack of access to studies that are relevant to the purpose of this study is limited. Because there are few studies that have already been done in this field of study. In addition, the greatest limitation in this study is related to the lack of access to real examples and data related to the number of medical tourists and the income from medical tourism.

8. List of Abbreviations

OCED : The Organization for Economic Co-operation and Development
NHS : National Health System
ACHS : Australian Council on Healthcare Standards
TAS : Trent Accreditation Scheme
ISO : International Organization for Standardization
HIPAA : Health Insurance Portability and Accountability Act
JCI : Joint International Accreditation
HIV : Human Immunodeficiency Virus
PRISMA : Preferred Reporting Items for Systematic Reviews and Meta-Analyses

References

Alsharif, M. J., Labonté, R. and Zuxun Lu. (2010). Patients beyond borders: A study of medical tourists in four countries. *Global Social Policy: An Interdisciplinary Journal of Public Policy and Social Development*, 10(3): 315–335.

Alvarez, M. M., Chanda, R. and Smith, R. D. (2011). The potential for bi-lateral agreements in medical tourism: A qualitative study of stakeholder perspectives from the UK and India. *Globalization and Health*, 7(1): 11.

Anifowose, O. N., Ghasemi, M. and Olaleye, B. R. (2022). Total quality management and Small and Medium-Sized Enterprises' (SMEs) performance: Mediating role of innovation speed. *Sustainability*, 14(14): 1–19.

Arueyingho, O., Ghasemi, M. and Bodamyali, T. (2022). Medical tourism in the Turkish Republic of Northern Cyprus: Strategies for improvement. *Journal of Multidisciplinary Academic Tourism*, 7(1): 1–18.

Arunanondchai, J. and Carsten, F. (2007). Trade in health services in the ASEAN region. World Bank Policy Research Working Paper 4147, Washington DC: World Bank, 2007.

Balaban, V. and Marano, C. (2010). Medical tourism research: A systematic review. *International Journal of Infectious Diseases*, 14: e135. doi:10.1016/j.ijid.2010.02.1784.

Beladi, H., Chao, C., Shan Ee, M. and Hollas, D. (2015). Medical tourism and health worker migration in developing countries. *Economic Modeling*, 46: 391–396.

Bezruchka, S. (2001). Medical tourism as medical harm to the third world: Why? For whom? *Wilderness Environ. Med.*, 11: 77–78.

Budeanu, A. (2005). Impacts and responsibilities for sustainable tourism: A tour operator's perspective. *Journal of Cleaner Production*, 13: 89–97.

Burkett, L. (2007). Medical tourism: Concerns, benefits, and the American perspective. *Journal of Legal Medicine*, 28: 223–245.

Caffarini, K. (2008). AMA meeting: Guidelines target safety of medical tourists. *American Medical News*, 51(25).

Carrera, P. and Lunt, N. (2010). A European perspective on medical tourism: The need for a knowledge base. *International Journal of Health Services*, 40: 469–84.

Carrera, P. and Bridges, J. (2006). Globalization and healthcare: Understanding health and medical tourism. *Expert Review of Pharmacoeconomics and Outcomes Research*, 6(Suppl 4): 447–454.

Centers for Medicare and Medicaid Services (US), and National Association of Insurance Commissioners. (2015). Choosing a Medigap Policy:... Guide to Health Insurance for People with Medicare (No. 2110). US Department of Health and Human Services, Centers for Medicare and Medicaid Services.

Chang, C. S., Chen, S. Y. and Lan, Y. T. (2013). Service quality, trust, and patient satisfaction in interpersonal–based medical service encounters. *BMC Health Services Research*, 13(1): 1–11.

Chantrapornchai, C. and Choksuchat, C. (2016). Ontology construction and application in practice case study of health tourism in Thailand. *SpringerPlus*, 5(1): 1–31.

Chee, H. (2010). Medical tourism and the state in Malaysia and Singapore. *Global Social Policy: An Interdisciplinary Journal of Public Policy and Social Development*, 10(3): 336–357.

Chen, Y. B. and Flood, C. M. (2013). Medical tourism's impact on health care equity and access in low-and middle-income countries: Making the case for regulation. *Journal of Law, Medicine and Ethics*, 41(1): 286–300.

Chinai, R. and Goswami, R. (2007). Medical visas mark growth of Indian medical tourism. *Bulletin of the World Health Organization*, 85: 164–165.

Cohen, E. H. (2008). *Explorations in Thai Tourism: Collected Case Studies*. Emerald Group Publishing.

Cohen, I. G. (2010). Medical tourism: The view from ten thousand feet. *Hastings Center Report*, 40(2): 11–12.

Connell, J. (2006). Medical tourism: Sea, sun, sand, surgery. *Tourism Management*, 27(6): 1093–1100.

Connell, J. (2008). Tummy tucks and the Taj Mahal? Medical tourism and the globalization of health care. In *Tourism Management: Analysis, Behaviour and Strategy* (pp. 232–244). Wallingford UK: CAB International.

Connell, J. (2013). Contemporary medical tourism: Conceptualisation, culture and commodification. *Tourism Management*, 34: 1–13.

Cormany, D. and Baloglu, S. (2011). Medical travel facilitator websites: An exploratory study of web page contents and services offered to the prospective medical tourist. *Tourism Management*, 32(4): 709–716.

Crooks, V. A., Turner, L., Snyder, J., Johnston, R. and Kingsbury, P. (2011). Promoting medical tourism to India: Messages, images, and the marketing of international patient travel. *Social Science and Medicine*, 72(5): 726–732.

Crooks, V. A., Kingsbury, P., Snyder, J. and Johnston, R. (2010). What is known about the patient's experience of medical tourism? A scoping review. *BMC Health Services Research*, 10(1): 1–12.

De Arellano, A. B. R. (2007). Patients without borders: The emergence of medical tourism. *International Journal of Health Services*, 37(1): 193–198.

Dunn, P. (2007). Medical tourism takes fight. *Hospitals and Health Networks*, 81(11): 40–44.

Edelheit, A. (2008). Ficino, Pico and Savonarola: The Evolution of Humanist Theology 1461/2–1498. Brill.

Essier, L. A. and Casken, J. (2013). Seeking health care through international medical tourism. *Journal of Nursing Scholarship*, 45(2): 177–184.

Fetscherin, M. and Stephano, R. M. (2016). The medical tourism index: Scale development and validation. *Tourism Management*, 52: 539–556.

Forgione, D. A. and Smith, P. C. (2007). Medical tourism and its impact on the US health care system. *Journal of Health Care Finance*, 34(1): 27–35.

Gan, L. L. and Frederick, J. R. (2011). Medical tourism facilitators: Patterns of service differentiation. *Journal of Vacation Marketing*, 17(3): 165–183.

Gan, L. L. and Frederick, J. R. (2011). Medical tourism in Singapore: A structure-conduct-performance analysis. *Journal of Asia-Pacific Business*, 12(2): 1–30.

Garud, A. (2005). Medical tourism and its impact on our healthcare. *Natl. Med. J. India*, 18: 318–319.

Ghasemi, M. and Hasanpoor, E. (2021). COVID-19 attack on the world's tourism industry. *Evidence Based Health Policy, Management and Economics*, 5(1): 1–3.

Ghasemi, M., Nejad, M. G., Alsaadi, N., Abdel-Jaber, M. T., Yajid, A., Shukri, M. and Habib, M. (2022). Performance measurment and lead-time reduction in epc project-based organizations: A mathematical modeling approach. *Mathematical Problems in Engineering*.

Ghasemi, M., Nejad, M. G. and Aghaei, I. (2021). Knowledge management orientation and operational performance relationship in medical tourism (overview of the model performance in the COVID-19 pandemic and post-pandemic era). *Health Services Management Research*, 34(4): 208–222.

Ghasemi, M., Sahranavard, S. A., Alola, U. V. and Hassanpoor, E. (2022). Can cost and quality management-oriented innovation enhance patient satisfaction in medical tourist destination? *Journal of Quality Assurance in Hospitality and Tourism*, 1–30.

Golabi, M. and Nejad, M. G. (2022). Intelligent and fuzzy UAV transportation applications in aviation 4.0. *In Intelligent and Fuzzy Techniques in Aviation 4.0* (pp. 431–458). Springer, Cham.

Gray, H. H. and Poland, S.C. (2008). Medical tourism: Crossing borders to access health care. *Kennedy Institute of Ethics Journal*, 18: 193–201.

Green, S. T. (2008). Medical tourism—A potential growth factor in infection medicine and public health. *Journal of Infection*, 57(5): 429.

Hall, C. M. (2011). Health and medical tourism: A kill or cure for global public health? *Tourism Review*.

Han, H. and Hwang, J. (2013). Multi-dimensions of the perceived benefits in a medical hotel and their roles in international travelers' decision-making process. *International Journal of Hospitality Management*, 35: 100–108.

Han, H. and Hyun, S. (2015). Customer retention in the medical tourism industry: Impact of quality, satisfaction, trust, and price reasonableness. *Tourism Management*, 46: 20–29.

Han, H. and Hyun, S. (2015). Image congruence and relationship quality in predicting switching intention: Conspicuousness of product use as a moderator variable. *Journal of Hospitality and Tourism Research*, 37(3): 303–329.

Hansen, F. (2008). A revolution in healthcare: Medicine meets the marketplace. *Public Affairs Review*, 59(4): 43–45.

HealthLeaders-InterStudy. (2011, May 3). Oklahoma City is quickly becoming the medical tourism destination for oncology treatment. Retrieved from http://hlisy.com/News-and-Events/Press-Releases/OKC-MO–050311.

Herrick, D. M. (2007). Medical tourism: Global competition in health care.

Heung, V. C. S., Kucukusta, D. and Song, H. (2011). Medical tourism development in Hong Kong: An assessment of the barriers. *Tourism Management*, 32: 995–1005.

Heung, V. C., Kucukusta, D. and Song, H. (2010). A conceptual model of medical tourism: Implications for future research. *Journal of Travel and Tourism Marketing*, 27(3): 236–251.

Horowitz, M. D. and Rosensweig, J. A. (2007). Medical tourism: Globalization of the health care market place. *Medscape General Medicine*, 9: 33.

Horowitz, M. D., Rosensweig, J. A. and Jones, C. A. (2007). Medical tourism: Globalization of the healthcare marketplace. *Medscape General Medicine*, 9(4): 33.

Hume, L. F. and Demicco, F. J. (2007). Bringing hotels to healthcare. *Journal of Quality Assurance in Hospitality and Tourism*, 8(1): 75–84.

Ibraghimova, I. (2006). Critical assessment of internet medical information. *American International Health Alliance (AIHA)*. Retrieved on October, 23, 2013.

Jaapar, M., Musa, G., Moghavvemi, S. and Saub, R. (2017). Dental tourism: Examining tourist profiles, motivation and satisfaction. *Tourism Management*, 61: 538–552.

Jagyasi, P. (2010). Medical Tourism impact: It's more than obvious. *Medical Tourism Magazine*, 17(1).

Joint Commission International, and Joint Commission on Accreditation of Healthcare Organizations. (2007). Joint Commission International accreditation standards for hospitals. Joint Commission Resources.

Kajzar, P. (2015). Selected destinations or health and wellness tourism in Europe. Masaryk university faculty of economics and administration department of regional economics and administration, 18th international colloquium on regional sciences (Hustopeče, June 17–19, 2015).

Keckley, P. H. and Underwood, H. R. (2008). Medical Tourism: Consumers in Search of Value. Washington USA: Deloitte Center for Health Solutions.

Khawand, S., Ghasemi, M. and Sahranavard, S. A. (2022). Employee involvement and socialization as an example of sustainable marketing strategy and organization's citizenship behavior: Empirical evidence from beirut hotel sector. *Sustainability*, 14(14): 8805.

Kmietowicz, Z. (2004). GPs to check on patients' residency status to stop "Health tourism". *Br. Med. J.*, 328–1217.

Law, E. (2000). Managing complexity and change in tourism: The case of a historic city. *International Journal of Health Tourism*, 2: 229–245.

Leahy, A. L. (2008). Medical tourism: The impact of travel to foreign countries for healthcare. *The Surgeon*, 6(5): 260–261.

Lee, J. Y., Kearns, R. A. and Friesen, W. (2010). Seeking affective health care: Korean immigrants' use of homeland medical services. *Health and Place*, 16(1): 108–115.

Legido-Quigley, H., Passarani, I. and Knai, C. (2011). Cross-border healthcare in the European Union: Clarifying patients' rights. *BMJ*, 342–296.

Lunt, N., Hardey, M. and Mannion, R. (2010). Nip, tuck and click: Medical tourism and the emergence of web-based health information. *The Open Medical Informatics Journal*, 4: 1–11.

Lunt, N., Horsfall, D. and Hanefeld, J. (2016). Medical tourism: A snapshot of evidence on treatment abroad. *Maturitas*, 88: 37–44.

Lunt, N. T., Mannion, R. and Exworthy, M. (2013). A framework for exploring the policy implications of UK medical tourism and international patient flows. *Social Policy and Administration*, 47(1): 1–25.

MacReady, N. (2007). Developing countries court medical tourists. *The Lancet*, 369(9576): 1849–1850.

Malekzadeh, R., Abedi, G., Abedini, E., Haghgoshayie, E., Hasanpoor, E. and Ghasemi, M. (2020). Use of ethical predictability in respect for human rights in Iranian hospitals with a 360-degree approach. *International Journal of Human Rights in Healthcare*, 13(5): 445–455.

Malekzadeh, R., Abedi, G., Hasanpoor, E., Ghasemi, M. and Mahmoodi, G. (2021). A hospital performance assessment model using the IPOCC approach. *Ethiopian Journal of Health Sciences*, 31(3).

Malekzadeh, R., Yaghoubian, S., Hasanpoor, E. and Ghasemi, M. (2021). Health system responsiveness in Iran: A cross-sectional study in hospitals of Mazandaran province. *International Journal of Human Rights in Healthcare*.

Marlowe, J. and Sullivan, P. (2007). Medical tourism: the ultimate outsourcing. *Human Resource Planning*, 30(2): 8–10.

Martinez, D., Ferriol, P., Tous, X., Cabrer, M. and Prats, M. (2008). Virtual health platform for medical tourism purposes. *Studies in Health Technology and Informatics*, 137: 269.

Mattoo, A. and Rathindran, R. (2005). Does health insurance impede trade in health care services? (Vol. 3667). World Bank Publications.

Mattoo, A. and Rathindran, R. (2006). How health insurance inhibits trade in health care. *Health Affairs*, 25(2): 358–368.

Mello, M. M. (2006). Understanding medical malpractice insurance: A primer.

Merrell, R. C., Boucher, D., Carabello, L., Herrick, D. M., Lazzaro, V., Ludwick, L., Toral, R., Woodman, J. and Doarn, C. R. (2008). Medical tourism. *Telemedicine and e-Health*, 14(1): 14–20.

Muller, H. and Kaufmann, E. (2001). Wellness tourism: Market analysis of a special health tourist segment and implications for the hotel industry. *Journal of Vacation Marketing*, 7(1): 5–17.

Muzaffar, F. and Hussain, I. (2007). Medical tourism: are we ready to take the challenge? *Journal of Pakistan Association of Dermatologists*, 17(4): 215–218.

Nejad, M. G. and Banar, M. (2018). Emergency response time minimization by incorporating ground and aerial transportation. *Annals of Optimization Theory and Practice*, 1(1): 43–57.

OECD. (2011). Medical tourism: Treatments, markets and health system implications: A scoping review. OECD, Directorate for Employment, Labour and Social Affairs.

Oorni, A. (2004). Consumer objectives and the amount of search in electronic travel and tourism markets. *Journal of Travel and Tourism Marketing*, 17(2/3): 3–14.

Ormond, M. (2011). Shifting subjects of healthcare: Placing 'medical tourism' in the context of Malaysian domestic healthcare reform. *Asia Pac. Viewpoint*, 52(3): 247–259.

Österle, A., Balazs, P. and Delgado, J. (2009). Travelling for teeth: Characteristics and perspectives of dental care tourism in Hungary. *British Dental Journal*, 206(8): 425–428.

Patterson, I. (2007). Information sources used by older adults for decision making about tourist and travel destinations. *International Journal of Consumer Studies*, 31: 528–533.

Pocock, N.S. and Phua, K.H. (2011). Medical tourism and policy implications for health systems: A conceptual framework from a comparative study of Thailand, Singapore and Malaysia. *Global Health*, 7: 12.

Ramanna, J. M. (2006). Medical tourism and the demand for hospital accreditation overseas. Retrieved March, 7, 2010.

Ratner, C. (2009). Marketing your hospital to medical tourists. *International Medical Travel Journal*.

Reed, C. M. (2008). Medical Tourism. *Medical Clinics of North America*, 92(6): 1433–1446.

Rhea, S. (2009). Still packing their bags: Health reform won't drastically alter the economics of medical tourism, but patients and providers can expect new opportunities, at home and abroad. *Modern Healthcare*, 39(30): 28–30.

Rooney, R. M., Cramer, E. H., Mantha, S., Nichols, G., Bartram, J. K., Farber, J. M. and Benembarek, P. K. (2004). A review of outbreaks of foodborne disease associated with passenger ships: Evidence for risk management. *Public Health Reports*, 119(4): 427–434.

Shirkhani, S. and Ghasemi, M. (2020). Medical Tourism Destination: Considering Distance, Cost and Quality with AHP Method. Asst. Prof. Dr. Mehmet Necati Cizreliogullari, Ph. D., 201.

Smith, R. D., Chanda, R. and Tangcharoensathien, V. (2009). Trade in health-related services. *Lancet*, 373(9663): 593–601.

Smith, M. and Puczko, T. (2009). Twenty-three districts in search of a city: Budapest—The capitaless capital? pp. 201–213. *In*: Maitland, R. and Ritchie, B. (eds.). *City Tourism: National Capital Perspectives*. Wallingford: CABI.

Smith, R., Álvarez, M. M. and Chanda, R. (2011). Medical tourism: A review of the literature and analysis of a role for bi-lateral trade. *Health Policy*, 103(2-3): 276–282.

Snepenger, D. (2006). Modeling Iso-Ahola's motivation theory in the tourism context. *Journal of Travel Research*, 45(2): 140–149.

Snepenger, D., King, J., Marshall, E. and Uysal, M. (2006). Modeling IsoAhola's motivation theory in the tourism context. *Journal of Travel Research*, 45(2): 140–149.

Sobo, E. J. (2009). Medical travel: What it means, why it matters. *Medical Anthropology*, 28(4): 326–335.

Timmermans, K. (2004). Developing countries and trade in health services: Which way is forward?. *International Journal of Health Services*, 34(3): 453–466.

Tourism Malaysia. (2007). Profile of tourist by selected markets 2007. Kuala Lumpur: Tourism Malaysia.

Turner, L. (2010). "Medical tourism" and the global marketplace in health services: US patients, international hospitals, and the search for affordable health care. *International Journal of Health Services*, 40(3): 443–467.

Turner, L. (2007). 'First world health care at third world prices': Globalization, bioethics and medical tourism. *BioSocieties*, 2(3): 303–325.

United Nations Economic and Social Commission for Asia and the Pacific Publications. (2007). EGM patients without borders: An overview of the medical travel industry in Asia, its challenges and opportunities. Bangkok, Thailand.

Vequist, D. and Valdez, E. (2008). Medical tourism economic report: Monterrey, Mexico. *Medical Tourism Magazine*, 4(21).

Vijaya, R. (2010). Medical tourism: Revenue generation or international transfer of healthcare problems? *Journal of Economic Issues*, 44(1): 53–70.

Waikar, A. M., Cappel, S. D. and Tate, U. S. (2011). Challenges and Opportunities for Developing Countries from Medical Tourism.

Weaver, D. and Oppermann, M. (2000). Tourism Management. Brisbane: Wiley.

Weber, W. (2002). Controversy leads to new Netherlands HIV-test guidelines. *Lancet*, 355 –475.

Whittaker, A. (2008). Pleasure and pain: Medical travel in Asia. *Global Public Health*, 3(3): 271–290.

Yap, J. (2007). Singapore: Medical travel and consumer driven health care. *AARP Journal*.

Yolal, M., Woo, E., Cetinel, F. and Uysal, M. (2012). Comparative research of motivations across different festival products. *International Journal of Event and Festival Management*, 3(1): 66–80.

Ytterhus, B. E. (2000). Supply chain management. Intermediary firms and sustainable developments: Some examples from the tourism and retailing sectors. *Norwegian School of Management, Sandvika, Norway*, 28.

Yu, J. Y. and Ko, T. G. (2012), A cross-cultural study of perceptions of medical tourism among Chinese, Japanese and Korean tourists in Korea. *Tourism Management*, 33(1): 80–88.

Chapter 17

Understanding the Importance of Communication in Romantic Relationships through the Framework SGD Goal 5

A Review

Abdulgaffar Olawale Arikewuyo,[1,*] *Tamar Haruna Dambo*[2] and *Olabola Taye Omisore*[3]

1. Introduction

The year 2015 marked the transformation of the then Millennium Development Goals (MDGs), which officially metamorphosed into the Sustainable Development Goals (SDG). Following this development, the United Nations (UN) initiated a resolution to birth sustainable development agendas that are expected to last till 2030, thereby establishing universal goals and 169 targets (Lim et al. 2016). Regarded as the most ambitious effort to improve human wellbeing to date (Agarwal 2018), the SDGs which are expected to last another fifteen years (2030) from commencement provided a more comprehensive framework to the MDGs (Liu et al. 2016). Because of the great potential that the SDGs were said to have to bring social, economic, and sustainable development, much enthusiasm accompanied the birth of the SDGs era as the campaign was kicked off with a motto that strongly suggested that no one would be left behind in this new era (Kim 2017).

[1] Birmingham City University International College, United Kingdom.
[2] Faculty of Communication and Media Studies, Eastern Mediterranean University, North Cyprus.
[3] Riveria College, Joensuu, Eastern Finland.
* Corresponding author: olawale7@yahoo.com

Amongst its targets, the SDGs intend to focus its development agenda on people, as such, consultations and deliberations were made so as to ensure its effectiveness. Quite interestingly, the UN engaged a number of NGOs, civil society organizations, scholars, and opinion leaders who made tangible inputs towards the actualization of the new plan.

SDGs have been affirmed as the goals that will make the world become a more habitable and sustainable place for its habitats (living and non-living) (Di Baldassarre et al. 2019). Indeed, with the aid of the SDGs, developing countries have been able to up their developmental strides, thereby improving the rate of global sustainable development (Sachs 2012). Furthermore, Swain and Yang-Wallentin (2020) asserted that developing countries are better beneficiaries of the SDGs as they helps them improve their social and environmental viability as these countries are still struggling to improve themselves economically and socially. Though country efficiency and performance according to the SDGs vary, the effectiveness or enforcement of the different SDGs themselves is not also at par. Some countries, for instance, have taken the issue of universal health coverage very seriously and continue to find better ways to make it work (Fullman et al. 2018). For example, while some countries are at their best in areas of quality education, health, and wellbeing, others are more attentive to gender equality, clean water, and sanitation. This therefore makes it pertinent for the continuous propagation of the seventeen different goals as the successful implementation of one could create more bandwidth to tackle other goals. As such, when countries are able to achieve success with a particular SDG, they will be encouraged to pick up more goals and continually achieve success, thereby creating an improved society.

Due to the urgent need for development, the SDG dream has further gained recognition in all areas and aspects of human endeavor (Sachs 2012). Though it is presumed that many countries may not meet the target of the SDGs, as what happened with the Millennium Development Goals (MDGs), it is advisable that the global frontliners do not relent in pursuing these goals (Lim 2016). This may be due to the obvious turnaround the MDGs and SDGs have achieved over their period of existence. These turnarounds have also provided a more creditable and reliable society globally (Liu et al. 2016).

The sustainable development programs have established a platform for continuous of growth and developments across the globe (Bowen et al. 2016). For instance, it is generally believed that if countries can focus solely on the sustainable development goals, by 2030, the world would be a more habitable and purpose driven place. Quite interestingly, the SDGs have also generated new partnerships while strengthening existing ones directed at enhancing social and human development. The SDGs have also influenced how countries in the global scene make decisions as well as facilitating a significant reduction of poverty (Kumar et al. 2016).

One of the strongest goals amongst the SDGs in the global scene is the SDG 5 which arguably wields great potential for generally improving quality of life of people (Agarwal 2018). SDG 5 is geared towards promoting gender equality. Several explorative and investigative studies have been conducted on the SDG 5 (Golding et al. 2017, Schwerhoff and Sy 2017, Cai and Choi 2020), these studies span through areas of health (Lim et al. 2016), economy (Lee et al. 2020), politics (Spann 2017), communication (ElAlfy et al. 2020), gender (Bhattacharya and Pal 2021), and environment (Alola et al. 2020), amongst others. Despite the remarkable successes in the efforts of government and non-governmental organizations, groups, and individuals in promoting gender balance in the society, the pertinent question is: To what extent does gender equality exist and get promoted at the macro level in the society? The family? How have romantic partners and the society boosted the SDG goal 5 in romantic relationships through communication? This study ultimately seeks to provide indepth clarification and justifications for the role of interpersonal communication in balancing the existing and threatening gender inequalities existing in romantic relationships.

2. Gender Inequity and Sustainable Development

Gender equality has been a global issue that has received significant attention across the globe. Due to the rigor in the discussion around gender equality globally, this issue has now received a significant improvement among nations (Elomäki 2015). In addition, research has continued to investigate the different dynamics associated with gender inequality (Fang and Munneke 2020). This cuts across many areas of human endeavor such as employment, relationships, communication, education, and other opportunities in human sustenance. In addition, among the sustainable goals, gender equality (which is referred to as the SDG 5) remains one of the most essential and crucial among others.

Gender inequality is a common phenomenon of concern in almost all countries (Carlsen and Bruggemann 2021). Psaki et al. (2018) observe that progress toward the SDG goals comprehensively require indepth understanding of global patterns in gender-based inequity as they happen and the systems that have given them room to fester. Eliciting these factors will enable policy makers to develop solutions that are capable of eliminating these obstacles to gender equality in contemporary society (Bertocchi and Bozzano 2019). This is because the ambitious goal of leaving no one behind cannot materialize until every gender is considered equally important in the development drive. Until gender equality is pushed to the front burner with clear vision and purposive efforts that inhibit its effective realization, societies may continue to witness uneven progress, which may in turn be counter-productive for them. Seeing as gender issues permeate every strata of society, it is crucial to consider the complex issues that are taken for granted as peeling back their layers might point societies in the right direction in tackling gender inequality. For example, while it can be easily taken for granted that farming is usually a male activity, Agarwal (2018) notes that the role of women in maintaining food security and sufficiency is highly underrated. This is because as it may be assumed that men may be more involved in farming activities, the role women play as well including value chain cannot be over emphasized. Gender inequity has been classified to occur in different physical and non-physical forms. For instance, Pirani et al. (2020) asserted that the way males and females react to emotional issues are not usually the same even when they pass through the same experience. As such, there is a need to increase motivating the female gender to participate in activities that may be stereotypically assigned to men in order to encourage other females (Denoncourt 2022).

While investigating the geographical perspectives for gender equity in sustainable cities and communities, de Miguel González and Sebastián-López (2022) asserted that men and women are not accorded equal opportunities in rural areas, thereby creating a level of inequality in their representation at different instances. For example, men are accorded more opportunities in terms or representation and their voices being heard. This fact is obvious in certain societies like the Gulf states where gender inequality is evident in the processes that undermine the national identity of women who are married to non-nationals—most women married to non-nationals cannot transfer their nationality to their husbands and children even though citizenship in most Arab states are by blood or land (Abu-Ras et al. 2022). Thus, citizenship rights are not equally distributed among men and women. In other developing nations, gender inequality is evident in the capital, infrastructural, business, economic, and environmental constraints that amplify gender inequality (Panda 2018). On a larger scale, gender inequality is also felt in the digital world where about 200 women fewer than men own mobile phones worldwide and just about 6% of women develop apps (Mariscal et al. 2019). Considering the fact that technology has become an integral part of societies today, this is a significant obstacle to SDG 5 as limited access to digital tools could also lead to other disadvantages in the labour markets, politics, healthcare, and even education since technology has permeated into every space that humans occupy. Such issues amplify the political, economic, cultural, social, and religious factors that can hinder the full actualization and implementation of gender parity, which in turn will delay the attainment of SDG 5 (Louis et al. 2021). However, poverty and lack of will by some governments still make gender inequity predominant, even in marriages.

3. Romantic Relationships

Romantic relationship involves an interpersonal interaction between two individuals where there exist physical, psychological, emotional, sexual, and other intimate interactions (Arikewuyo et al. 2021). As such, there exists a form of bound between the individuals who are intimately involved in each other and pursue a common goal. It is important to note that intimacy is an essential part of a romantic relationship, thereby making it a determining factor for a relationship to become romantic (Luo et al. 2019). Intimacy may involve sexual, physical, or non-physical as individuals involved in the relationship see each other as an important element of their livelihood. Individuals also become dependent on each other in a romantic relationship, thereby making each other a determining factor of their activities and decision-making (Arikewuyo et al. 2020). Generally speaking, romantic relationships are expected to be built on trust.

Romantic relationships help in further boosting any form of mutual relationship between individuals, thereby facilitating a bond. In addition, individuals in a relationship enter into it with their specific goals, desires, wants, and needs (Arikewuyo et al. 2019). These may range from sexual, financial, or psychological gratification where the partner makes efforts to ensure they are happy in the relationship. Quite interestingly, it could be stated that there is no relationship without its peculiar problems and challenges. This may range from economic, social or gender inequality, lack of trust, and infidelity amongst others. (Garbinsky et al. 2020) asserted that infidelity has been shown to be one of the most common forms of relational instability. However, infidelity may not necessarily be sexual but include financial infidelity, which occurs when an individual does not openly disclose or reveal their financial status to their romantic partner.

Before a relationship is built, the individuals involved pass through a stage called relationship initiation. The is the early part of the relationship where the individuals involved tries to study each other's behavioral pattern as well as upgrade their attraction and intimacy (Mahmut and Croy 2019). As such, anyone is capable of maintaining a stable and beneficial relationship where individuals involved will experience one form of satisfaction or the other (Doyle and Molix 2015). However, it is not impossible that individuals in a relationship may have dialectic goals in the relationship. In addition, romantic relationships are found to be more purposeful when individuals involved set out purposeful and achievable goals (Strunz et al. 2017). This will help in enhancing satisfaction in the relationship based on the similarity in purpose.

Just like every other human activity, romantic relationships are not without challenges. These challenges range from inequality, sustainability, accountability, and other life events. However, commitment and dedication to a relationship are essential in achieving a purposeful and successful dyadic relationship. Successful romantic relationships are also dependent largely on the individuals' ability to support each other's dreams and aspirations (Shulman and Connolly 2013). As such, partner support will help build a lasting and purposeful relationship. However, Chen (2020) asserted that the attitude of individuals towards their romantic partner is based on their parental influence. As such, individuals in a relationship will most likely behave based on their parental experience. This may come in form of the parents' interaction/communication with each other, as well as their ability to erode gender imbalance in such relationship (Lemay et al. 2015).

Relationship civility has, furthermore, been described as the hallmark of romantic relationships. As such, individuals must engage on different affectionate activities that will help promote romantic involvement and thereafter, relationship satisfaction (Soller 2014). In addition, romantic power also has the capability to influence affection in relationships. This means romantic power must be able to distinguish when there is the need for compromise in decisions that will help promote stability. As such, it will give way for interdependence in the relationship, thereby promoting stability (Gordon and Chen 2013). In all of these activities in a romantic relationship, the question that arises amongst individuals, relationship experts as well researchers is that with the existing pattern of interaction in a relationship, how pertinent is gender balance between the individuals involved therein?

4. Communication in Romantic Relationships

Communication is the most important aspect of the social world. In fact, humans cannot do without communication. This signals the role that communication plays in our human experiences. Considering the role it plays in our daily living, communication also plays a crucial role in romantic relationships. Since the basic foundation of the social world is predicated on communication, every relationship is built on communication, thereby making it the core of the relationship. Also, romantic relationships exist based on the interpersonal relations and communication that exists between two individuals (Strunz et al. 2017). The ability to establish and maintain social interaction and communication in romantic relationships is highly important. To achieve a healthy relationship, individuals must establish effective communication between themselves.

Generally, romantic relationships are regarded as a relevant source of emotional bonding that can increase self-positivity as well as heighten social integration (Gómez-López et al. 2019). In addition to this, communication between couples in romantic relationships is considered to have significant impact on a couple's wellbeing (Gómez-López et al. 2019). There are many reasons for this; communication creates opportunities to resolve conflict, it can heighten the system that couples need to feel supported, and it can also build trust. Overall, communication or lack of it in romantic relationships can affect intimacy among couples. Thus, communication in romantic relationships is an essential part of sustaining and maintaining such a relationship.

Due to the emergence of communication technology facilities, virtual communication has now become the focal point of communication (Luo and Tuney 2015). For instance, individuals do not need to establish face-to-face contact before they can effectively communicate. This is largely due to the emergence of cellphones which birthed text messaging and social media interactions (Arikewuyo et al. 2021). However, texting has been found to be a frequent means of communication among romantic partners (Morey et al. 2013). This may occur through the usual text messaging as well as through instant messaging platforms. As such, stability in communication can be said to be a means through which individuals improve the quality of their relationships, as well as the quality of relational attachments.

One of the problems romantic relationships face such as barrier in communication is fueled by long-distance relationships. This involves the process where both romantic partners are not resident at the same place as a result of certain factors like work, education, and other social activities (Arikewuyo et al. 2020). However, due to social media and instant messaging platforms, individuals are now able to process audio visual communication which enhances meaningful communication and thereafter add to relational stability (Arikewuyo et al. 2021). In addition, Janning et al. (2018) assert that audio-visual form of communication further enhances intimacy among romantic partners rather than paper communication, which enhances truthfulness.

Social media platforms have become pertinent channels of communication in romantic relationships (Ruppel 2015). Indeed, with the day-to-day advancement in technology, communication online has further been boosted with the aid of pictures, videos, and other intimate online patterns (Arikewuyo et al. 2020). Social media platforms have also made expression of love and affection a seamless activity. This is further boosted with the use of smileys, emojis, stickers, gifs, and many other online features that have made communication in romantic relationships easy. Mediated communication has therefore been largely accepted by users as a reliable means of communication (Len-Ríos et al. 2016).

5. Gender Equality in Romantic Relationships

Generally, gender is classified as typical attitudes and patterns attributed to a male or female based on time and culture with individuals behave based on the gender they represent (Dewinter et al. 2017). Gender equality occurs when both parties have equal rights and privileges to all forms of physical and non-physical resources. As such, with the propagation of gender equality, which is a vital part of the SDGs, romantic relationships are expected to attain more success. Gender imbalance may be said to occur in a relationship over a period of time, however, it will most likely reduce as the relationship gets better or as romantic partners age (Uslu and Erenoğlu 2022). Family patterns and dynamic have been categorized as factors that influence gender relation in romantic relationships. As such, if a male child comes from a family that is male dominant, this may influence his ideology, perception, attitude. and general way of life (Tyrell et al. 2016).

Gender imbalance in romantic relationships can be said to occur right from the stage of conception (Liu et al. 2022). For instance, it is widely believed that a male should be the one to woo the female. This has over the years created heated arguments and debates, however, till date the situation has not significantly changed. This stereotype is largely promoted by cultural orientation where the male is generally assumed to be the romantically strong individual; as such it may portray the woman negatively if she assumes the position of the romantically strong individual (Hartley et al. 2014). Though, it is generally assumed that the males are more sexually active, other points of view believe there should be equity in sexual dominance (De Meyer et al. 2017). Indeed Stavropoulos et al. (2018) affirmed that gender is not a determining factor to the level of attachment that is experienced in the romantic relationship. Invariably, it may be assumed that any of the individuals could be weaker in terms of romantic attachment to the other. Interestingly, attachment and anxiety in individuals have shown to be associated with poor communication among romantic partners (Weger and Tang 2022).

Also, gender may not be largely associated with the extent of love that is inherent in the relationship. As such, the male could love more or be more romantically attached in the relationship (Van Lankveld et al. 2018). Furthermore, it has generally been observed that the male gender is more sexually attached to the females in a romantic relationship (Akers et al. 2021). For instance, Thomeer et al. (2020) asserted that men have a higher level of sexual urge to women. As such, it may be inferred that women are able to control their sexual urge and desires more. Conversely, it was discovered that males are more self-reliant, emotionally stable, and less reliant in romantic relationships (Stavropoulos et al. 2018). As such, the male gender will less likely become destabilized in case of challenges or obstacles in the relationship. This may account for the notion behind the stereotypical referral to the males as the dominant gender in the relationship (Pulice-Farrow et al. 2020).

6. Conclusions

The ultimate goal of this study was to provide an indepth understanding to the role of interpersonal communication in balancing the existing and threatening gender inequalities existing in romantic relationships. Findings show that existing literature understand that there exists to a certain extent, communication and relational imbalance among individuals in a romantic relationship. As such, this has affected so many areas of sustainability of these relationships. Our study adds new perspectives to understanding interpersonal communication in romantic relationships and how gender imbalance, a sustainable development goal occurs and is being addressed in romantic relationships.

Gender inequality is still highly visible and present in human communication and interaction and has somewhat influenced romantic relationships. As such, it has further transcended into other areas of these relationships and the lives of the individuals involved even though some relationships allow a considerable degree of balance in their relationships. This is done through financial transparency as well as managing the societal, traditional, and economic activities of both the romantic partners are involved. It is also pertinent to note that gender balance in romantic relationships promotes friendliness, and helps build stronger and balanced family ties, and orientation, as such, the joy and happiness that radiates from the relationship and family is further transmitted to the society. Among all forms of perceived inequalities in a relationship which largely benefits the male gender, it can generally be argued that where females have enough self-awareness with regards to sexual consent, this further gives them an opportunity to share power in the relationship.

Quite importantly, the study concluded that communication is key to the sustainability of any relationship as well as promoting healthy relationships. This study hereby suggests that individuals entering into a romantic relationship should consider the personality and compatibility of their partner before entering into the lifelong partnership. Dyadically, romantic partners should also be ready to support their partner's dreams and work together to achieve success. No doubt, to ensure gender equity, romantic partners need to promote effective and purposeful interpersonal communication, thereby helping to further achieve SDG 5. Furthermore, individuals need to care for themselves as well as their romantic partners; this is where equity will set in without any rancor. Though romantic partners may experience turbulence at the early stage of their bond, the ability to persevere and jointly break through challenges will help in keeping the relationship.

References

Abu-Ras, W., Elzamzamy, K., Burghul, M. M., Al-Merri, N. H., Alajrad, M. and Kharbanda, V. A. (2022). Gendered citizenship, inequality, and well-being: the experience of cross-national families in qatar during the gulf cooperation council crisis (2017–2021). *International Journal of Environmental Research and Public Health*, 19(11): 6638.

Agarwal, B. (2018). Gender equality, food security and the sustainable development goals. *Current Opinion in Environmental Sustainability*, 34: 26–32.

Akers, W. P., Cashwell, C. S. and Blake, S. D. (2021). Images of resilience: Outness in Same-gender romantic relationships. *Journal of LGBTQ Issues in Counseling*, 15(2): 168–194.

Alola, A. A., Arikewuyo, A. O., Ozad, B., Alola, U. V. and Arikewuyo, H. O. (2020). A drain or drench on biocapacity? Environmental account of fertility, marriage, and ICT in the USA and Canada. *Environmental Science and Pollution Research*, 27(4): 4032–4043.

Arikewuyo, A. O., Efe-Özad, B. and Owolabi, A. (2020). I U': A semiotic analysis of romantic relationship bitmojis on social media. *It Happened on Tinder: Reflections and Studies on Internet-Infused Dating*, 134–144.

Arikewuyo, A. O., Efe-Özad, B., Dambo, T. H., Abdulbaqi, S. S. and Arikewuyo, H. O. (2021). An examination of how multiple use of social media platforms influence romantic relationships. *Journal of Public Affairs*, 21(3): e2240.

Arikewuyo, A. O., Eluwole, K. K. and Özad, B. (2021). Influence of lack of trust on romantic relationship problems: The mediating role of partner cell phone snooping. *Psychological Reports*, 124(1): 348–365.

Arikewuyo, A. O., Eluwole, K. K., Dambo, T. H. and Abdulbaqi, S. S. (2021). Do low self-esteem, relationship dissatisfaction and relationship insecurity exacerbate the intention to break up in romantic relationships? *Current Psychology*, 1–12.

Arikewuyo, A., Ozad, B. and Lasisi, T. T. (2019). Erotic use of social media pornography in gratifying romantic relationship desires. *The Spanish Journal of Psychology*, 22.

Bali Swain, R. and Yang-Wallentin, F. (2020). Achieving sustainable development goals: Predicaments and strategies. *International Journal of Sustainable Development and World Ecology*, 27(2): 96–106.

Bertocchi, G. and Bozzano, M. (2019). Origins and implications of family structure across Italian provinces in historical perspective. *Cliometrics of the Family*, 121–147.

Bhattacharya, A. and Pal, M. (2021). Fifth sustainable development goal gender equality in India: Analysis by mathematics of uncertainty and covering of fuzzy graphs. *Neural Computing and Applications*, 33(21): 15027–15057.

Bowen, K. J., Cradock-Henry, N. A., Koch, F., Patterson, J., Häyhä, T. and Vogt, J. (2017). Implementing the "Sustainable Development Goals": Towards addressing three key governance challenges—collective action, trade-offs, and accountability. *Current Opinion in Environmental Sustainability*, 26: 90–96.

Bowen, P. K., Shearier, E. R., Zhao, S., Guillory, R. J., Zhao, F., Goldman, J. and Drelich, J. W. (2016). Biodegradable metals for cardiovascular stents: From clinical concerns to recent Zn-Alloys. *Advanced Healthcare Materials*, 5(10): 1121–1140.

Cai, Y. J. and Choi, T. M. (2020). A United Nations' Sustainable Development Goals perspective for sustainable textile and apparel supply chain management. *Transportation Research Part E: Logistics and Transportation Review*, 141: 102010.

Carlsen, L. and Bruggemann, R. (2021). Gender Equality in Europe: The Development of the Sustainable Development Goal No. 5 Illustrated by exemplary cases. *Social Indicators Research*, 158(3): 1127–1151.

Chen, W. W., Yuan, H., Yang, X. and Lai, S. K. (2020). Parenting, self-concept, and attitudes about romantic relationships. *Journal of Adolescence*, 82: 41–49.

De Meyer, S., Kågesten, A., Mmari, K., McEachran, J., Chilet-Rosell, E., Kabiru, C. W. and Michielsen, K. (2017). Boys should have the courage to ask a girl out: Gender norms in early adolescent romantic relationships. *Journal of Adolescent Health*, 61(4): S42–S47.

de Miguel González, R. and Sebastián-López, M. (2022). Education on Sustainable Development Goals: Geographical perspectives for gender equality in sustainable cities and communities. *Sustainability*, 14(7): 4042.

Denoncourt, J. A. (2022). Supporting Sustainable Development Goal 5 Gender Equality and entrepreneurship in the tanzanite mine-to-market. *Sustainability*, 14(7): 4192.

Dewinter, J., De Graaf, H. and Begeer, S. (2017). Sexual orientation, gender identity, and romantic relationships in adolescents and adults with autism spectrum disorder. *Journal of Autism and Developmental Disorders*, 47(9): 2927–2934.

Di Baldassarre, G., Sivapalan, M., Rusca, M., Cudennec, C., Garcia, M., Kreibich, H. and Blöschl, G. (2019). Sociohydrology: Scientific challenges in addressing the sustainable development goals. *Water Resources Research*, 55(8): 6327–6355.

Doyle, D. M. and Molix, L. (2015). Social stigma and sexual minorities' romantic relationship functioning: A meta-analytic review. *Personality and Social Psychology Bulletin*, 41(10): 1363–1381.

ElAlfy, A., Darwish, K. M. and Weber, O. (2020). Corporations and sustainable development goals communication on social media: Corporate social responsibility or just another buzzword? *Sustainable Development*, 28(5): 1418–1430.

Elomäki, A. (2015). The economic case for gender equality in the European Union: Selling gender equality to decision-makers and neoliberalism to women's organizations. *European Journal of Women's Studies*, 22(3): 288–302.

Fang, L. and Munneke, H. J. (2020). Gender equality in mortgage lending. *Real Estate Economics*, 48(4): 957–1003.

Fullman, N., Barber, R. M., Abajobir, A. A., Abate, K. H., Abbafati, C., Abbas, K. M. and Dubey, M. (2017). Measuring progress and projecting attainment on the basis of past trends of the health-related Sustainable Development Goals in 188 countries: An analysis from the Global Burden of Disease Study 2016. *The Lancet*, 390(10100): 1423–1459.

Fullman, N., Yearwood, J., Abay, S. M., Abbafati, C., Abd-Allah, F., Abdela, J. and Chang, H. Y. (2018). Measuring performance on the Healthcare Access and Quality Index for 195 countries and territories and selected subnational locations: A systematic analysis from the Global Burden of Disease Study 2016. *The Lancet*, 391(10136): 2236–2271.

Garbinsky, E. N., Gladstone, J. J., Nikolova, H. and Olson, J. G. (2020). Love, lies, and money: Financial infidelity in romantic relationships. *Journal of Consumer Research*, 47(1): 1–24.

Golding, N., Burstein, R., Longbottom, J., Browne, A. J., Fullman, N., Osgood-Zimmerman, A. and Hay, S. I. (2017). Mapping under-5 and neonatal mortality in Africa, 2000–15: A baseline analysis for the Sustainable Development Goals. *The Lancet*, 390(10108): 2171–2182.

Gómez-López, M., Viejo, C. and Ortega-Ruiz, R. (2019). Psychological well-being during adolescence: Stability and association with romantic relationships. *Frontiers in Psychology*, 10: 1772.

Gordon, A. M. and Chen, S. (2013). Does power help or hurt? The moderating role of self–other focus on power and perspective-taking in romantic relationships. *Personality and Social Psychology Bulletin*, 39(8): 1097–1110.

Hartley, J. E., Wight, D. and Hunt, K. (2014). Presuming the influence of the media: Teenagers' constructions of gender identity through sexual/romantic relationships and alcohol consumption. *Sociology of Health and Illness*, 36(5): 772–786.

Janning, M., Gao, W. and Snyder, E. (2018). Constructing shared "space": Meaningfulness in long-distance romantic relationship communication formats. *Journal of Family Issues*, 39(5): 1281–1303.

Kim, Y. Y. (2017). Cross-cultural adaptation. *Oxford Research Encyclopedia of Communication.*

Kumar, S., Kumar, N. and Vivekadhish, S. (2016). Millennium development goals (MDGS) to sustainable development goals (SDGS): Addressing unfinished agenda and strengthening sustainable development and partnership. *Indian Journal of Community Medicine: Official Publication of Indian Association of Preventive and Social Medicine*, 41(1): 1.

Lee, K. H., Noh, J. and Khim, J. S. (2020). The Blue Economy and the United Nations' sustainable development goals: Challenges and opportunities. *Environment International*, 137: 105528.

Lemay Jr, E. P., Lin, J. L. and Muir, H. J. (2015). Daily affective and behavioral forecasts in romantic relationships: Seeing tomorrow through the lens of today. *Personality and Social Psychology Bulletin*, 41(7): 1005–1019.

Len-Ríos, M. E., Streit, C., Killoren, S., Deutsch, A., Cooper, M. L. and Carlo, G. (2016). US Latino adolescents' use of mass media and mediated communication in romantic relationships. *Journal of Children and Media*, 10(4): 395–410.

Lim, S. S., Allen, K., Bhutta, Z. A., Dandona, L., Forouzanfar, M. H., Fullman, N. and Chang, J. C. (2016). Measuring the health-related Sustainable Development Goals in 188 countries: A baseline analysis from the Global Burden of Disease Study 2015. *The Lancet*, 388(10053): 1813–1850.

Liu, L., Oza, S., Hogan, D., Chu, Y., Perin, J., Zhu, J. and Black, R. E. (2016). Global, regional, and national causes of under-5 mortality in 2000–15: An updated systematic analysis with implications for the Sustainable Development Goals. *The Lancet*, 388(10063): 3027–3035.

Liu, Y., Gao, J., Wang, H., Semba, S., Gu, C. and Yang, H. (2022). Families' influence on romantic relationship and its reconstruction. *Chaos, Solitons and Fractals*, 155: 111662.

Louis, N., Shyleen, C. and Tirivangasi, H. M. (2021). Barriers towards Sustainable Development Goal (Sdg 5) on achieving gender equality and empowerment of women and girls in Nyanga rural district of Manicaland Province, Zimbabwe. *African Journal of Gender, Society and Development*, 10(3): 33.

Luo, B., Fang, W., Shen, J. and Cong, X. F. (2019). Gift–image congruence and gift appreciation in romantic relationships: The roles of intimacy and relationship dependence. *Journal of Business Research*, 103: 142–152.

Luo, S. and Tuney, S. (2015). Can texting be used to improve romantic relationships?—The effects of sending positive text messages on relationship satisfaction. *Computers in Human Behavior*, 49: 670–678.

Mahmut, M. K. and Croy, I. (2019). The role of body odors and olfactory ability in the initiation, maintenance and breakdown of romantic relationships–A review. *Physiology and Behavior*, 207: 179–184.

Mariscal, J., Mayne, G., Aneja, U. and Sorgner, A. (2019). Bridging the gender digital gap. *Economics*, 13(1).

Morey, J. N., Gentzler, A. L., Creasy, B., Oberhauser, A. M. and Westerman, D. (2013). Young adults' use of communication technology within their romantic relationships and associations with attachment style. *Computers in Human Behavior*, 29(4): 1771–1778.

Panda, S. (2018). Constraints faced by women entrepreneurs in developing countries: Review and ranking. *Gender in Management: An International Journal*.

Pirani, N., Ricker, B. A. and Kraak, M. J. (2020). Feminist cartography and the United Nations Sustainable Development Goal on gender equality: Emotional responses to three thematic maps. *The Canadian Geographer/Le Géographe Canadien*, 64(2): 184–198.

Pulice-Farrow, L., McNary, S. B. and Galupo, M. P. (2020). "Bigender is just a Tumblr thing": Microaggressions in the romantic relationships of gender non-conforming and agender transgender individuals. *Sexual and Relationship Therapy*, 35(3): 362–381.

Ruppel, E. K. (2015). Use of communication technologies in romantic relationships: Self-disclosure and the role of relationship development. *Journal of Social and Personal Relationships*, 32(5): 667–686.

Sachs, J. D. (2012). From millennium development goals to sustainable development goals. *The Lancet*, 379(9832): 2206–2211.

Schwerhoff, G. and Sy, M. (2017). Financing renewable energy in Africa–Key challenge of the sustainable development goals. *Renewable and Sustainable Energy Reviews*, 75: 393–401.

Shulman, S. and Connolly, J. (2013). The challenge of romantic relationships in emerging adulthood: Reconceptualization of the field. *Emerging Adulthood*, 1(1): 27–39.

Soller, B. (2014). Caught in a bad romance: Adolescent romantic relationships and mental health. *Journal of Health and Social Behavior*, 55(1): 56–72.

Spann, M. (2017). Politics of poverty: The post-2015 sustainable development goals and the business of agriculture. *Globalizations*, 14(3): 360–378.

Stavropoulos, V., Mastrotheodoros, S., Burleigh, T. L., Papadopoulos, N. and Gomez, R. (2018). Avoidant romantic attachment in adolescence: Gender, excessive internet use and romantic relationship engagement effects. *PloS One*, 13(7): e0201176.

Strunz, S., Schermuck, C., Ballerstein, S., Ahlers, C. J., Dziobek, I. and Roepke, S. (2017). Romantic relationships and relationship satisfaction among adults with Asperger syndrome and high-functioning autism. *Journal of Clinical Psychology*, 73(1): 113–125.

Thomeer, M. B., Umberson, D. and Reczek, C. (2020). The gender-as-relational approach for theorizing about romantic relationships of sexual and gender minority mid-to later-life adults. *Journal of Family Theory and Review*, 12(2): 220–237.

Tyrell, F. A., Wheeler, L. A., Gonzales, N. A., Dumka, L. and Millsap, R. (2016). Family influences on Mexican American adolescents' romantic relationships: Moderation by gender and culture. *Journal of Research on Adolescence*, 26(1): 142–158.

Uslu, N. and Erenoğlu, R. (2022). Does the gender course have any effects on young people's gender equality perceptions, dating violence, and problem-solving behaviours in romantic relationships? A mixed-method study. *Perspectives in Psychiatric Care*, 58(2): 471–481.

Van Lankveld, J., Jacobs, N., Thewissen, V., Dewitte, M. and Verboon, P. (2018). The associations of intimacy and sexuality in daily life: Temporal dynamics and gender effects within romantic relationships. *Journal of Social and Personal Relationships*, 35(4): 557–576.

Weger, H. and Tang, C. Y. (2022). Attachment avoidance and anxiety as predictors of confirming and disconfirming communication in romantic relationships. *Southern Communication Journal*, 87(1): 28–43.

Index

N

O

P

R

S

T

U

V

About the Editors

Andrew Adewale Alola (PhD in Economics) is an Associate Professor of Sustainability at CREDS-Centre for Research on Digitalization and Sustainability, Inland Norway University of Applied Science, Norway. As a Research Fellow at Adnan Kassar School of Business, LAU, Lebanon, he's on leave as Assistant Professor at *Nişantaşı* University, Turkey. Having authored over 200 scientific studies in top tier peer-reviewed journals, he has been named among Standard University's Top 2% Global Scientists since 2020.

Festus Victor Bekun is an Assistant professor at Istanbul Gelisim University and a senior research fellow at Lebanese American University. He has published over 250 scientific papers in top-tier outlets and has high indices in Google Scholar, Scopus, and Web of Science. He is recognized as one of the top 2% of influential scientists by Stanford University. Dr. Bekun serves on the editorial boards of various scientific journals and is currently the editor in chief of Open Agriculture.

Uju Violet Alola earned her PhD degree in Tourism Management from Eastern Mediterranean University, Famagusta, Northern Cyprus. She also holds an MSc in Marketing Management from the same university and a Bachelor's degree from Anambra State University, Nigeria. Presently, she is a faculty member at Istanbul Gelisim University, Turkey. She has contributed to various reputable journals such as Tourism Management Perspectives, International Journal of Tourism Research, Current Issues in Tourism.

For Product Safety Concerns and Information please contact our EU representative GPSR@taylorandfrancis.com Taylor & Francis Verlag GmbH, Kaufingerstraße 24, 80331 München, Germany